T0181254

IFIP Advances in Information and Communication Technology 578

Editor-in-Chief

Kai Rannenberg, Goethe University Frankfurt, Germany

Editorial Board Members

IFIP – The International Federation for Information Processing

IFIP was founded in 1960 under the auspices of UNESCO, following the first World Computer Congress held in Paris the previous year. A federation for societies working in information processing, IFIP's aim is two-fold: to support information processing in the countries of its members and to encourage technology transfer to developing nations. As its mission statement clearly states:

> IFIP is the global non-profit federation of societies of ICT professionals that aims at achieving a worldwide professional and socially responsible development and application of information and communication technologies.

IFIP is a non-profit-making organization, run almost solely by 2500 volunteers. It operates through a number of technical committees and working groups, which organize events and publications. IFIP's events range from large international open conferences to working conferences and local seminars.

The flagship event is the IFIP World Computer Congress, at which both invited and contributed papers are presented. Contributed papers are rigorously refereed and the rejection rate is high.

As with the Congress, participation in the open conferences is open to all and papers may be invited or submitted. Again, submitted papers are stringently refereed.

The working conferences are structured differently. They are usually run by a working group and attendance is generally smaller and occasionally by invitation only. Their purpose is to create an atmosphere conducive to innovation and development. Refereeing is also rigorous and papers are subjected to extensive group discussion.

Publications arising from IFIP events vary. The papers presented at the IFIP World Computer Congress and at open conferences are published as conference proceedings, while the results of the working conferences are often published as collections of selected and edited papers.

IFIP distinguishes three types of institutional membership: Country Representative Members, Members at Large, and Associate Members. The type of organization that can apply for membership is a wide variety and includes national or international societies of individual computer scientists/ICT professionals, associations or federations of such societies, government institutions/government related organizations, national or international research institutes or consortia, universities, academies of sciences, companies, national or international associations or federations of companies.

More information about this series at http://www.springer.com/series/6102

Aravindan Chandrabose ·
Ulrich Furbach · Ashish Ghosh ·
Anand Kumar M. (Eds.)

Computational Intelligence in Data Science

Third IFIP TC 12 International Conference, ICCIDS 2020
Chennai, India, February 20–22, 2020
Revised Selected Papers

Springer

Editors
Aravindan Chandrabose (iD)
SSN College of Engineering
Chennai, India

Ashish Ghosh (iD)
Indian Statistical Institute
Kolkata, India

Ulrich Furbach (iD)
University of Koblenz-Landau
Koblenz, Germany

Anand Kumar M. (iD)
National Institute of Technology Karnataka
Mangalore, India

ISSN 1868-4238   ISSN 1868-422X   (electronic)
IFIP Advances in Information and Communication Technology
ISBN 978-3-030-63469-8   ISBN 978-3-030-63467-4   (eBook)
https://doi.org/10.1007/978-3-030-63467-4

This Springer imprint is published by the registered company Springer Nature Switzerland AG
The registered company address is: Gewerbestrasse 11, 6330 Cham, Switzerland

Preface

We are excited to present the proceedings of the IFIP Third International Conference on Computational Intelligence in Data Science (ICCIDS 2020) which was held at Sri Sivasubramaniya Nadar College of Engineering (SSN), Chennai, India, during February 20–22, 2020. The conference was organized by the Department of Computer Science and Engineering and the Machine Learning Research Group (MLRG) of SSN. In particular, D. Thenmozhi, J. Bhuvana, and P. Mirunalini, meticulously worked on all the finer details towards the successful organization of this conference. The first two editions of ICCIDS were conducted during June 2017 and February 2019 at SSN.

The aim of this series of conferences is to provide an international platform to discuss, exchange, disseminate, and cross-fertilize the innovative ideas that solicit experimental, theoretical work, and methods in solving problems and applications of current issues of computational intelligence and data science. The enthusiastic response received for this third edition of the conference was overwhelming. The papers presented in the conference were across a spectrum of areas such as machine learning, deep learning, artificial intelligence, data analytics, text processing, image processing, and Internet of Things (IoT).

We reached out to more than 120 experts in the fields of computational intelligence and data science to review the papers received. Each submitted paper went through comprehensive reviews by at least three reviewers and their review comments were communicated to the authors before the conference. Out of 94 papers received through the EasyChair submission management system, only 31 papers were accepted. 27 papers were registered and presented by authors from India and abroad. All the presented papers were again reviewed by the session chairs and their comments were communicated to the authors for revising their work. The final submissions were checked by the Program Committee to ensure that all the comments had been addressed.

Two pre-conference workshops were scheduled on February 20, 2020, on the topics "Text analysis and Information Extraction & Retrieval — TIER 2020" and "Image and Video Analysis — IVA 2020." TIER 2020 had 30 participants and the sessions were handled by Suriyadeepan and Kamal Raj of Saama Technologies, India. Demonstrations on GPU based computations for deep learning were held by Laxmi Nageswari, Global Lead, AI Education & Solutions, Boston Limited, India, and Krishna Mouli, Nvidia certified deep learning faculty, Boston Limited, India. Hands-on training on transformer models was given by D. Thenmozhi, B. Senthil Kumar, and S. Kayalvizhi of SSN. In IVA 2020, which had 27 participants, technical sessions were handled by T. T. Mirnalinee of SSN and Hemalatha of Madras Institute of Technology, Anna University, Chennai, India. Hands-on training on U-Net was given by P. Mirunalini and T. T. Mirnalinee of SSN.

The conference proceedings were inaugurated on February 21, 2020, with two keynote presentations: "Machine Learning and Big Data Analysis" by Ashish Ghosh of the Indian Statistical Institute, Kolkata, India, and "Industry 4.0 — AI, Robotics, and

Automation" by Prahlad Vadakkepat of the National University of Singapore, Singapore. Presentations of the accepted papers were organized as four tracks:

- Computational Intelligence for Text and Speech Analysis
- Computational Intelligence for Sensor and Other Data
- Computational Intelligence for Image and Video Analysis
- Data Science

Two parallel tracks, with seven presentations each, were scheduled on the first day of the conference. The sessions started with an invited talk on "Code Variants Retrieval" by Venkatesh Vinayak Rao of Chennai Mathematical Institute, Chennai, India.

The second day of the conference had two keynote talks, the first one on "The assessment of Information Credibility in the Social Web" by Gabriella Pasi, Professor, Department of Informatics, Systems, and Communication (DISCo), University of Milano-Bicocca, Italy, followed by a talk on "Transfer Learning for Natural Language Processing" by Anand Kumar M. of the National Institute of Technology, Suratkal, India. Raj Ramesh, Chief AI Officer, DataFoundry, USA, delivered an invited talk on "How Data Scientists can help the Business?" Two parallel tracks were scheduled on the second day of the conference, one with seven presentations and the other with six presentations.

This volume contains revised versions of 26 papers presented at the conference. We sincerely hope that these papers provide significant research contributions and advancements in the fields of computational intelligence and data science. We thank the International Federation of Information Processing (IFIP) for their support and encouragement, and Springer for publishing these proceedings.

We would like to thank and acknowledge the financial support provided by SSN, HCL, Boston, Frontier Systems, Tata Motors, CoMMNet, USAM, and Petrofac. We thank all the Program Committee members, reviewers, session chairs, Organizing Committee members, and the participants for their contributions towards the success of the conference.

July 2020   Aravindan Chandrabose
                                                            Ulrich Furbach
                                                            Ashish Ghosh
                                                        Anand Kumar M.

Organization

General Chair

Aravindan Chandrabose   SSN College of Engineering, India

Program Committee Chairs

Ulrich Furbach   Universität Koblenz-Landau, Germany
Ashish Ghosh   Indian Statistical Institute, Kolkata, India

Steering Committee

Chitra Babu   SSN College of Engineering, India
Aravindan Chandrabose   SSN College of Engineering, India
Ulrich Furbach   Universität Koblenz-Landau, Germany
Ashish Ghosh   Indian Statistical Institute, Kolkata, India
Eunica Mercier Laurent   Lyon 3 University, France
Anand Kumar M.   NITK Suratkal, India
Gabriella Pasi   University of Milano-Bicocca, Italy

Program Committee

Vallavaraj A.   National University of Science and Technology, Oman
Thenmozhi D.   SSN College of Engineering, India
Biju Issac   University of Northumbria, UK
Bhuvana J.   SSN College of Engineering, India
Mostafa Hajiaghaei Keshteli   Mazandaran University of Science and Technology,
        Iran
Sri Krishnan   Ryerson University, Canada
Sobha L.   AU-KBC, MIT, India
Mirunalini P.   SSN College of Engineering, India
Milton R. S.   SSN College of Engineering, India
Srinivasa Rao   IIT Kharagpur, India
Claudia Schon   Universität Koblenz-Landau, Germany
Mirnalinee T. T.   SSN College of Engineering, India

Additional Reviewers

Beulah A.
Chamundeswaria A.
Sigappi A. N.
Bharathi B.
Monica Jennifer B.
Prabavathy B.
Rajesh Kanna B.
Senthil Kumar B.
Charulatha B. S.
Annie Bennat
Prabir Kumar Biswas
Saroj Biswas
Birla Bose
Saranya Jyothi C.
Suba Lalitha C. N.
Aroul Canessane
Jonathan Hoyin Chan
Venkata Vara Prasad D.
Satchidananda Dehuri
Arthi
Jayabhadhuri
Golden Julie E.
Damodar Reddy Edla
Shyni Emilin
Felix Enigo
Pradeep G.
Raghuraman G.
Shahul Hameed H.
Joe Luis Paul I.
Kavitha J.
Raja Sekar J.
Suresh J.
Femilda Josephin J. S.
Sheeba J. I.
Babu K.
Jayanthi K.
Lekshmi K.
Muneeswaran K.
Saruladha K.
Sarathchandran K. R.
Ramya L.
Anwesha Law
Janaki Meena M.

Saritha M.
Sathyavani M.
Sujaritha M.
Thachayani M.
Thenmozhi M.
Kavitha M. G.
Sandhya M. K.
Arun
Bemesha
Pranab K. Muhuri
Imon Mukherjee
Selvakumar Murugan
Bhalaji N.
Prabhu Ram N.
Radha N.
Sripriya N.
Sujaudeen N.
Venkateswaran N.
Arjun P.
Salini P.
Sanju P.
Vidhya P
Sreeja P. S.
Rajarshi Pal
Lavanya Palani
Latha Parthiban
Ahila Priyadharshini
Deepa R.
Jayabhadhuri R.
Juliana R.
Kanchana R.
Priyadharsini R.
Sandanalakshmi R.
Vidhya R.
Ponmagal R. S.
Reshma Rastogi
Rahul Roy
Swarup Roy
Angel Deborah S.
Chithra S.
Devi Mahalakshmi S.
Ithaya Rani S.
Karthika S.

Kavitha S.	Jaishakthi S. M.
Lakshmipriya S.	Malaikannan Sankarasubbu
Manisha S.	Vijay Sunder Ram
Mohanavalli S.	Debadatta Swain
Padma Priya S.	Sreesharmila T.
Poornima S.	Veena T.
Rajalakshmi S.	Anithaa Shri T. P.
Rajkumar S.	Ananthalakshmi V.
Saraswathi S.	Balasubramanian V.
Sasirekha S.	Nandini V.
Sathya Priya S.	Sundararaman V.
Sheerazuddin S.	Thanikachalam V.
Sumathi S.	Vijayalakshmi V.
Uma Rani S.	T. Veerakumar
Vanitha Sivagami S.	Harold Robinson Y.
Jansi Rani S. V.	Lokeswari Y. V.

Sponsors

Contents

Data Science

Computational Intelligence for Text Analysis

Emotion Recognition in Sentences - A Recurrent Neural Network Approach

N. S. Kumar⬤, Mahathi Amencherla⬤, and Manu George Vimal(✉)⬤

PES University, Bangalore, India
nskumar@pes.edu, mahathiamencherla15@gmail.com, manugeorge004@gmail.com

Abstract. This paper presents a Long short-term memory (LSTM) model to automate the tagging of sentences with a dominant emotion. LSTM is a type of artificial recurrent neural network (RNN) architecture. A labeled corpus of news headlinesc [1] and textual descriptions [3] are used to train our model. The data set is annotated for a set of six basic emotions: Joy, Sadness, Fear, Anger, Disgust and Surprise and also takes into consideration a 7th emotion, Neutral, to adequately represent sentences with incongruous emotions. Our model takes into account that one sentence can represent a conjunct of emotions and resolves all such conflicts to bring out one dominant emotion that the sentence can be categorized into. Furthermore, this can be extended to categorize an entire paragraph into a particular emotion. Our model gives us an accuracy of 85.63% for the prediction of emotion when trained on the above mentioned data set and an accuracy of 91.6% for the prediction of degree of emotion for a sentence. Additionally, every sentence is associated with a degree of the dominant emotion. One can infer that a degree of emotion means the extent of the emphasis of an emotion. Although, more than one sentence conveys the same emotion, the amount of emphasis of the emotion itself can vary depending on the context. This feature of determining the emphasis of an emotion, that is, degree of an emotion, is also taken care of in our model.

Keywords: Sentiment analysis · Sentence level emotion analysis · Natural Language Processing · Recurrent neural networks · Long Short-Term Memory · Emotion recognition · Degree of emotion

1 Introduction

A single distinct element of speech or writing is a word. Every word carries a meaning, a string of words can be combined to form a sentence that is not only meaningful but also expressive and can be associated with a particular state of mind, that is, emotion.

A lot of work has been carried out for the sentiment analysis of text wherein the text is classified into one of the three categories: positive, negative and neutral. This has been mostly achieved by the use of lexical resources such as the

© IFIP International Federation for Information Processing 2020
Published by Springer Nature Switzerland AG 2020
A. Chandrabose et al. (Eds.): ICCIDS 2020, IFIP AICT 578, pp. 3–15, 2020.
https://doi.org/10.1007/978-3-030-63467-4_1

SentiWordNet that assigns each word with a positive, negative or neutral tag. This approach only considers the individual words that constitute the sentence and not the relative positioning of the words amongst themselves, or to the context of the sentence. This process analysis text at a rudimentary level and also does not identify the particular emotion. As a sentence tagged negative encompasses a large variety of emotions such as sorrow, anger and fear and the above methodology does not differentiate amongst them and does not help us to discern the emotion being conveyed in a sentence. Although the sentiment analysis of text has its own relevance as a market analysis tool we opine that we could further benefit by an emotion annotation which has the potential to further strengthen the analysis and increase the effectiveness of these applications.

A similar lexical resource-based approach was taken to achieve emotional analysis of text using the WordNet-Affect List. The WordNet-Affect list contains words annotated with six emotional categories: joy, fear, anger, sadness, disgust and surprise. This too suffered a similar drawback wherein the context of the sentence and the overall situation expressed in the sentence were not taken into consideration.

The emotion in a sentence need not necessarily be conveyed plainly, with the use of explicit words. For example, "I am happy I passed my exams" can clearly be identified as a sentence that expresses "Joy" due to the presence of the word, "appy". Whereas the sentence, "I finally passed my exam" although does not contain any words that express a single emotion explicitly; it can clearly be categorized as a sentence that conveys "Joy". This information was not extracted from the sentence based on the cue provided by a single word but rather by analyzing the situation expressed in the sentence through ones own past experience. Therefore, it can clearly be observed that the emotion of a sentence cannot be determined merely based on the words that form it, but also by analysing the underlying context evinced by a sentence. Furthermore, the extent to which an emotion is expressed can be understood by associating each emotion with a degree. We have given each emotion four levels characterized by one of the following values, 0.25, 0.50, 0.75 and 1; 0.25 being the lowest degree and 1 being the highest. Using these values, our model predicts the degree of emotion.

Neural Networks have been employed to widely and achieved fairly significant results in the field of Natural Language Processing, such as translation, text summarization etc. A Recurrent Neural Network (RNN) is a type of Neural Network that remembers its previous states. But, a significant disadvantage of RNN is that it works only on a short-term basis. Meaning, when context comes into play, the desired outcome is not achieved. This issue has been resolved by the Long Short-Term Memory (LSTM) Neural Network. A typical/vanilla RNN modifies the entire information as a whole and does not take into consideration important and non-important information. Whereas, LSTMs have a mechanism known as cell states in which information is modified with multiplications and additions. Hence, LSTM is used to predict emotion in sentences because it can selectively retrieve or drop information.

2 Data Sets

2.1 SemEval 2007 Affect Sensing Corpus

The SemEval 2007 Affect Sensing Corpus data set [1] contains 1250 news headings along with a list of their corresponding emotions from the set of emotions-Joy, Sadness, Fear, Anger, Disgust, and Surprise. This data set is considered to be our gold standard because it not only complies with Ekman's basic emotion model [2] but also has been annotated previously by humans. An advantage of this data set is that each sentence also comes with a valence/degree of emotion underlying in the sentence. Sentences in this data set have contextual meaning rather than an obvious emotion. For example, the sentence "Resolution approved for international games" is tagged as "Joy". But, this data set is skewed in the sense that sentences tagged as "Surprise" are very less compared to sentences tagged as "Joy". Hence, all emotions do not have equal number of sentences. In addition, each sentence is annotated with a degree varying between 0 and 1 for each of the six emotions, with 0 stating that the sentence does not carry the particular emotion and 1 stating that it definitely carries the particular emotion.

2.2 ISEAR

The ISEAR data set [3] contains 7666 situational sentences with a list of their corresponding emotions from the set of emotions- Joy, Sadness, Fear, Anger, Disgust, Shame and Guilt. For example, the sentence "When I pass an examination which I did not think I did well" is tagged as "Joy". Unlike the above mentioned data set, this data set is not skewed. All seven of these emotions have almost equal number of sentence examples. Having a symmetrical data set is important so as to have a uniform feature extraction in our Neural Network (LSTM). In addition to the Ekman's basic emotion model [2], where the emotions are Joy, Sadness, Fear, Anger, Disgust and Surprise, this data set has two emotions-Shame and Guilt. However, for our model, we decided against using these two emotions because they are not included in Ekman's basic emotion model [2]. Ekman proposed that only these six emotions are strongly expressed across all the cultures in the world. Since these sentences are based on situations, emotion tagging can be done based on context as well.

3 Literature Review

This section covers an extensive survey on the approaches previously used to detect the emotion of a sentence.

Dipankar Das and Shivaji Bandyopadhyay [4] employ a machine learning approach using Conditional Random Field (CRF). Their method is a two step process wherein, the first step is to create an emotion for each word in a sentence using WordNet's Affect List and the next step is to find the dominant emotion of each sentence using weight scores of each word in the sentence. The first step uses

CRF for word level annotation which searches for words in the sentence that are present in the Affect list and returns word level tagging of emotions. The second step is done using these word level emotions to find the overall sentence level emotion based on weight scores of each word in a sentence. Although this paper produces a high accuracy of 87.65%, it is limited by the Synonyms set (SynSet) in the Affect list. This model fails to account for words that do not exist in the said list. For example, "Serena misses Bangalore tournament" is a sentence in the SemEval 2007 data set [1]. If this model were to find the emotion of this sentence, it would fail to do so because the word "misses" indicates Sadness in this context but Sadness Affect list does not contain this word.

The authors in Paper [5] do an experimental analysis on different approaches to tackle the problem of sentence level emotion tagging. They use five approaches, which consist of four knowledge-based ideas and one corpus-based idea. The first approach in knowledge-based ideas, WordNet-Affect Presence, deals with annotating the emotions in a text simply based on the presence of words from the Word-Net Affect lexicon. The second approach, Latent Sentiment Analysis (LSA), the LSA similarity between the given text and each emotion is calculated and each emotion is defined as a vector of the word expressing that particular emotion. The third method, LSA emotion SynSet, is a method in which in addition to the word denoting an emotion, its synonyms from the WordNet SynSet are also used. Finally, the fourth method, LSA all emotion words, which aids the previous set by adding the words in all the SynSets labeled with a given emotion, as found in WordNet Affect List. What can be inferred from this paper [5], is that knowledge-based approaches are constrained by unknown words. Moreover, the corpus-based approach used in this paper [5] uses a machine learning classifier, Naive Bayes that is trained on Blog Posts to classify emotion in a labelled data set. This approach is a more practical approach and has been employed in a similar fashion in our model.

In paper [6], the authors discuss a method to find emotion labels using two methods: keyword spotting and lexical affinity. These methods use a ready-made lexical corpus to find words pertaining to a particular emotion of the above stated set of emotions. This paper [6] does not take into consideration negation words like not, neither, and never, which can give the polarizing emotion of the sentence. However, it tags an emotion to sentence based on the context rather than the word level emotion weights. Their process assigns an emotion to the sentence by weighing the relation between the different words and emotions present in it. Each word is looked up in the lists of emotional words (LEW) and if found, its value is obtained. Else, the ANEW list (Bradley and Lang 1999) is consulted, and if found, its value is obtained. If the words are not found in either of the lists, the hypernyms of the word are retrieved from WordNet. The model for the above example, "Serena misses Bangalore tournament" successfully classifies it as "Sadness".

The model built by the authors in paper [7] employs a rule-based approach for the classification of the sentences with a particular emotion. A data set consisting of user reviews is used. A particularly novel approach explored in this paper is to improve the performance of the categorization of the sentences by including additional emotion-related signals, such as emoticons, emotion words, polarity sifters, slang and negations, to detect and classify emotions efficiently in user reviews. One possible limitation is that it cannot incorporate domain-specific words because it is dependant on SentiWordNet. Another disadvantage of this model, is the fact that a rule-based approach requires an extensive research and possible linguistic experts.

The authors in paper [8] discuss Disgust. They find relations between the words used in the sentence to the emotion depicted by it and further the deeper contextual meaning imparted by each of the words to the sentence. Once these terms and the relations are found, they are generalized to construct a set of rules knows as emotion recognition rules (ERRs). These rules are then used to recognize the emotion from any given text. However a possible drawback is that they use WordNet and ConceptNet to generalize their lexicon, hence any words not present in either of them would not be identified and classified and rules relating to them would not be constructed.

4 Model Architecture

4.1 Hybrid Data Set

As discussed in Sect. 2, the two aforementioned data sets have their own pros and cons. An amalgam of the two data sets was used to produce a hybrid data set that is uniform and has equal sentences of each emotion. The data set now not only consists of news headlines but also situational descriptions, which enables the model to have an all round robust learning. The SemEval data set [2], which has sentences like "Bombers kill shoppers", does not train the model well because its sentences are vague. Therefore, combining it with the ISEAR data set [3] allows the model to understand context and implement emotion extraction and therefore, contextual learning as well as word level feature extraction is achieved. The advantage of the proposed Hybrid Data set is twofold- one being data from Semeval data set and ISEAR data set were chosen such that the final number of sentences in each category of emotion is the same thereby tackling the issue of skewedness; the other being, since now the data set consists of single sentence news headlines [1] which allow learning at the word level and multiple sentence situational descriptions [3] which allow learning on a contextual level. This data set was split into training and testing data, where 80% was used in training and 20% was used in testing.

4.2 Pre-processing

The first step before any work can be carried out is to pre-process the data to bring about uniformity. Since our data set contains sentences ranging from news headlines to situational descriptions, it is essential to conform these various sentences to a single template. The quality of data directly affects any models' ability to learn. Hence, we convert all the text to lower case and remove punctuation marks such as colons and quotations. This also provides uniformity in one hot encoding of the words.

4.3   Sentence Splitter

The next step is to check every sample for conjunctions, commas and full stops and split it accordingly. The common conjunctions include and, but, or, nor, for, so and yet. When a sample sentence is broken down into its several pieces, each piece can carry an emotion and hence, each emotion of each piece contributes to the overall emotion of the whole sentence. Taking the example, "I am sad that I didn't study for my exam but I am happy I passed", it can be seen that when the sentence is split into "I am sad that I didn't study for my exam" and "I am happy I passed" based on the conjunction "but", each of these pieces carry a different emotion, namely, Sadness and Joy. Therefore, this influences the final emotion of the entire sentence. The method used for this, is to check for the presence of a conjunction, comma (,) or full stop (.) and send these individual sentences to the subsequent step.

4.4   Obtaining Emotion Words

A very important aspect of emotion detection is the Part Of Speech (POS) tag. It can be concluded that most words which are either Nouns, Adjectives, Verbs or Adverbs have an underlying emotion. The sentence, "When I began school at UC, the pre-enrollment, the classes and the question of success scared me" has many words like the and when that do not contribute to the overall emotion of the sentence. It can be inferred that the words with a POS tag of Nouns, Adjectives, Verbs or Adverbs aids the sentence's emotion. In the above case, "question", "success" and most importantly, "scared" contributes to the overall emotion of "Fear" to the sentence. Hence, we decided to drop all words which have POS tags of anything other than the ones mentioned above.

4.5   Model

Prediction of emotion and degree is carried out by the use of a Long Short-Term Memory (LSTM) model. In the LSTM architecture there are three gates, forget gate, input gate and an output gate and a memory cell. The forget gate decides which information should be discarded and which should be retained. The input gate handles the addition of relevant information to the cell state. The output gate deals with selecting useful information from the current cell state of all the available information to show as an output. LSTM in its core, preserves information from inputs that have already passed through it using the hidden state. More formally each gate can be modelled as follows (Fig. 1)

$$i_t = \sigma(w_i[h_{t-1}, x_t] + b_i) \tag{1}$$

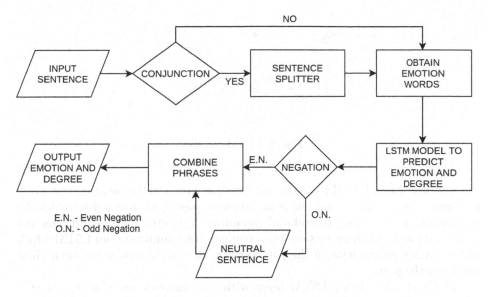

Fig. 1. System architecture

$$f_t = \sigma(w_f[h_{t-1}, x_t] + b_f) \tag{2}$$

$$o_t = \sigma(w_o[h_{t-1}, x_t] + b_o) \tag{3}$$

where,

i_t represents input gate, f_t represents forget gate, o_t represents output gate, σ represents sigmoid function, w_x represents weight of the neurons for the respective gate(x), h_{t-1} represents output of LSTM block at previous timestamp, x_t represents input at current timestamp, b_x represents biases for the respective gate(x).

Equations for the cell state, candidate state and the final output are as follows

$$c'_t = tanh(w_c[h_{t-1}, x_t] + b_c) \tag{4}$$

$$c_t = f_t * c_{t-1} + i_t * c'_t \tag{5}$$

$$h_t = o_t * tanh(c^t) \tag{6}$$

where,

c_t represents the cell state at current timestamp, c'_t represents the candidate for the cell state at current timestamp, h_t output after passing through the activation function.

The cell state considers what information it needs to discard from the previous state $(f_t * c_{t-1})$ and what it needs to include from the current candidate state $(i_t * c'_t)$. Finally the state is passed through the activation function to obtain h_t which is the output of the current LSTM at timestamp t. This is then passed through the last softmax layer to get the predicted output (Fig. 2).

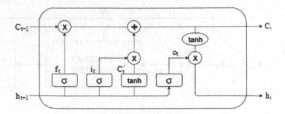

Fig. 2. LSTM block

A bidirectional LSTM has been employed, so as to run the inputs in two ways, one from past to future and one from future to past. This method proved to be advantageous by relating to and understanding the context better, and showed a stark improvement in prediction when compared to unidirectional LSTM which only retained information of the past because the only inputs it has been given are from the past.

The first layer is an LSTM layer with 100 memory units which receives sequences of data rather than scattered data. A dropout layer is implemented to prevent overfitting of the model. The last layer is a fully connected layer with a softmax activation function.

The model has been trained using back propagation and the objective loss function used is the cross-entropy loss. The model is fit over 20 epochs with a batch size of 64.

The high accuracy attained by our model compared to standard machine learning and knowledge-based approaches can be attributed to the sequential learning nature of LSTM and its ability to discard and retain information to maximize output accuracy. In the emotion classification task, the input words are processed by LSTM networks sequentially and the last output of the LSTM represents the meaning of sentence, which is finally used to predict the emotion.

4.6   Post-processing

After categorizing the sentence under one emotion, the penultimate step is to check for the presence of negation words in the sentence. Given a sentence, "I am not happy", we see that the sentence is classified as Joy due to the presence of the word "happy". However, we know that this is not the case as the occurrence of "not" before the "happy" essentially means the opposite of happy. Therefore, it is imperative to check for negation words. Keeping in mind that most of the emotions of a text are subjective to the reader, if a sentence contains odd number of negation words then the predicted emotion is changed to a seventh emotion- Neutral because it is to be noted that, "not happy" does not directly correlate to another emotion in the Ekman's model [2] but rather depends on context again. The sentence "I am not happy" could imply disappointment, anger or sadness. The Neutral emotion is used to tackle such conflicts by making the overall emotion of a sentence neutral. But, if a sentence contains a double negation, such as, "It is not that I'm not happy", it is implied that the negations cancel each other out, keeping the original emotion of the sentence intact. Hence, if a sentence has even number of negation words, the original emotion is not replaced with the Neutral emotion. If multiple sentences carrying varying set of emotions is considered, for example, "I am happy I reached the flight on time. But I am sad I spent a lot of money to go to the airport", we tackle the emotion of the sentences individually first and use Table 1 to get the overall emotion of the set of sentences. In the example, sentence 1 is Joy and sentence 2 is Sadness, so the overall emotion is Neutral. In general, if a set of sentences has conflicting emotions, the overall emotion for the set of sentences is categorized as Neutral as shown in Table 1.

Conflicting emotions are resolved as a post processing step and not within the prediction model itself because of two main reasons. Firstly, the data set does not include sentences tagged with a neutral emotion. So, the LSTM model does not learn what makes a sentence as neutral whereas since the data set has six other emotions, the model is able to learn what each of those emotions' mean. Secondly, as mentioned above if the sentence contains an even number of negation words then the sentence is left intact with its predicted emotion, only in the case wherein the sentence contains an odd number of negation words, it is classified as neutral. Therefore, it is essential that the sentence is first classified into one of Ekman's six emotions before being checked for negation, so that in the event of an even negation the sentence can be left untouched and classified under its predicted emotion.

Table 1. Dominant emotion

Emotion	Joy	Sadness	Fear	Anger	Disgust	Surprise	Neutral
Joy	Joy	Neutral	Neutral	Neutral	Neutral	Surprise	Joy
Sadness	Neutral	Sadness	Fear	Anger	Disgust	Neutral	Sadness
Fear	Neutral	Fear	Fear	Anger	Fear	Fear	Fear
Anger	Neutral	Anger	Anger	Anger	Anger	Anger	Anger
Disgust	Neutral	Disgust	Fear	Anger	Disgust	Surprise	Disgust
Surprise	Joy	Neutral	Fear	Anger	Surprise	Surprise	Surprise

If a sentence is found to be composed of multiple phrases joined by a conjunction, then each phrase is classified with a particular emotion. Since we require the entire sentence to be tagged with a single emotion, it is essential to merge the emotions of the phrases to produce a single dominant emotion. To achieve such a result, we have come up with a method to resolve ties between emotions. Table 1 shows the resulting emotion on combining two incongruous emotions. This table was obtained on careful observation and analysis of different conjunct sentences that expressed two or more contrary emotions.

Another interesting feature has been considered in our model known as degree of an emotion. Along with the dominant emotion, a sentence will have the extent of the emotion conveyed as well. The difference in the following sentences, "I am happy I passed my exam", "I am very happy I passed my exam" and "I'm ecstatic that I passed my exam!" is apparent in the sense that each sentence conveys varying degrees of the emotion "Joy". So, the sequence is least to greatest degree of emotion in the above three sentences. Degree of every emotion has been categorised into four values, 0.25, 0.50, 0.75, 1.0 where the greatest value of 1.0 indicates highest degree and least value of 0.25 indicates the lowest degree. The LSTM model successfully categorised every sentence of the data set into one of these four categories. The SemEval 2007 data set [1] comes with a list of valences/degrees for every emotion and consequently, every sentence. To begin with, the valences have been rounded off to the aforementioned categories of degree. This was used to train the Neural Network (LSTM) which achieved an accuracy of 91.6%. Examples from the tested data set are as follows-"Mother and daughter stabbed near school" received a degree of 1.0 of the Sadness emotion, "Hurricane Paul nears Category 3 status" received a degree of 0.75 of the Fear emotion, "UK announces immigration restrictions." received a degree of 0.50 of the Anger emotion and "Messi makes Barcelona squad return" received a degree of 0.25 of the Joy emotion.

5   Evaluation and Results

A comprehensive analysis has been conducted on the following data sets:- SemEval 2007 data set [1], ISEAR data set [3] and a hybrid of the previous two, to produce emotions. Four metrics are used to compare the model's performance on these data sets, namely, Accuracy, Precision, Recall and F1 score.

Accuracy can be given by,

$$(TP + TN)/(TP + FP + FN + TN) \tag{7}$$

Precision can be given by,

$$TP/(TP + FP) \tag{8}$$

Recall can be given by,

$$TP/(TP + FN) \tag{9}$$

F1 Score can be given by,

$$2 * (Recall * Precision)/(Recall + Precision) \tag{10}$$

where TP is number of true positives, TN is number of true negatives, FP is number of false positives and FN is the number of false negatives.

The following metrics have been tabulated and displayed in Table 2. It is inferred from the table that the combination of the SemEval 2007 data set and ISEAR data set improves the overall working of the LSTM model because it predicts not only based on situations but also on context. Table 3 shows how our deep learning model fairs with the state of the art classical models: Naïve Bayes

Table 2. Results

Data set	Accuracy	Precision	Recall	F1 score
SemEval	68.45%	0.681	0.459	0.548
ISEAR	63.74%	0.729	0.356	0.478
Hybrid	85.63%	0.752	0.524	0.618

Table 3. Model comparison

Model	SemEval	ISEAR	Hybrid
Deep learning (ours)	68.45%	63.74%	85.63%
Naïve Bayes	63.16%	67.72%	71.23%
Knowledge Based Technique	51.13%	53.6%	56.18%

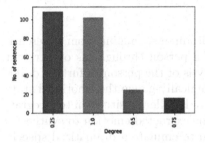

Fig. 3. Annotated degree   Fig. 4. Model degree

Classifier and Knowledge Based Technique, in terms of accuracy. It is evident that the model we have built performs more precisely than the classical models.

Additionally, the degree of emotion was predicted for each emotion and had an accuracy of 91.6%. Figure 3 is a bar graph which indicates the data set's annotated degree and Fig. 4 is a bar graph that indicates our model's predicted degree. Taking a few examples, "I am happy" is given a degree of 0.25 and "I am very happy" is given a degree of 0.50.

6 Conclusion

In this paper three novelties are presented. The first is that a sentence can be a conjunct of emotions and it has to be amicably resolved. For example, in the sentence "I passed the examination but I am disappointed by my below par performance", although clearing the exam makes the speaker happy the general expression is that of sorrow as the speaker is disappointed with his/her performance.

The second is that, not all sentences can be classified into an emotion and some may express a neutral emotion. For example, "The sun sets in the west" is not a sentence that bursts with a particular emotion but is a rather tame expression that just declares a fact in a neutral tone.

The third novelty is that our model also takes into consideration that a sentence belonging to a particular emotion can show varying levels of the emotion. For example, anger has two sub categories- hot anger and cold anger; where hot anger can be of the form rage and cold anger can be of the form passive aggressiveness. This is expressed by an additional output, that is, degree of the emotion that describes the extent of the emotion. The output degree has a set of four discrete values (0.25, 0.50, 0.75, 1).

This we conclude that our deep learning model fairs better than other state of the art classical models such as the Naïve Bayes Classifier and Knowledge Based Technique, and gives us a maximum accuracy of 85.63% to predict the emotion of text and an accuracy of 91.6% to predict the degree of the sentence.

7   Future Work

The application of such an analysis is multitudinous, ranging from differentiating the mental and psychological state of a person through his or her text messages to helping in the psychological analysis of the person to further developing an empathetic chat bot, that can dynamically gauge the emotional state of the customer/user and be better equipped to handle the situation to integrating a text-to-speech engine that can utilize the extracted emotion to synthesize speech that can incorporate the said emotion to emulate a dramatized speech. This dramatized speech can be used to generate more realistic audio books with varying emotion so as to mimic a human tone.

Furthermore, emotion recognition has helped opinion mining in social media of various reviews and political discussions progress one step further than sentiment analysis. Now, with the help of varying degree of emotion, progress can be made further on classifying what opinions social media users have and digitally analysing these opinions.

Our model has been developed which can effectively categorize sentences into one of the six basic emotions (Joy, Sadness, Fear, Anger, Disgust, Surprise) and an additional emotion category: Neutral. This additional information obtained can be passed along with the sentence to construct emotional speech. The model can also be expanded to include more emotions other than the above mentioned six. Rather than representing degree of an emotion by a set of discrete values, fine grained values, that is a continuous set of values that gives a greater control over the degree [0, 100] too should be explored to further strengthen the emotional analysis of text.

References

1. Strapparava, C., Mihalcea, R.: SemEval-2007 task 14: affective text. In: Proceedings of the Fourth International Workshop on Semantic Evaluations (SemEval-2007), pp. 70–74 (2007)

2. Ekman, P.: An argument for basic emotions. Cogn. Emot. **6**(3–4), 169–200 (1992)
3. International Survey on Emotion Antecedents and Reactions data set. https://www. unige.ch/cisa/index.php/download_file/view/395/296/
4. Das, D., Bandyopadhyay, S.: Sentence level emotion tagging. In: 2009 3rd International Conference on Affective Computing and Intelligent Interaction and Workshops, pp. 1–6. IEEE (2009)
5. Strapparava, C., Mihalcea, R.: Learning to identify emotions in text. In: Proceedings of the 2008 ACM Symposium on Applied Computing, pp. 1556–1560 (2008)
6. Francisco, V., Gervás, P.: Exploring the compositionality of emotions in text: word emotions, sentence emotions and automated tagging. In: AAAI-06 Workshop on Computational Aesthetics: Artificial Intelligence Approaches to Beauty and Happiness (2006)
7. Asghar, M.Z., Khan, A., Bibi, A., Kundi, F.M., Ahmad, H.: Sentence-level emotion detection framework using rule-based classification. Cogn. Comput. **9**(6), 868–894 (2017)
8. Shaheen, S., El-Hajj, W., Hajj, H., Elbassuoni, S.: Emotion recognition from text based on automatically generated rules. In: IEEE International Conference on Data Mining Workshop, pp. 383–392 (2014)

Context Aware Contrastive Opinion Summarization

S. K. Lavanya[1(✉)] and B. Parvathavarthini[2]

[1] Jerusalem College of Engineering, Chennai, India
sklavanyasambath@gmail.com
[2] St. Joseph's College of Engineering, Chennai, India
parvathavarthini@gmail.com

Abstract. Model-based approaches for context-sensitive contrastive summarization depend on hand-crafted features for producing a summary. Deriving these hand-crafted features using machine learning algorithms is computationally expensive. This paper presents a deep learning approach to provide an end-to-end solution for context-sensitive contrastive summarization. A hierarchical attention model referred to as Contextual Sentiment LSTM (CSLSTM) is proposed to automatically learn the representations of context, feature and opinion words present in review documents of each entity. The resultant document context vector is a high-level representation of the document. It is used as a feature for context-sensitive classification and summarization. Given a set of summaries from positive class and a negative class of two entities, the summaries which have high contrastive score are identified and presented as context-sensitive contrastive summaries. Experimental results on restaurant dataset show that the proposed model achieves better performance than the baseline models.

Keywords: Opinion summarization · Attention model · Pointer-generator network · Context-aware sentiment classification

1 Introduction

Accumulating opinions from review documents is termed opinion summarization. There are two types of opinion summarization techniques: extraction- and abstraction-based summarization. A summary is developed by selecting certain prominent input sentences in extraction-based techniques. Abstractive summarization techniques produce novel summaries by rephrasing sentences which are rather abstractive in nature. A feature-based summarization is an abstraction-based method that presents the opinion distribution of each feature separately and helps users to make better decisions.

Though different summarization techniques have been proposed, there is still a need for summarizing different opinions about a particular feature of different products. Contrastive opinion summarization methods produce a summary consisting of a set of contrastive sentence pairs by choosing the most representative

© IFIP International Federation for Information Processing 2020
Published by Springer Nature Switzerland AG 2020
A. Chandrabose et al. (Eds.): ICCIDS 2020, IFIP AICT 578, pp. 16–29, 2020.
https://doi.org/10.1007/978-3-030-63467-4_2

and comparable sentences from two sets of positively and negatively opinionated sentences. Such a summary helps users easily interpret contrastive opinions
about a product or service. Most of the present works on contrastive summarization select contrastive sentences only based on the explicit opinion words
present. The context information is also necessary to understand the implicit
opinion present in the sentences. For example, the sentence "city gas mileage
is horrible", does not convey a negative sentiment always. It conveys that for
city traffic, mileage is not good. It would be helpful for the users if the context
(reason) is also considered while producing summaries.

Context refers to the circumstances in terms of which an event or entity
can be fully understood. Text data is associated with rich context information.
Meaning of an unknown word can be guessed by looking at the context words
surrounding that word [7]. Deriving these context words using a machine learning
algorithm is computationally expensive [3,6]. Model-based approaches rely on
hand-crafted features produced by machine learning algorithm for producing
a contrastive summary. In this work, an end-to-end deep learning framework is
proposed for the problem of context-sensitive contrastive opinion summarization.

2 Related Work

The emergence of deep neural networks has motivated the researchers to model
the abstractive summarization task using different network architectures. A
sequence-to-sequence attention model [16] generates a short summary conditioned on the input sentence. A novel convolutional attention-based decoder [5]
generates a summary by focusing on the appropriate input words while preserv-
ing the meaning of the sentence. A Feature-rich encoder [14] captures linguistic
features in addition to the word embeddings of the input document. To address
the problem of out-of-vocabulary words, pointer-generator model was used by
the decoder.

A general single-document summarization framework [4] applies attention
directly to select sentences or words similar to pointer networks for producing a
summary. A read-again mechanism based model [24] computes the representation
of each word by taking into account the entire sentence not just considering
the history of words before the target word. A hierarchical document encoder
and an attention-based extractor model [10] incorporated the latent structure
information of summaries to improve the quality of summaries generated using
abstractive summarization.

A selective encoding model [26] selects the encoded information using selective gate from encoder to improve encoding effectiveness and control the information flow to the decoder. A graph-based attention neural model [19] discovers
the salient information of a document. In addition to the saliency problem, flu-
ency, non-redundancy and information correctness are addressed using a unified
framework. A hierarchical model [11] improves text summarization using significant supervision by the sentiment classifier. The summarizer captures the
sentiment orientation of the text thus improving the coherence of the generated summary. An end-to-end framework [25] integrates the sentence selection

strategy with the scoring model to predict the importance of sentence based on the previously selected sentences. The keywords extracted from the text using an extractive model is encoded using a Key Information Guide Network (KIGN) [9]. These keywords guides the decoder in summary generation. Thus the method combines the advantage of extractive and abstractive summarization.

A multi-layered attentional peephole convolution LSTM network [15] automatically generates summary from a long text. Instead of traditional LSTM, convolution LSTM was used which allows access to the content of previous memory cell through a peephole connection. An LSTM-CNN based model [18] produces abstractive text summarization by taking semantic phrases rather than words as input. The model was able to produce natural sentences as output summary.

Most of the existing techniques for summary generation differ in network structure, inference mechanism, and decoding method. In this paper, a neural network model has been proposed for improving the contrastive summary generation process by selecting and encoding essential keywords.

3   Proposed Framework

The proposed context-sensitive contrastive summary model is a sequence-to-sequence model with attention and a pointer-generator mechanism. The input of the proposed model is documents of entities taken for comparison. These are a sequence of words passed to an attentive encoder, Contextual Sentiment LSTM Encoder. The encoder's outputs are then passed to an attentive decoder which generates a summary using pointer-generator mechanism. The document summaries are then passed to a multi-layer perceptron which computes a contrastive score. The summaries which have high contrastive score are then presented as contrastive summaries. The architecture of the proposed context-sensitive contrastive summary model is shown in Fig. 1. The components and variables in the figure are explained in detail in the forthcoming section.

3.1   Contextual Sentiment LSTM Encoder

An attention-based Long short-term memory (LSTM) model takes a sequence of words as input and produces a probability distribution quantifying the importance of the target word in each position of the input sequence. An attentive network, the Sentic LSTM was proposed in [12] which adds common sense knowledge concepts from the Senticnet through the recall gate to control information flow in the network based on the target. A Coupled Multi-layer Attention network for the automatic co-extraction of feature and opinion words was proposed in [20]. The authors furnished an end-to-end solution by providing a couple of attention with a tensor operator in each layer of the network.

The proposed Contextual Sentiment LSTM model is a hierarchical attention model which automatically learns the representation of context, feature and opinion words present in review sentences. The attention mechanism of the proposed Contextual Sentiment LSTM model is shown in Fig. 2.

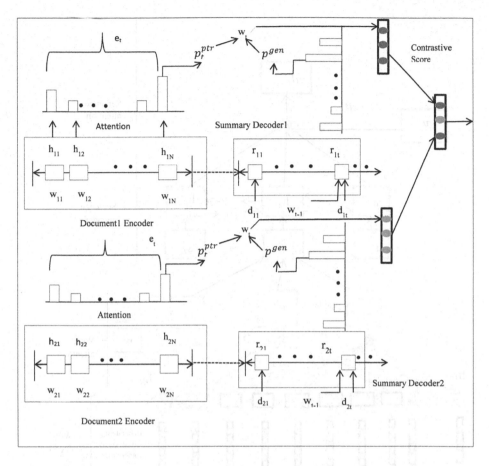

Fig. 1. Architecture of context-sensitive contrastive summary model.

It consists of a word encoder, a word attention, a sentence encoder, and a sentence attention layer. Word embeddings are given as input to the word encoder in order to avoid the assignment of random weights to the input words during training. The dependency-based word embeddings proposed in [8] is used to generate word embeddings corresponding to the input words. Dependency-based word embeddings are generated based on dependency relation between words and hence best suitable for the problem of opinion summarization.

The word encoder transforms the dependency-based word embeddings into a sequence of hidden outputs. Multi-factor attention is built on top of the hidden outputs: feature words attention, opinion words attention and context words attention. Each attention takes as input the hidden outputs found at the position of the feature, opinion or context words and computes an attention weight over these words by finding the similarity between the hidden vectors and vectors from knowledge bases using a bilinear operator. A sentence context vector is

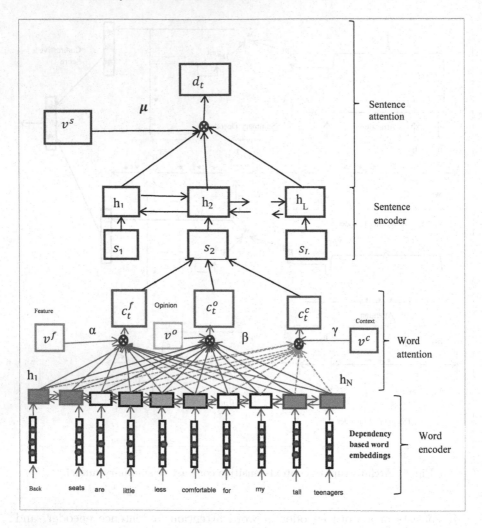

Fig. 2. Attention mechanism of the proposed Contextual Sentiment LSTM Model

generated by aggregating the context vectors of feature, opinion and context words produced using attention mechanism.

The document encoder transforms the sentence vectors into a sequence of hidden outputs. A document context vector is generated using sentence level attention. The document context vector captures the important information present in the input document.

4   Context-Sensitive Contrastive Summary Model

Given a review document, a Skip-Gram model generates dependency-based word embeddings for each word in the document. The dependency-based word embeddings are given as input to the bi-directional LSTM. The sequence of hidden outputs is produced by concatenation of both the forward and backward hidden outputs. The information flow at the current time step is controlled by input, output and forget gates of the LSTM cell as given by Eq. (1–6)

$$i_t = \sigma(W_i[h_{t-1}; w_t] + b_i) \tag{1}$$

$$f_t = \sigma(W_f[h_{t-1}; w_t] + b_f) \tag{2}$$

$$o_t = \sigma(W_o[h_{t-1}; w_t] + b_o) \tag{3}$$

$$\tilde{c}_t = \tanh(W_c[h_{t-1}; w_t] + b_c) \tag{4}$$

$$c_t = f_t * c_{t-1} + i_t * \tilde{c}_t \tag{5}$$

$$h_t = o_t * \tanh(c_t) \tag{6}$$

where i_t, f_t, o_t and c_t are the input gate, forget gate, output gate and memory cell respectively. W_i, W_f and W_o are the weight matrices, b_i, b_f, b_o and b_c are the bias for the input gate, forget gate, output gate and memory cell respectively, and w_t is the word vector for the current word. The sequence of hidden outputs produced is denoted by H = $\{h_1, h_2, ..., h_N\}$, where each hidden element is a concatenation of both the forward and backward hidden outputs.

4.1   Word Attention

It is essential to learn the embedding vector for each feature, opinion and context word present in a sentence for the problem of context-sensitive summarization. Hence three attentions: Feature attention, opinion attention and context attention are explored.

The ConceptNetNumberbatch embeddings proposed in [19] is used to guide the network to attend to the feature words present in a document for feature attention. The AffectiveSpace embeddings proposed in [2] is used as a source of attention for opinion words. The context hints gathered from different user surveys are utilized to build attention source vector for context words.

$$c_t^f = \sum \alpha_t h_t \tag{7}$$

$$c_t^o = \sum \beta_t h_t \tag{8}$$

$$c_t^c = \sum \gamma_t h_t \tag{9}$$

The sentence context vector is acquired by fusing together the vectors of feature, opinion and context. The context vector captures the most important features of a sentence with respect to feature, opinion and context words.

$$s_t = (c_t^f + c_t^o + c_t^c) \tag{10}$$

4.2 Sentence Attention

Sentence level attention guides the network to attend the most important sentences in a document that serve as clues for sentiment classification. A sentence attention weight is calculated by introducing a sentence attention source vector.

$$d_t = \sum \mu_t h_t \tag{11}$$

4.3 Context-Sensitive Sentiment Classification

The document context vector is fed into a unidirectional LSTM which converts it into a vector whose length is equivalent to the quantity of classes {positive, negative}. The softmax layer changes the vector to a conditional probability distribution.

$$p_d^r = softmax(W^r d_t + b_d^r) \tag{12}$$

4.4 Context-Sensitive Summary Generation

For constructing a context-sensitive summary of the input document a unidirectional LSTM is used as a decoder. The hidden state of the decoder is updated using the document context vector and the previous decoder state.

$$r_t = LSTM(r_{t-1}, d_t, w_{t-1}) \tag{13}$$

The pointer-generator network proposed in [17] is adopted on top of LSTM for generating summaries. The pointer-generator network generates summary by copying important words from the input document via pointing and generating words from a fixed vocabulary.

At each time step, the generator selects and outputs an important word from a distribution of words computed through a softmax layer. The pointer network directly copies a word based on attention mechanism from the input document to the output summary.

$$p(w_t) = p_t^{ptr} p^{gen}(w_t) + (1 - p_t^{ptr}) \sum_i d_{t,i} \tag{14}$$

4.5 Context-Sensitive Contrastive Summary Generation

Given a set of summaries from two entities, the summaries which have high contrastive score are identified and presented as contrastive summaries. To compute contrastive score, the element-wise product of summary vectors is given as input to a Multi-layer perceptron.

$$Score_{i,j} = \pi_{i=1}^N \pi_{j=1}^N sigmoid(MLP([s_i s_j]; \theta)) \tag{15}$$

5   Experimental Results

Experiments are conducted on SemEval 2014 task 4 restaurant reviews dataset in [13]. The restaurant dataset consists of 1,125,457 documents with a vocabulary size of 476,191. Since the restaurant dataset is large, a sample of 50,000 documents is taken for experimentation to reduce the training time of the model. The sentences are annotated with feature, opinion and context labels for word attention as given in Table 1. For context-sensitive sentiment classification, the feature words are annotated with their corresponding polarities as shown in Table 2. The statistics of the datasets used for context-sensitive summary generation is shown in Table 3. Initially, sentences are tokenized using sentiment-aware tokenizer and words that occur less than 100 times are filtered. All the word vectors are initialized by Dependency-based word embeddings in [8]. The dimension of word embeddings is set to 300. The dependency-based word embeddings capture arbitrary contexts based on dependency relation between words which helps the proposed CSLSTM model to learn better weights.

The dependency-based word embeddings are converted into hidden representations through CSLSTM model implemented with Theano library stated in [1]. The number of hidden units is set to 50. For training, a batch size of 30 samples

Table 1. Dataset description for word attention.

Training	Testing	Total
3,041	800	3,841

Table 2. Feature distribution per sentiment class.

Features	Positive		Negative	
	Train	Test	Train	Test
Food	867	302	209	69
Price	179	51	115	28
Service	324	101	218	63
Ambience	263	76	98	21
Others	546	127	199	41
Total	2179	657	839	222

Table 3. Statistics of the dataset for a context-sensitive summary generation.

Statistics	Training	Testing
Number of documents	45,000	5000
Number of document-summary pairs	283,326	11,360
Maximum document length	90	90
Maximum summary length	14	14

was used. AdaGrad optimizer is used for minimizing the training loss with a learning rate of 0.01.

5.1 Context-Sensitive Sentiment Classification Using Soft-Max Classifier

The sentiment classifier takes the feature, opinion and context words identified using CSLSTM as input and determines the polarity of each feature word. From the sample sentence shown in Fig. 3, it can be seen that the classifier detects a multi-word phrase "delivery times" as a feature with the help of feature attention source "service" and determines its polarity as "positive" based on context "city" detected using context attention source. The classifier predicts the correct polarity of a feature even if the sentence is long and has a complicated structure.

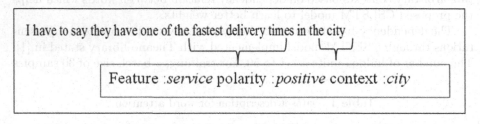

Fig. 3. Context-sensitive classification of sample sentences from restaurant dataset.

5.2 Context-Sensitive Summary Generation Using Pointer-Generator Network

For the pointer-generator network, the generator vocabulary size is set to 20k based on most attended words using an attention mechanism. A sample context-sensitive summary generated for restaurant dataset is shown in Table 4. It can be seen that the model repeats some of the phrases in the reference summary. Rarely the model generates summary consisting of words that is not truthful with respect to the input document.

Table 4. Sample context-sensitive summary generated using pointer-generator network for restaurant dataset.

DocumentAll the appetizers and salads were fabulous, the steak was mouth watering and the pasta was delicious!!! The sweet lassi was excellent as was the lamb chettinad and the garlic naan but the rasamalai was forgettable..... **Gold summary**Food was excellent with a wide range of choices and good services. It was a bit expensive though. **Context-sensitive summary**Excellent food with a wide range of choices and good services.

5.3   Summary Generation by Selecting Sentences Using Contrastive Score

The summaries that have high contrastive score are identified and chosen for a summary generation. Table 5 shows the sample contrastive summary pairs of Restaurant X and Restaurant Y.

Table 5. Sample contrastive summary pairs of Restaurant X and Restaurant Y.

Restaurant X	Restaurant Y
My fiance surprise 30th birthday dinner here could not be happier	Magnificent restaurant to celebrate my birthday the other night with my boyfriend
Great food at reasonable prices	Definitely not worth the price
Great place to dine for dinner	Place I would reconsider revisiting for dinner

6   Performance Analysis of Proposed Context-Sensitive Summary Method with Existing Methods

The performance of the proposed method is evaluated in terms of F_1 score, accuracy and ROUGE metrics.

6.1   Word Attention

The attention results of the proposed CSLSTM model is compared with two existing methods, RNCRF [21] and CMLA [20]. Attention in RNCRF model is based on pre-extracted syntactic relations. The CMLA model uses multi-layer attentions with tensors to learn the interaction between features and opinions.

F_1 Score. F_1 score is used to measure the accuracy of attention. It is based on the precision and recall score of the test dataset. It is represented by Eq. (16):

$$F_1 score = (1 + \beta^2) + (P * R)/(\beta^2.P) + R \tag{16}$$

A comparison of F_1 scores of word attention by CSLSTM and existing methods for restaurant dataset is shown in Table 6. The RNCRF and CMLA models attend only to the feature and opinion words in a sentence. The CSLSTM model attends to the context words in addition to feature and opinion words. The proposed model outperforms existing models as the model attends to important words based on attention sources.

Table 6. Comparison of F_1 score of word attention by CSLSTM with existing models for restaurant dataset.

Model	Feature	Opinion	Context
RNCRF [21]	84.05	80.93	–
CMLA [20]	85.29	83.18	–
Proposed CSLSTM	85.70	82.16	78.12

6.2   Context-Sensitive Sentiment Classification

The sentiment classification results are compared with two existing methods, Sentic LSTM [12] and ATAE-LSTM [22]. The Sentic-LSTM model replaces the encoder with a knowledge embedded LSTM for filtering data that does not coordinate with ideas in the knowledge base. ATAE-LSTM model incorporates feature embeddings into the representation of a sentence.

Accuracy. Accuracy is the ratio of the correctly classified sentences to the total number of sentences in the dataset. It is represented by Eq. (17):

$$Accuracy = Number\ of\ correct\ predictions/Total\ number\ of\ predictions$$
$$(17)$$

Figure 4 illustrates the comparison of the accuracy of sentiment classification by proposed CSLSTM with existing methods for restaurant dataset with respect to sentence length.

The ATAE-LSTM performs better than the Sentic LSTM as it captures important information pertaining to a given feature for sentiment classification. The Sentic LSTM performs slightly lower level than the ATAE-LSTM, as knowledge base ideas may at times delude the network to manage words irrelevant to features. The CSLSTM network achieves the best performance of all, as it discriminates information in response to given feature, opinion and context inputs.

6.3   Context-Sensitive Summary Generation

The summary generated using the proposed CSLSTM model is compared with two existing models, MARS model [23] and pointer-generator model [17]. MARS is a sentiment-aware abstractive summarization system which leverages on text-categorization to improve the performance of summarization. Pointer-generator model introduces coverage based network to avoid repetition of phrases in the summary.

ROUGE. The results of summarization are evaluated using three different metrics provided by ROUGE. In addition to the two metrics ROUGE-1 and ROUGE-2, ROUGE-L is taken as the third metric for evaluation.

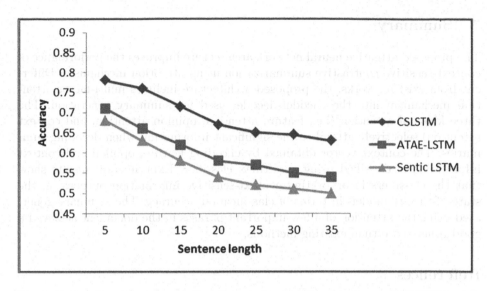

Fig. 4. Comparison of accuracy of sentiment classification by proposed CSLSTM with existing methods for restaurant dataset with respect to sentence length.

ROUGE-1: The number of unigrams in common between human summary and model summary.

ROUGE-2: Captures bigram overlap, thus measuring the readability of summaries.

ROUGE-L: The longest common sequence between the reference summary and model summary.

Table 7 shows the comparison of $ROUGE_{F1-score}$ of the summary produced by CSLSTM and existing methods. The attention of important phrases in the document by the proposed CSLSTM model results in great $ROUGE-1_{F1-Score}$ of around 36 in restaurant dataset. The proposed model also achieves a high $ROUGE-L_{F1-Score}$ compared to existing models.

Table 7. Comparison of $ROUGE_{F1-score}$ of summarization by CSLSTM with existing models for restaurant dataset.

Model	$ROUGE-1$	$ROUGE-2$	$ROUGE-L$
Pointer-generator model [17]	31.83	11.70	28.3
MARS model [23]	35.66	15.2	32.42
Proposed CSLSTM	36.34	17.6	33.57

7   Summary

The proposed attentive neural network architecture improves the performance of context-sensitive contrastive summarization using attention mechanism. Different from existing works, the proposed architecture includes multi-factor attention mechanism into the encoder-decoder used for summary generation. The three kinds of attention (i.e., feature attention, opinion attention, and context attention) selectively attend to the significant information when decoding summaries. The context vector obtained by encoding feature, opinion and context information is classified using a softmax classifier. Experimental results show that the classifiers incorporating context-sensitive information outperform the state-of-the-art models in terms of classification accuracy. The summary generated using the attention of a few important phrases in the document achieved a good rouge score than existing methods.

References

1. Bastien, F., et al.: Theano: new features and speed improvements. arXiv preprint arXiv:1211.5590 (2012)
2. Cambria, E., Poria, S., Bajpai, R., Schuller, B.: SenticNet 4: a semantic resource for sentiment analysis based on conceptual primitives. In: Proceedings of COLING 2016, the 26th International Conference on Computational Linguistics: Technical Papers, pp. 2666–2677 (2016)
3. Chen, G., Chen, L.: Augmenting service recommender systems by incorporating contextual opinions from user reviews. User Model. User-Adapt. Interact. **25**(3), 295–329 (2015). https://doi.org/10.1007/s11257-015-9157-3
4. Cheng, J., Lapata, M.: Neural summarization by extracting sentences and words. arXiv preprint arXiv:1603.07252 (2016)
5. Chopra, S., Auli, M. Rush, A.M.: Abstractive sentence summarization with attentive recurrent neural networks. In: Proceedings of the 2016 Conference of the North American Chapter of the Association for Computational Linguistics: Human Language Technologies, pp. 93–98 (2016)
6. Hariri, N., Mobasher, B., Burke, R., Zheng, Y.: Context-aware recommendation based on review mining. In: Proceedings of the 9th Workshop on Intelligent Techniques for Web Personalization and Recommender Systems, p. 30 (2011)
7. Lahlou, F.Z., Benbrahim, H., Mountassir, A., Kassou, I.: Inferring context from users' reviews for context aware recommendation. In: Bramer, M., Petridis, M. (eds.) Research and Development in Intelligent Systems XXX, pp. 227–239. Springer, Cham (2013). https://doi.org/10.1007/978-3-319-02621-3_16
8. Levy, O., Goldberg, Y.: Dependency-based word embeddings. In: Proceedings of the 52nd Annual Meeting of the Association for Computational Linguistics, pp. 302–308 (2014)
9. Li, C., Xu, W., Li, S., Gao, S.: Guiding generation for abstractive text summarization based on key information guide network. In: Proceedings of the 2018 Conference of the North American Chapter of the Association for Computational Linguistics: Human Language Technologies, Volume 2, pp. 55–60 (2018)
10. Li, P., Lam, W., Bing, L., Wang, Z.: Deep recurrent generative decoder for abstractive text summarization. arXiv preprint arXiv:1708.00625 (2017)

11. Ma, S., Sun, X., Lin, J., Ren, X.: A hierarchical end-to-end model for jointly improving text summarization and sentiment classification. arXiv preprint arXiv:1805.01089 (2018)
12. Ma, Y., Peng, H., Cambria, E.: Targeted aspect-based sentiment analysis via embedding commonsense knowledge into an attentive LSTM. In: Proceedings of 32nd Association for the Advancement of Artificial Intelligence, pp. 5876–5883 (2018)
13. Manandhar, S.: SemEval-2014 task 4: aspect based sentiment analysis. In: Proceedings of the 8th International Workshop on Semantic Evaluation, pp. 27–35 (2014)
14. Nallapati, R., Zhou, B., Gulcehre, C., Xiang, B.: Abstractive text summarization using sequence-to-sequence RNNs and beyond. arXiv preprint arXiv:1602.06023 (2016)
15. Rahman, M., Siddiqui, F.H.: An optimized abstractive text summarization model using peephole convolutional LSTM. Symmetry 11(10), 1290 (2019)
16. Rush, A.M., Chopra, S., Weston, J.: A neural attention model for abstractive sentence summarization. arXiv preprint arXiv:1509.00685 (2015)
17. See, A., Liu, P. J., Manning, C. D.: Get to the point: summarization with pointer-generator networks. arXiv preprint arXiv:1704.04368 (2017)
18. Song, S., Huang, H., Ruan, T.: Abstractive text summarization using LSTM-CNN based deep learning. Multimed. Tools Appl. 78(1), 857–875 (2018). https://doi.org/10.1007/s11042-018-5749-3
19. Speer, R., Havasi, C.: ConceptNet 5: a large semantic network for relational knowledge. In: Gurevych, I., Kim, J. (eds.) The People's Web Meets NLP. Theory and Applications of Natural Language Processing, pp. 161–176. Springer, Heidelberg (2013). https://doi.org/10.1007/978-3-642-35085-6_6
20. Wang, W., Pan, S.J., Dahlmeier, D., Xiao, X.: Coupled multi-layer attentions for co-extraction of aspect and opinion terms. In: Proceedings of Association for Advancement of Artificial Intelligence, pp. 3316–3322 (2017)
21. Wang W., Pan, S.J., Dahlmeier D., Xiao, X.: Recursive neural conditional random fields for aspect-based sentiment analysis. arXiv preprint arXiv:1603.06679 (2016)
22. Wang, Y., Huang, M., Zhu, X., Zhao, L.: Attention-based LSTM for aspect-level sentiment classification. In: Proceedings of the 2016 Conference on Empirical Methods in Natural Language Processing, pp. 606–615 (2016)
23. Yang, M., Qu, Q., Shen, Y., Liu, Q., Zhao, W., Zhu, J.: Aspect and sentiment aware abstractive review summarization. In: Proceedings of the 27th International Conference on Computational Linguistics, pp. 1110–1120 (2018)
24. Zeng, W., Luo, W., Fidler, S., Urtasun, R.: Efficient summarization with read-again and copy mechanism. arXiv preprint arXiv:1611.03382 (2016)
25. Zhou, Q., Yang, N., Wei, F., Huang, S., Zhou, M., Zhao, T.: Neural document summarization by jointly learning to score and select sentences. arXiv preprint arXiv:1807.02305 (2018)
26. Zhou, Q., Yang, N., Wei, F., Zhou, M.: Selective encoding for abstractive sentence summarization. arXiv preprint arXiv:1704.07073 (2017)

Tamil Paraphrase Detection Using Encoder-Decoder Neural Networks

B. Senthil Kumar[✉], D. Thenmozhi, and S. Kayalvizhi

Department of CSE, Sri Sivasubramaniya Nadar College of Engineering,
Chennai, India
{senthil,theni_d,kayalvizhis}@ssn.edu.in

Abstract. Detecting paraphrases in Indian languages require critical analysis on the lexical, syntactic and semantic features. Since the structure of Indian languages differ from the other languages like English, the usage of lexico-syntactic features vary between the Indian languages and plays a critical role in determining the performance of the system. Instead of using various lexico-syntactic similarity features, we aim to apply a complete end-to-end system using deep learning networks with no lexico-syntactic features. In this paper we exploited the encoder-decoder model of deep neural network to analyze the paraphrase sentences in Tamil language and to classify. In this encoder-decoder model, LSTM, GRU units and gNMT are used as layers along with attention mechanism. Using this end-to-end model, there is an increase in f1-measure by 0.5% for the subtask-1 when compared to the state-of-the-art systems. The system was trained and evaluated on DPIL@FIRE2016 Shared Task dataset. To our knowledge, ours is the first deep learning model which validates the training instances of both the subtask-1 and subtask-2 dataset of DPIL shared task.

Keywords: Tamil paraphrase detection · Deep learning ·
Encoder-decoder model · Sequence-to-sequence (Seq-2-Seq)

1 Introduction

Recent advances in deep neural network architecture has motivated many researchers in natural language processing to apply for different tasks such as PoS, language modeling, NER, relation extraction, paraphrase detection, semantic role labeling etc., Paraphrase detection is one of primary and important task for many NLP applications. The two sentences which convey the same meaning in a language are said to be semantically correct or paraphrases. Paraphrases can be detected, extracted and can be generated. Paraphrase detection is an important task in paraphrase generation and extraction system. In paraphrase generation, the paraphrase detection improves the quality by picking up the best semantic equivalent paraphrase from the list of generated sentences. In paraphrase extraction, the detection of paraphrase plays a vital role in validating

© IFIP International Federation for Information Processing 2020
Published by Springer Nature Switzerland AG 2020
A. Chandrabose et al. (Eds.): ICCIDS 2020, IFIP AICT 578, pp. 30–42, 2020.
https://doi.org/10.1007/978-3-030-63467-4_3

the extracted sentences. Apart from this, paraphrase detection is also utilized in document summarization, plagiarism detection, question-answering system etc.

Paraphrases in sentences are detected in earlier systems using different traditional machine learning techniques. Recently paraphrase detection was explored in some regional Indian languages. One of the shared task which was first of its kind for Indian languages – Detecting Paraphrases in Indian languages – DPIL@FIRE2016 [9,10] highlighted the performance of different systems using traditional machine learning techniques. The issues associated with these systems are: (1) they are dependent on the hand-crafted feature engineering which used pure lexical, syntactic and semantic features (2) the features are language-dependent (3) the morphological variations in Indian languages increase the complexity in detection and (4) the lack of resources such as WordNet, word2vec pre-trained embeddings for Indian languages deterred the systems from detecting the sentencial paraphrases. These limitations can be addressed by using the deep neural networks which require less or no features to build the model and is language independent.

Our main contributions in this paper are: (a) the use of encoder-decoder model to build paraphrase detection system for Tamil language which require no lexico-syntactic features. This is an end-to-end model. (b) the system is evaluated in a monolingual setting to detect the paraphrases in Tamil using DPIL-FIRE2016 shared task dataset. (c) we compared the performance of this system with the state-of-the-art paraphrase detection system for Tamil using deep neural networks.

2   Related Work

The shared task on Detecting Paraphrases in Indian Languages DPIL@FIRE 2016 [9,10] was a good effort towards creating a benchmark data for paraphrases in Indian Languages – Hindi, Tamil, Malayalam and Punjabi. Paraphrase detection for Indian languages are more challenging and complex due to its morphological variations and Tamil, Malayalam belongs to Dravidian language family which are more agglutinative in nature than Hindi and Punjabi from Indo-Aryan language family. Totally eleven teams submitted their results out of which only four teams evaluated the results for all the four Indian languages. In general most of teams used traditional machine learning techniques such as Logistic Regression, Random Forest, Gradient Tree Boosting, Multinomial Logistic Regression, Maximum Entropy and Probabilistic NN models. The commonly used lexical features are PoS, Lemma, Stopwords, word overlaps and the similarity features are Cosine, Jaccard coefficient, Levenshtein distance measures and the semantic feature like synonyms are used.

It was observed from the four teams which submitted their results for all the four languages that, there was a huge difference in the f1-score and accuracy across the category or family – Dravidian and Indo-Aryan – of languages. This revealed that the morphological inflections and agglutinative characteristics of Dravidian language family deter the performance of the system. HIT2016

[7] which scored top result for all languages other than Hindi, submitted their results based on the Cosine similarity, jaccard coefficient, METEOR – a machine translation-based metric for the character n-grams. The accuracy of the system was 89% and 94% for Hindi and Punjabi respectively. For the Dravidian languages, the accuracy obtained was 81% and 79% for Malayalam and Tamil respectively. This clearly indicate that the features to be used in detecting the paraphrases are dependent on the language characteristics. Similarly the three teams JU_NLP [15], KS_JU [17] and NLP_NITMZ [17] which submitted their results for all the four languages had showed the difference in accuracy across the family of languages for their lexico-syntactic features. In general, the accuracy of Tamil and Malayalam are lesser when compared to the accuracy obtained by the Hindi and Punjabi languages for the similar lexico-syntactic features.

Another team Anuj [16] submitted his result only for Hindi language without using any similarity measures but by using only the lexical features such as lemma, stop word removal and synonym, yielded the better result than the team which used the similarity-based features. This clearly highlights that the similarity features, lexical or semantic features used to detect the paraphrases are dependent on the natural language characteristics. Moreover there is a lack of resources such as WordNet, word2vec pre-trained embeddings for Indian languages. To overcome the shortcomings of traditional machine learning techniques which depends on hand-crafted feature engineering, some attempts were made to develop models using deep neural networks for Indian languages. This greatly reduced the need of any lexical or syntactic resources for the regional languages. One such attempt for Tamil was by Mahalakshmi et al. [13], Bhargava et al. [2] and Senthil Kumar et al. [18]. Mahalakshmi et al. [13] used the recursive auto-encoders (RAE) to compute the word representations of the paraphrase sentences.

The model then used the euclidean distances to measure the similarity and then softmax classifier is used to detect the paraphrases. They evaluated the training dataset and obtained the accuracy of 50%. Bhargava et al., attempted to use three deep neural networks model for detecting paraphrases in four Indian languages – Tamil, Malayalam, Hindi, Punjabi – and English. They used CNN, CNN using WordNet and stacked LSTM models. It was found that all the systems performed well for English than the Indian languages. Among the Indian languages Hindi, Punjabi paraphrase sentences were detected better than Tamil and Malayalam. They reported the evaluation of subtask-1 dataset of all the Indian languages. For Tamil, the CNN model outperformed the other models and scored 74.3% of f1-measure. Senthil Kumar et al., used the deep Bi-LSTM-layered networks and evaluated the model with two attention mechanisms. The model was evaluated for 5-fold cross validation of training dataset and obtained the accuracy of 65.2%.

3   Dataset Description

The dataset used to train and evaluate our system is the dataset released by the DPIL@FIRE 2016 shared task for four Indian languages. The shared task required participants to identify the sentential paraphrases in four Indian languages, namely Hindi, Tamil, Malayalam, and Punjabi. The corpora mainly contains the news articles from various web-based news sources. The shared task organized two subtasks: Subtask 1-to classify the given paraphrases into Paraphrase (P) or Non-Paraphrase (NP). Subtask 2-to classify the given paraphrases into Paraphrase (P), Semi-Paraphrase (SP) or Non-Paraphrase (NP). The corpora contains two subsets, one for each subtask. For subtask-1, the Tamil corpus was classified into one of binary class for the 2500 sentencial paraphrases and the 900 test pairs. For the subtask-2, the corpus contain 3500 sentential Tamil paraphrases and the 1400 test pairs. The analysis on the corpus showed that the average number of words used in the paraphrase sentences are less for the aggultinative languages. This is because of the more morphological inflections in the agglutinative languages. This lead to more unique words and hence larger vocabulary size. These morphological variations and agglutinative characteristics play vital role in determining the performance of the paraphrase detection system for agglutinative language like Tamil.

Table 1. Data population for subtask-1

Sub task-1	Training	Testing	Total
Paraphrase	1000	400	1400
Non-paraphrase	1500	500	2000
Total	2500	900	3400

Table 2. Data population for subtask-2

Sub task-2	Training	Testing	Total
Paraphrase	1000	400	1400
Non-paraphrase	1500	500	2000
Semi-paraphrase	1000	500	1500
Total	3500	1400	4900

```
<Paraphrase pID="TAM0071">
<Sentence1>
    இ.எஸ்.டி. மசோதாவை நிறைவேற்ற எதிர்க்கட்சிகளின் ஆதரவை பெற முயற்சிப்போம் என்றார்
    வெங்கையா நாயுடு.
</Sentence1>
<Sentence2>
    இ.எஸ்.டி. மசோதாவை நிறைவேற்ற எதிர்க்கட்சிகளின் ஆதரவைப் பெற திட்டமிட்டுள்ளதாக
    வெங்கையா நாயுடு தெரிவித்துள்ளார்.
</Sentence2>
<Class>P</Class>
</Paraphrase>
```

Fig. 1. A pair of Tamil paraphrase sentences in the dataset

Figure 1 shows an example of paraphrase sentences from the DPIL@ FIRE2016 dataset which is tagged as P category. The dataset was annotated with one of the paraphrase category (P, SP, NP) and in the XML format. Each paraphrase was allocated with a unique ID and the pair of sentences are separated and annotated with the corresponding paraphrase tag.

4   Preprocessing

The pair of sentences from the dataset are extracted and presented to the network as shown in Fig. 2. The pair of sentences are appended together. The two sentences are delimited with the tag <eol> which acted as separator between the pair of sentences and <s>, </s> are used to denote the boundary of the pair of input sentences. These sentences are given as input to the encoder-decoder model as input sequences.

<s> ஜி.எஸ்.டி. மசோதாவை நிறைவேற்ற எதிர்க்கட்சிகளின் ஆதரவை பெற முயற்சிப்போம் என்றார் வெங்கையா நாயுடு . <eol> ஜி.எஸ்.டி. மசோதாவை நிறைவேற்ற எதிர்க்கட்சிகளின் ஆதரவைப் பெற திட்டமிட்டுள்ளதாக வெங்கையா நாயுடு தெரிவித்துள்ளார் . </s>

Fig. 2. A preprocessed input sentences $w_1, ... w_n$

5   Encoder-Decoder Model

To detect the Tamil paraphrases using encoder-decoder model we have adopted the Neural Machine Translation architecture [11] which was based on the sequence-to-sequence learning [3, 19]. The sequence-to-sequence (seq-2-seq) models were pioneered by Ilya Sutskever [15], applied that model for machine translation (MT) task. This is the most promising model in natural language processing (NLP) tasks such as sequence labeling [4, 20], machine translation [11], syntactic parsing [8, 21], semantic parsing [5]. In our approach we treated the encoder-decoder model as the classifier to classify the source sequence of words to the corresponding paraphrase label as the target. This encoder-decoder model or sequence to sequence (seq-2-seq) model is the most popular model in learning the target sequence conditioned on the source sequence. This model typically uses the encoder to encode the given input sequence into a fixed vector representation which is then fed into the decoder. The decoder then predicts the next word given the source sequence representation, the previous output and the current hidden state. The decoder predicts the next word conditioned on the previous input sequences. We treated the paraphrase detection as the classification problem. Given the pair of paraphrases as input, the model has to predict the class label. The encoder and decoder uses a multilayered RNN variant units to map the input sequence to a vector of a fixed dimensionality, and then decode that to the target label from that vector.

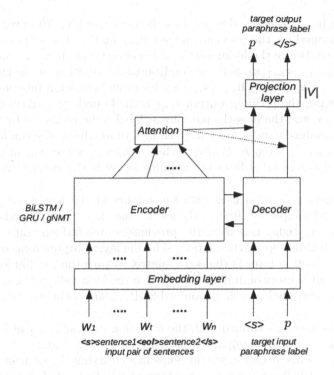

Fig. 3. Encoder-Decoder model for paraphrase detection

The model has an embedding layer, an encoder, a decoder and a projection layer. The encoder and the decoder receives the following input: first the source sentence, then a boundary marker which indicates the transition from the encoding to the decoding mode, then the target input and target output class label. Here the source sentence is the input sequence which is a pair of paraphrase sentences with a delimiter and the target is the corresponding class label. The system is feed with those sentences in tensors, which are in time-major format. Once the source sentences and target label are given, the model first look up its word embeddings to retrieve the corresponding word representations. The vector representation for each word is derived by choosing a fixed vocabulary of size V for source and target sequences. The embedding weights are usually learned during training. Since the input is time major, the encoder receives the input till the boundary marker. At each time step t, the hidden state $h_{<t>}$ of the encoder is updated by, $h_{<t>} = f(h_{<t-1>}, w_t)$, where f is a non-linear activation function, w_t is the source sentences. Here in our case w_t is the word sequences from the pair of input sentences.

Once the source sentences are encoded, the last hidden state of the encoder is used to initialize the decoder. Once the boundary marker reached, it marked the beginning of the decoder. The hidden state of the encoder is a vector of fixed dimensionality **s** – which is a summary of the source sentences. Given the hidden state $\mathbf{h}_{<t>}$, the decoder predicts the next word $\mathbf{v}_t$. However $\mathbf{h}_{<t>}$ and v_t are conditioned on the previous output $\mathbf{v}_{t-1}$ and on the summary s of the input sequence. Hence the hidden state of the decoder at time t is computed by, $\mathbf{h}_{<t>} = f(\mathbf{h}_{<t-1>}, \mathbf{v}_{t-1}, \mathbf{s})$, and the conditional distribution of the next symbol is, $P(v_t|\mathbf{v}_{t-1}, ..., \mathbf{v}_1, \mathbf{s}) = g(\mathbf{h}_{<t>}, \mathbf{v}_{t-1}, \mathbf{s})$, for given activation functions f and g. Here in our case, the previous output v_{t-1} is the boundary marker of the paraphrase sentence and the v_t is the paraphrase label to be predicted by the model. Hence the Encoder-Decoder are jointly trained to maximize the conditional log-likelihood, max_θ $1/N$ log p_θ $(\mathbf{v}_n|\mathbf{w}_n)$ where $(\mathbf{w}_n, \mathbf{v}_n)$ is the pair of paraphrase sentences and target label from the training set, θ is the varying model hyper parameters.

The projection layer is feed with the tensors of the target output words, these are decoder inputs shifted to the left by one step with a boundary marker appended on the right. Based on the previous source hidden state and target input, for each timestep on the decoder side, the layer outputs a maximum logit value. The projection layer is the dense matrix to turn the top hidden states of decoder, to logit vectors of dimension V. Given the logit values, the training loss can be easily computed by using standard SGD/Adam optimizer by varying the learning rate.

The model was also trained with the attention mechanism, which computes the attention weight by comparing the current decoder hidden state with all encoder states. This attention weight helps in identifying the sequence of source input words that helps in predicting the target label. Instead of discarding all the source hidden states, attention mechanism provides direct connection between source sentences and target label by paying attention to the relevant source sequences. We have used both scaled-luong [12] and normed-Bahdanau [1] attention mechanisms. There are several key differences between the attention model proposed by Bahdanau et al. [1] and Luong et al. [12]. The primary difference is that the Luong attention simply uses hidden states at the top LSTM layers in both the encoder and decoder, where as the Bahdanau attention use the concatenation of the forward and backward source hidden states in the bi-directional encoder and target hidden states in their decoder. For M:N sequence mapping tasks such as sequence labelling, machine translation task, Luong attention mechanism performs better as the word alignment is required during the translation by the decoder from the encoder. Since our task is to map the n word sequences to the paraphrase label which is N:1 mapping, Bahdanau attention produces better result as only the encoder attention weight is required to predict the paraphrase label.

The encoder-decoder model basically was built by using RNN units in the stacked layers. It can also be built based on the variants of RNN units such as LSTM, Bi-LSTM, GRU and gNMT units [11]. As shown in Fig. 3, we used GRU and gNMT units in the encoder-decoder layers to detect the Tamil paraphrases.

6   Experiment

There were earlier attempts using LSTM, Bi-LSTM as stacked layers in the encoder-decoder models to predict the Tamil paraphrase sentences [2,17]. For some of the tasks such as machine translation [11], GRU and gNMT model performed better than the RNN variants – LSTM, Bi-LSTM. This motivated us to use the other RNN-variant GRU and gNMT in the stacked layers of encoder-decoder model. On fine tuning the hyper parameters of the model, we found that 2-layered GRU and 5-layered gNMT – 1 Bi-LSTM with 3 residual layers – at both encoder-decoder performs better than the other for the subtask-1. These two models were trained and evaluated on the DPIL@FIRE 2016 dataset for the Tamil language. Both the systems were trained and validated for the training data of both subtask-1 and subtask-2. The subtask-1 contains 2500 pairs of training sentences and 900 pairs of testing sentences. Out of which only 40% are paraphrase pairs, the remaining 60% pairs of sentences are non-paraphrase pairs. The subtask-2 contains 3500 paraphrase pairs and 1400 testing instances. For the subtask-2, we experimented with all the variations in encoder-decoder models using Bi-LSTM, GRU and gNMT model. The models are fine-tuned with various options and hyperparameters. The gNMT model for the subtask-1 performed better with the SGD as optimizer, 128 LSTM units, learning rate 1.0, initial weight was set to 0.1 with forget_bias 1.0 to guard against the vanishing gradient, batch size 128, dropout is 0.2, with the normed-bahdanau attention mechanism. The model was trained and tested on using single GPU.

7   Results

The system was evaluated for the training instances of the subtask-1 and subtask-2. The existing deep learning systems for Tamil paraphrase detection [11,13,18] reported their performance only for the subtask-1 training instances. From the conventional state-of-the-art systems using DPIL dataset, the HIT2016 [7] scored high across all the languages which used Cosine similarity as one of its feature, the Indo-Aryan family languages obtained an average accuracy of 91.5% whereas the Dravidian languages obtained 80% only. The cosine similarity feature used for Indo-Aryan family could not perform well for the Dravidian languages, since it is agglutinative. Hence there is a need for system which identifies the feature or model specific to Dravidian languages.

To our knowledge, ours is the first end-to-end deep learning model which validates the training instances of both the subtask-1 and subtask-2 dataset. We developed two variations with respect to attention techniques on the deep neural network – system using scaled-luong (SL), normed bahdanau (NB) as attention mechanisms. We performed 5-fold cross validation on the training instances of both the subtask-1 and subtask-2 datasets. In subtask-1, for the given 2500 instances, each fold the data is split into 2000 pairs of sentences as the training instances and 500 pairs of sentences as testing instances. Our split for the subtask-1 is similar to Mahalakshmi et al. [13]. For subtask-2, each fold contains

3000 instance pairs as the training and 500 pairs of sentences as the testing instances. Our model for subtask-1 was treated as the binary classifier which predicts P/NP class. For the subtask-2 our model was treated as the multi-class classifier which predicts P/SP/NP class. The performance of the models for both the subtask-1 and subtask-2 was measured by precision, recall and f1-measure metrics. We compared our different models – GRU, gNMT variants – with the existing deep tamil paraphrase detection systems. For the subtask-1, gNMT-layered encoder-decoder model obtained 0.5% improvement than the state-of-the-art systems.

Table 3. Cross-validation performance of subtask-1

Subtask-1 models	Precision	Recall	Accuracy	F1-score
GRU-NB	74.46	52.24	67.16	61.40
GRU-SL	72.78	54.96	67.2	62.3
gNMT-SL (1 Bi-LSTM + 1 residual)	75.84	61.76	71.04	68.08
gNMT-NB (1 Bi-LSTM + 1 residual)	72.98	67.84	71.36	70.32
gNMT-NB (1 Bi-LSTM + 3 residual)	**78.14**	**71.76**	**75.04**	**74.81**
Mahalakshmi et al. [13]	–	–	50.0	–
Bhargava et al. [2]	–	–	–	74.3
Senthil Kumar et al. [18]	–	–	65.2	–

Mahalakshmi et al. [13] reported the overall accuracy of the system detecting the Tamil paraphrases by 50% and Senthil Kumar et al. [18] used Bi-LSTM-layered encoder-decoder model to detect the Tamil paraphrases and performance in accuracy was reported as 65.2%. Bhargava et al. [2] used the three models and the best CNN-based deep neural network model was measured in f-measure of 74.3%. Table 3, shows the comparison of subtask-1 dataset evaluation of the existing deep learning models with our encoder-decoder models.

For the subtask-2, we experimented with all the variants of RNN units – BiLSTM, GRU and gNMT units in the encoder-decoder layers with scaled-luong and normed-bahdanau attention mechanisms. 2-layered BiLSTM with scaled-luong attention has higher accuracy than the other encoder-decoder models for subtask-2 as shown in Table 4. When comparing with the subtask-1 models which is binary classification, gNMT models are not performing well for the multi-class classification.

For the test data, HIT2016 - a traditional learning system scored top result in the shared task obtained 82.11% for the subtask-1, our gNMT-based deep learning system obtained 55.6%. For the subtask-2, HIT2016 obtained 75.5% whereas our GRU-based obtained 49.2% which is low compared with the traditional machine learning model. Data sparsity is one of the main reason behind the low performance of deep neural networks models.

Table 4. Cross-validation performance of subtask-2

Subtask-2 models	Precision	Recall	Accuracy	F1-score
2 BiLSTM-SL	**64.52**	**62.74**	**75.42**	61.76
2 BiLSTM-NB	61.98	61.06	74.45	60.58
GRU-SL	63.56	62.54	75.05	**61.81**
GRU-NB	63.91	62.37	74.66	61.71
gNMT-SL (1 Bi-LSTM + 1 residual)	60.55	58.98	72.66	58.11
gNMT-NB (1 Bi-LSTM + 1 residual)	61.39	60.16	73.33	59.48
gNMT-NB (1 Bi-LSTM + 3 residual)	60.55	60.5	73.39	60.49

8 Error Analysis

From Table 1, the number of non-paraphrase sentence pairs are 60% whereas the paraphrase sentence pairs are only 40%. Table 5 shows the class-wise prediction of the sentence pairs by the different models. There is a huge difference in predicting the paraphrase sentences by the different models from 65.3–89.7%. Where as the non-paraphrase sentence pairs prediction by all the models are approximately between 65–68%. Eventhough the number of non-paraphrase sentences are higher, the system performance depends only on predicting the paraphrase sentence pairs. There is a drastic increase around 24% in detecting the paraphrase sentence pairs across the models when compared with the non-paraphrase sentence pairs detection with only around 3%. From the Table 2, it is clear that 43% of sentences are non-paraphrase sentences where as around 28.5% of each are paraphrase and semi-paraphrase sentence pairs in the training dataset. Table 6 shows the class-wise f1-score across all the models. To calculate the class-wise performance, the multiclass data will be treated as if binarized under a one-vs-rest transformation [6]. In this subtask-2 also, even though the number of non-paraphrase sentence pairs are more when compared to the paraphrase and semi-paraphrase pairs, the performance of the system highly depends on the detection of paraphrase and semi-paraphrase sentence pairs. The average detection of P and SP sentence pairs are 63.39% and 63.82% f1-score respectively, which are more when compared to the average detection of NP sentence

Table 5. Class-wise prediction (in %) for subtask-1

Subtask-1 models	P	NP
GRU-NB	65.3	**68.40**
GRU-SL	68.7	66.2
gNMT-SL (1Bi-LSTM + 1 residual)	77.2	66.93
gNMT-NB (1Bi-LSTM + 1 residual)	84.8	62.40
gNMT-NB (1Bi-LSTM + 3 residual)	**89.7**	65.27

pairs with f1-score of 54.65% only, which is almost 9% less when compared to the P and SP detection.

Table 6. Class-wise prediction (in f-score) for subtask-2

Subtask-2 model	P	NP	SP	F1-score
2 BiLSTM-SL	64.43	**56.70**	64.16	61.76
2 BiLSTM-Nb	**65.45**	53.56	62.71	60.58
GRU-SL	64.47	54.75	**66.20**	**61.80**
GRU-NB	63.51	55.68	65.93	61.71
gNMT-SL (1Bi-LSTM + 1 residual)	60.53	52.19	61.58	58.11
gNMT-NB (1Bi-LSTM + 1 residual)	62.34	52.99	63.10	59.48
gNMT-NB (1Bi-LSTM + 3 residual)	63.01	56.69	63.62	60.49
Average	63.39	54.65	63.82	

Hence the above analysis indicates that the performance of the system depends only on the P, P & SP sentence pairs in subtask-1 and subtask-2 dataset respectively, but not on the NP sentence pairs. Eventhough the NP sentences are more in the dataset when compared with the other class(es) of sentences in both the subtasks, the performance of the system does not depends on the NP sentence pairs. The lack in detecting the NP sentence pairs may be due to the encoder-decoder model characteristic, since the model learn word representations in continuous space, it tends to map the source sentence to target label that occur frequently in the context [14].

9   Conclusions

We used the encoder-decoder deep neural network model to detect the Tamil paraphrases from the DPIL@ FIRE2016 corpus. We evaluated our systems for 5 folds on the both the subtasks. We developed two variations with respect to attention techniques on the deep neural network – system using scaled-luong (SL), normed bahdanau (NB) as attention mechanisms. We compared our different models – GRU, gNMT variants – with the existing deep tamil paraphrase detection systems. For the subtask-1, gNMT-layered encoder-decoder model obtained 0.5% improvement in f1-score than the state-of-the-art systems. For the subtask-2, GRU-SL encoder-decoder model obtained overall f1-score of 61.81%. To our knowledge, ours is the first end-to-end deep learning model which validated the training instances of both the subtask-1 and subtask-2 dataset of DPIL@ FIRE2016. Eventhough the state-of-the-art systems for Tamil paraphrase detection have used different neural networks using CNN, MLP and Bi-LSTM, ours is the complete end-to-end encoder-decoder model with less or no preprocessing and lexico-syntactic features. The performance of this system can be improved further with more amount of training instances. Since the deep neural network system require more data to be trained, the DPIL training instances for Tamil paraphrase – 2500 and 3500 – was not enough for the deep neural network model to capture and learn the syntactic feature of the language from the given instances. Another way to improve the performance of the systems is to apply the recent deep neural network architectures that is suitable for the paraphrase detection.

References

1. Bahdanau, D., Cho, K., Bengio, Y.: Neural machine translation by jointly learning to align and translate. arXiv preprint arXiv:1409.0473 (2014)
2. Bhargava, R., Sharma, G., Sharma, Y.: Deep paraphrase detection in Indian languages. In: Proceedings of the 2017 IEEE/ACM International Conference on Advances in Social Networks Analysis and Mining 2017, pp. 1152–1159. ACM (2017)
3. Cho, K., et al.: Learning phrase representations using RNN encoder-decoder for statistical machine translation. arXiv preprint arXiv:1406.1078 (2014)
4. Daza, A., Frank, A.: A sequence-to-sequence model for semantic role labeling. arXiv preprint arXiv:1807.03006 (2018)
5. Dong, L., Lapata, M.: Language to logical form with neural attention. arXiv preprint arXiv:1601.01280 (2016)
6. Haghighi, S., Jasemi, M., Hessabi, S., Zolanvari, A.: PyCM: multiclass confusion matrix library in python. J. Open Source Softw. **3**(25), 729 (2018). https://doi.org/10.21105/joss.00729
7. Kong, L., Chen, K., Tian, L., Hao, Z., Han, Z., Qi, H.: HIT 2016@ DPIL-FIRE2016: detecting paraphrases in Indian languages based on gradient tree boosting. In: FIRE (Working Notes), pp. 260–265 (2016)
8. Konstas, I., Iyer, S., Yatskar, M., Choi, Y., Zettlemoyer, L.: Neural AMR: sequence-to-sequence models for parsing and generation. arXiv preprint arXiv:1704.08381 (2017)
9. Kumar, M.A., Singh, S., Kavirajan, B., Soman, K.: DPIL@ FIRE 2016: overview of shared task on detecting paraphrases in Indian languages (DPIL), vol. 1737, pp. 233–238 (2016)
10. Anand Kumar, M., Singh, S., Kavirajan, B., Soman, K.P.: Shared task on detecting paraphrases in Indian languages (DPIL): an overview. In: Majumder, P., Mitra, M., Mehta, P., Sankhavara, J. (eds.) FIRE 2016. LNCS, vol. 10478, pp. 128–140. Springer, Cham (2018). https://doi.org/10.1007/978-3-319-73606-8_10
11. Luong, M.T., Brevdo, E., Zhao, R.: Neural machine translation (seq2seq) tutorial (2017). https://github.com/tensorflow/nmt
12. Luong, M.T., Pham, H., Manning, C.D.: Effective approaches to attention-based neural machine translation. arXiv preprint arXiv:1508.04025 (2015)
13. Mahalakshmi, S., Anand Kumar, M., Soman, K.: Paraphrase detection for Tamil language using deep learning algorithm. Int. J. Appl. Eng. Res **10**(17), 13929–13934 (2015)
14. Nguyen, T.Q., Chiang, D.: Improving lexical choice in neural machine translation. arXiv preprint arXiv:1710.01329 (2017)

15. Saikh, T., Naskar, S.K., Bandyopadhyay, S.: JU_NLP@ DPIL-FIRE2016: para-phrase detection in Indian languages-a machine learning approach. In: FIRE (Working Notes), pp. 275–278 (2016)
16. Saini, A., Verma, A.: Anuj@DPIL-FIRE2016: a novel paraphrase detection method in Hindi language using machine learning. In: Majumder, P., Mitra, M., Mehta, P., Sankhavara, J. (eds.) FIRE 2016. LNCS, vol. 10478, pp. 141–152. Springer, Cham (2018). https://doi.org/10.1007/978-3-319-73606-8_11
17. Sarkar, K.: KS_JU@ DPIL-FIRE2016: detecting paraphrases in Indian languages using multinomial logistic regression model. arXiv preprint arXiv:1612.08171 (2016)
18. Senthil Kumar, B., Thenmozhi, D., Aravindan, C., Kayalvizhi, S.: Tamil para-phrase detection using long-short term memory networks. In: Proceedings of Tamil Internet Conference - TIC2019, Chennai, India, pp. 4–10 (2019). ISSN 2313–4887
19. Sutskever, I., Vinyals, O., Le, Q.: Sequence to sequence learning with neural net-works. In: Advances in NIPS (2014)
20. Thenmozhi, D., Kumar, S., Aravindan, C.: SSN_NLP@ IECSIL-FIRE-2018: deep learning approach to named entity recognition and relation extraction for con-versational systems in Indian languages. In: FIRE (Working Notes), pp. 187–201 (2018)
21. Vinyals, O., Kaiser, Ł., Koo, T., Petrov, S., Sutskever, I., Hinton, G.: Grammar as a foreign language. In: Advances in Neural Information Processing Systems, pp. 2773–2781 (2015)

Trustworthy User Recommendation Using Boosted Vector Similarity Measure

Dhanalakshmi Teekaraman[1] , Sendhilkumar Selvaraju[1]([⊠]) ,
and Mahalakshmi Guruvayur Suryanarayanan[2]

[1] Department of Information Science and Technology, Anna University, Chennai 600025, India
ssk_pdy@yahoo.co.in
[2] Department of Computer Science and Engineering, Anna University, Chennai 600025, India

Abstract. An online social network (OSN) is crowded with people and their huge number of post and hence filtering truthful content and/or filtering truthful content creator is a great challenge. The online recommender system helps to get such information from OSN and suggest the valuable item or user. But in reality people have more belief on recommendation from the people they trust than from untrusted sources. Getting recommendation from the trusted people derived from social network is called Trust-Enhanced Recommender System (TERS). A Trust-Boosted Recommender System (TBRS) is proposed in this paper to address the challenge in identifying trusted users from social network. The proposed recommender system is a fuzzy multi attribute recommender system using boosted vector similarity measure designed to predict trusted users from social networks with reduced error. Performance analysis of the proposed model in terms of accuracy measures such as precision@k and recall@k and error measures, namely, MAE, MSE and RMSE is discussed in this paper. The evaluation shows that the proposed system outperforms other recommender system with minimum MAE and RMSE.

Keywords: Recommender system · Provenance · Trust · Social network · Fuzzy classifier

1 Introduction

The Social network is congested by a large number of posts such as blog, reviews, opinions, image, video, etc. Extracting the required information from such a congested network is very difficult and time consuming task. An online recommender system helps to retrieve the desired information from this crowded network. For example, in Amazon's recommender system, item-to-item collaborative filtering approach is used for item recommendation. Similarly, Facebook, LinkedIn and other social networking sites to examine the network of connections between a user and their friends to suggest a new group based on interest. The downside of this online recommender system is that, the recommendations are generated based on anonymous people similar to the target user. This recommendation does not guarantee that the recommendation generated is from

© IFIP International Federation for Information Processing 2020
Published by Springer Nature Switzerland AG 2020
A. Chandrabose et al. (Eds.): ICCIDS 2020, IFIP AICT 578, pp. 43–58, 2020.
https://doi.org/10.1007/978-3-030-63467-4_4

trusted people. Therefore, people tend to rely more on trusted person's recommendation than online recommendation [21]. The recommender system designed for the trust network is called trust-based recommender system.

When the trust model becomes potentially vulnerable then the transparency of the trust rating is lost [13]. The critical analysis of content or web resources makes the trust rating transparent, which is made possible only by provenance. Provenance provides Meta information about the creation and processing of content. Thus, in this model, the trust rating is computed using the provenance data derived from W7 model. Also, the trust ratings derived in the models [1, 2, 8–10, 14] are single rating or single preference. For example, in five rating scale, the trust value '4' represents high trust while trust value '1' represents very low trust. With a single rating or preference, the multiple aspects of the user or item cannot be expressed which will either directly or indirectly reduce the recommendation quality. Therefore, if the trust rating is derived using multiple criteria or features such as 'Originality of the content', 'Timeliness of the post', and 'Relevancy of the content' as (4, 3, 2), then evidently the quality of recommendation is improved.

The issues discussed above are handled by the proposed recommender system. One is multi-dimensional or multi attributes based trust evaluation than single dimension or a single attribute. If multiple aspects of users are analyzed for trust computation, then the impact of recommendation is stronger and positive. Next the attribute information gain is used as weight component and weighted similarity measure is computed. This multiple dimensions are easily represented using vector and hence vector similarity measures such as Jaccard, Dice and Cosine are used. Then this similarity is boosted by users trust degree or trust level and recommendation is made. The contributions of the proposed recommender system are as follows.

- Modeling the user
- Formation of fuzzy vector space
- Finding preference and recommending top-k users

The structure of the paper is as follows. Section 2 briefs about the related research. The proposed recommender system is elaborately discussed in Sect. 3. Performance evaluation is discussed in Sect. 4. Finally, conclusion and future works are stated in Sect. 5.

2 Related Work

The trust-based recommendation techniques depend on two important components, namely recommendation techniques and representation of trust models.

2.1 Trust-Enhanced Recommendation Techniques

The trust enhanced recommendation algorithms are generally an enhancement of standard recommendation techniques such as Simple mean; Pearson weighted mean, Pearson collaborative filtering. The former method receives recommendation from trusted peers, whereas the latter method received recommendation from normal users. The most common trust enhanced recommender strategy is asking the users to explicitly mention the trust statements about other users. For instance the Moleskiing recommender system [3] uses FOAF files that contain trusted information scale ranging from 1 to 9. The Trust model proposed by A. Abdul Rahman and S. Hailes [1] for virtual communities grounded in real-world social trust characteristics, reputation or word-of-mouth. Falcone et al. proposed a fuzzy cognitive map model [8] to derive the trust based on belief value of an agent. This model shows how different component (belief) may change and how their impact can change depending on the specific situation and from the agent personality. The aim of a Golbeck's trust model [9] is, to determine how much one person in the network should trust another person to whom they are not directly connected. This algorithm accurately analyses the opinions of the people in the system. TidalTrust algorithm works based on trust-based weighted mean which uses the trust value of users as a weight for the ratings of other users.

Hang et al. [10] used a graph-based approach to recommend a node in a social network using similarity in trust networks. Massa and Aversani [14] proposed a trust-based recommendation system where it is possible to search for trustable users by exploiting trust propagation over the trust network. Andersen et al. [2] explored an axiomatic approach for trust-based recommendation and propose several recommendation models, some of which are incentive compatible. In MoleTrust method the similarity weight attributed to ratings by user. A trust-filtered collaborative filtering technique is used by O'Donovan and Smith in [4]. Here the trust value is used as a filtering mechanism to choose only, the item raters who are trusted above a certain threshold. An Ensemble trust technique is proposed by victor et al. [17] aims to take into account all possible ways to obtain a positive weight for a rater of an item while favoring trust over similarity.

2.2 Trust Model Representation

Trust representations can be classified from three different perspectives, namely (i) Probabilistic vs. gradual trust (ii) Single vs. multi-dimensional trust and (iii) Trust vs. distrust. Probabilistic representations use probabilities to indicate how much trust is placed by a user to another [17] Stronger trust corresponds to a higher probability. Gradual representations [17] use continuous values to represent trust. The values can be any values so they cannot be explained as probabilities. The values directly indicate trust strengths. Here, (u, v, t) denotes that the trust value from u to v is t. Trust is a complex concept with multiple dimensions (i) Multifaceted trust and (ii) Trust evolution. It is an extension of single trust representations of multi-dimensional trust representations [11]. Trust is context dependent. Trusting someone on one topic does not necessarily mean he will be trusted by others. The trust value is represented with <u, v, f, p>, where u trust v with probability p in the facet f. Also author suggests that trust evolves as humans interact over time T. Josang's subjective logic explores the probabilistic model [12] that considers both trust and distrust simultaneously. A gradual trust model for both trust and distrust can be found in [5, 8, 16]. Guha et al. use a pair (t, d) [18] with trust degree t and distrust degree d and final suggested trust value is obtained by subtracting d from t i.e. t-d.

3 Proposed Recommender System

The proposed recommender system is built to recommend the top-k reviewers in a book-based social network. For this, the data about the reviewer and the review is collected from Goodreads, Google Books and Amazon using ad hoc-API and scrapping HTML pages. The fields collected from the social network are given in Table 1. More than 61,000 reviews and associated reviewer's data available from 2007 to 2015 is collected and details of the dataset collected is given in Table 2. Number of reviews and number of reviewers are not same always. But this dataset has single review from each reviewer.

Table 1. Fields collected from the social network

S. no	Fields	S. no	Fields
1	Review_Text	9	Reviewer_ID
2	Year_Month_of_Joining	10	Review_ID
3	Review_Postdate	11	Reviewer_URL
4	Rate_of_Review	12	Review_URL
5	Number_of_Likes	13	Commenter_URL
6	Month_Last_Active	14	Date_of_Comment
7	Number_of_Comments	15	Comments
8	Number_of_Reply	16	Profile_Status

Table 2. Dataset details

Title of the book	No. of reviews	Overall rating	No. of years of review
Adventures of Huckleberry Finn	10,153	3.78	8 (2008–2015)
Good Night, Mr. Tom	1,055	4.25	9 (2007–2015)
Harry Potter and the Sorcerer's Stone	49,955	4.40	9 (2007–2015)

The collected data is preprocessed and from this the trust score of each reviewer is computed using W7 provenance model [22]. Then, using DoT pruned Fuzzy Decision Tree (FDT) classifier [7] the reviewers are classified and fuzzy rules were generated. Finally, fuzzy rules are combined with a target user's request to perform recommendation. The major components of the proposed recommender system are as follows.

1. Provenance Based Trust Assessment
2. Fuzzy Decision Tree Based Classifier
3. Trust-Boosted Recommender System

3.1 Provenance Based Trust Assessment

This section briefs about how trust assessment is made using W7 provenance model. For the experiment the data are collected from Goodreads.com book-based social network. The provenance elements defined in W7 model are a 7-tuple: (WHAT, WHEN, WHO, HOW, WHY, WHICH, WHERE). The description of provenance elements in the context of trust is shown in Table 3.

Table 3. Description of provenance elements

Provenance element	Description
P_{WHAT}	Describes the review content that is relevant to the topic
P_{WHEN}	Represents the effective time spent by the reviewer
P_{WHERE}	Refers to the location (IP_Address, Domain_name) from where review is posted
P_{WHO}	Refers to the reviewer who is an author or creator of the review
P_{HOW}	Describes how review content is deviated from the rating given by the reviewer
P_{WHICH}	Refers to the application or device used to post the review
P_{WHY}	Describes the intention behind the post of review content

Since, the data for P_{WHERE} and P_{WHICH} is not provided by the domain, these two elements cannot be modeled. Therefore, the core provenance elements taken for trust quantification are P_{WHAT}, P_{WHEN}, P_{WHO}, P_{HOW} and P_{WHY}. Trust assessment algorithm quantifies these five provenance elements. This trust value is then given to the learning model to classify the users with various levels.

3.2 Fuzzy Decision Tree Based Classification

The learning model takes the quantified provenance value obtained using W7 model as a trust input. This is fuzzified using Triangular Membership Function (TMF) and rule base is constructed using Mamdani's 'If... Then' interpretation. Fuzzy Decision Tree (FDT) [19] takes the rule base and generates decision trees using a fuzzy ID3 [6] algorithm. To construct FDT, two criteria need to be evaluated, one is splitting criterion and the other is stopping criterion. The former one helps to choose the root node and child nodes. The latter one controls the growth of the tree.

In FDT, provenance element having highest information gain is assigned as the root node and leaf node denotes trust decision. Each distinct path from root to a leaf produces distinct rule. Each generated rule is assigned Degrees of Truth (DoT) [15] to state that how much truth value it holds. If $DoT = 1$, then the rule is absolutely true and if $DoT = 0$ then the rule is absolutely false. Sample fuzzy rules are shown in Fig. 1. Here, reviewers are classified into 5 different trust levels as VHGT (Very High Trust), HT (High Trust), MT (Moderate Trust), LT (Low Trust) and VLWT (Very Low

Trust). The abbreviation for the linguistic terms present in Fig. 1 is as follows. MSM (Moderately Same), HD (Highly Deviated), HSM (Highly Same), MD (Moderately Deviated), MITM (Moderately Ineffective Time Spent), HITM (Highly Ineffective Time Spent), HUTR (Highly Untruthful), HR (Highly Relevance), NTR (Neutrally Truthful), MIR (Moderately Irrelevance) and MDSML (Moderately Dissimilar).

Rule 1: If P_{HOW} is MSM and P_{WHEN} is MITM and P_{WHAT} is HR Then Trust is HT (DTH=1)

Rule 2: If P_{HOW} is HD and P_{WHY} is HUTR Then Trust is LT (DTH=0.93)

Rule 3: If P_{HOW} is HSM and P_{WHAT} is HR Then Trust is VHGT (DTH=1)

Rule 4: If P_{HOW} is MD and P_{WHY} is NTR and P_{WHEN} is MITM Then Trust is MT (DTH=0.98)

Rule 5: If P_{HOW} is MD and P_{WHY} is NTR and P_{WHEN} is HITM and P_{WHAT} is MIR and P_{WHO} is
         MDSML Then Trust is VLWT (DTH=1)

Fig. 1. Fuzzy rules

In order to get better accuracy with minimum number of rules, the stopping criterion (β) is used. The value of β chosen is 0.9 and 1 and lengths of rules ranges from 2 to 5. The Table 4 shows the number of rules generated. For example, rule #2 and #3 in Fig. 1 has length 2 and rule #5 has length 5. These fuzzy rules are taken as input to build a trust-boosted recommender system.

Table 4. Rules generated

β	Total rules generated	No. of rules			
		2	3	4	5
0.9	721	16	161	343	201
1	477	8	61	207	201

3.3 Trust-Boosted Recommender System

The proposed trust-boosted recommender system recommends the trustworthy users to the target user U_T is shown in Fig. 2. The major components of this recommender system model are:

- User Profile Learning
- Recommendation

The target user or requester (U_T) sends a query as a request (R_q) asking for recommendations from the trust network. This query is sent to the trusted network and it

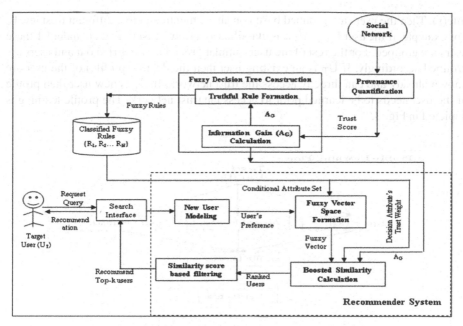

Fig. 2. Proposed trust boosted recommender system

checks whether the U_T is new user or not. If U_T is the existing user, then recommend the highly trusted users. Otherwise the request is sent to the profile learner where profile data (P_{data}) are updated based on the query and existing profile information. Then this updated P_{data} is sent to the trust network. In the trust network, each reviewer is grouped based on trust levels <VHGT, HT, MT, LT, VLWT>. From the set of fuzzy rules, extract the conditional attribute and the decision attributes. For each conditional attribute, generate fuzzy vector space (FVSP). The FVSP consists of a tuple <Attribute, Preference based Fuzzy Number>. The vector similarity measures such as Jaccard, Dice and Cosine is carried out to find how much target user is similar to the others in trust network.

The gain value of each attribute (A_G) is assigned as a weight component and it is applied to the above mentioned three similarity measures and weighted similarity value is calculated. Then this similarity is boosted by the corresponding decision attribute's trust degree (trust level). Then, based on the boosted similarity value, the trusted users are ranked from highest to lowest. Finally, top-k users are recommended to the target user (U_T). After a recommendation, the target user's feedback is collected and profile learner will update the P_{data} accordingly. To collect the feedback, set of feedback query (FD_{qry}) is formulated based on five attributes P_{WHAT}, P_{WHEN}, P_{WHO}, P_{HOW} and P_{WHY}. For each FD_{qry}, users are asked to provide a quantitative value in the scale 0 to 1. This recommendation process is repeated for each user request with the updated profile.

User Profile Learning Phase
In user modeling phase, the fuzzy rules are extracted from the rule database derived using fuzzy decision tree (discussed in Sect. 3.2). Each user may have one or more than one rule as shown in Fig. 1. Using rule matching algorithm, each users is assigned matched

rule(s). Then the users are grouped based on above mentioned five different trust levels. For example, if the user U_{109}; U_{169} is classified as a Low Trust (LT) then under LT these users are grouped. For the rest of the users similar procedure is carried out and users are grouped accordingly. If U_T is an existing user then the details (profile) of the user are known already and can directly access the trust network. If U_T is new user then profile of the user needs to be learned prior to access the trust network. The profile learning is depicted in Fig. 3.

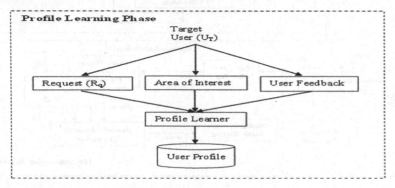

Fig. 3. Profile learning model

Initially U_T's field of interest and training examples or already labeled items are collected and this forms a basic profile (P_b) of the user. The set of feedback (F_b) provided by the U_T for the items is also collected. Finally the (P_b, F_b, R_q) are combined and sent to the profile learner. Then the learned user profile is given as input to the recommendation phase.

Recommendation Phase

The recommendation is carried out in two steps. One is creation of FVSP and the other is recommendation of the user.

Formation of FVSP

The rules extracted from the trust network are partitioned into conditional attributes set (CAS) and decision attributes set (DAS). The CAS consists of all the trust attributes $<P_{WHAT}, P_{WHEN}, P_{WHO}, P_{HOW}, P_{WHY}>$. The DAS consists of trust decision $<$VLWT, LT, MT, HT, VHGT$>$.

Step 1: For each attribute in the conditional attribute set, assign attribute grade. This is based on the position of the triangular fuzzy function and is given in Table 5. The linguistics space of each attribute is given below.

$$Linguistic\ Space = \begin{pmatrix} P_{WHAT} = [HIR, MIR, NR, MR, HR] \\ P_{HOW} = [HSM, MSM, NSM, MD, HD] \\ P_{WHEN} = [HITM, MITM, NETM, METM, HETM] \\ P_{WHY} = [HTR, MTR, NTR, MUTR, HUTR] \\ P_{WHO} = [HDSML, MDSML, NDSML, MSML, HSML] \end{pmatrix}$$

Table 5. Fuzzy number for the attribute P_{WHAT}

Grade	Fuzzy number
1	(0.0, 0.0, 0.25)
2	(0.0, 0.25, 0.50)
3	(0.25, 0.50, 0.75)
4	(0.50, 0.75, 1.0)
5	(0.75, 1.0, 1.0)

For example, in P_{WHAT} attribute the position of 'HIR' has low grade, i.e. 1 and 'HR' has high grades, i.e. 5. Similarly, in P_{WHY} attribute the position of 'MTR' (Moderately Truthful") has medium grade, i.e. 3.

Step 2: Assign the fuzzy number for each linguistic term based on the grade. Since it follows the triangular fuzzy logic, the fuzzy number assigned for each grade is shown below. For example, the fuzzy number for the linguistic term for the attribute P_{WHAT} is shown in Table 6. For other attributes, fuzzy number is same as that of shown in Table 5.

Table 6. Fuzzy number for the attribute P_{WHAT}

P_{WHAT} linguistic term	Fuzzy number
HIR (Highly Irrelevant)	(0.0, 0.0, 0.25)
MIR (Moderately Irrelevant)	(0.0, 0.25, 0.50)
NR (Neutrally Relevant)	(0.25, 0.50, 0.75)
MR (Moderately Relevant)	(0.50, 0.75, 1.0)
HR (Highly Relevant)	(0.75, 1.0, 1.0)

Step 3: The fuzzy number for each attribute is now represented as a vector in FVSP. The FVSP for each rule is represented as a pair $\{<A_K, FN_{AK}>\}$. where,

- K refers to a number of attributes, here K = 5.
- A_K represents the current attribute and
- FN_{AK} refers to the fuzzy number for the specified attribute A_K. That is FVSP = $\{<A_1, (a_{11}, a_{12}, a_{13})>, <A_2, (a_{21}, a_{22}, a_{23})> \ldots <A_5, (a_{51}, a_{52}, a_{53})>\}$. Here (a_{11}, a_{12}, a_{13}) is a triplet used in TMF to define the fuzzy number where $0 \leq a_{11} \leq a_{12} \leq a_{13} \leq 1$.

For example FVSP for the rule1 shown in Fig. 1 is given below. $\{<P_{HOW}, (0.25, 0.50, 0.75)>, <P_{WHY}, (0.50, 0.75, 1.0)>, <P_{WHEN}, (0.50, 0.75, 1:0), <P_{WHAT}, (0.50, 0.75, 1.0)>, <P_{WHO}, (0.75, 1.0, 1.0)>\}$. This FVSP is taken as input to calculate the vector similarity and to suggest the top-k trustworthy users.

Recommendation of the Top-k Users

In the vector space there are some similarity measures between two vectors which have been successfully applied in fields such as pattern recognition, classification of complex objects and other decision making problems. The vector similarity measures chosen in the proposed recommendation system are the Cosine similarity. Using this FVSP, above vector based similarity measure is carried out to find how much U_T is similar to the other users in trust network. The gain value of each attribute (A_G) is taken as a weight component and it is applied to the above mentioned measures and similarity value is calculated.

Let $X = U_T = (a_1, a_2, a_3)$ and $Y = U_N = (b_1, b_2, b3)$ is the fuzzy number of the target user (U_T) and the other user (U_N) from the trust network respectively, then the cosine similarity measure is given in Eq. (1) is as follows.

$$S = Cosine(U_T, U_N) = \sum_{k=1}^{5} A_{Gk} \frac{\sum_{f=1}^{3}\left(FN_{A_{Tkf}} \cdot FN_{A_{Nkf}}\right)}{\sqrt{\sum_{f=1}^{3}\left(FN_{A_{Tkf}}^2\right)} \cdot \sqrt{\sum_{f=1}^{3}(FN_{A_{Nkf}}^2)}} \tag{1}$$

where,

- A_G represents the attribute gain,
- f represents the fuzzy number of values in each fuzzy number,
- a_1, a_3, b_1 and b_3 are the endpoints of fuzzy numbers,
- a_2 and b_2 are the peak point of fuzzy numbers

After finding the similarity (S), boost this value by a corresponding trust score of the user U_N given in Eq. (2). Boosting is linear, since it is done with associated trust level. Using this boosted similarity (S_b), prediction of the target user's trust score is carried out. The prediction formula is given in Eq. (3).

$$S_b = S * S^{T_{wt}} \tag{2}$$

$$Pred\left(U_T, I_j\right) = \begin{cases} tr_{U_T}, & if\ S_b = 0\ or\ if\ tr_{U_N, I_j} = \overline{tr_{U_N}} \\ tr_{U_T} + \dfrac{\sum_{U_N \in NB} S_b(U_N, U_T) \times \left(tr_{U_N, I_j} - \overline{tr_{U_N}}\right)}{\sum_{U_N \in NB} |S_b(U_N, U_T)|}, & else \end{cases} \tag{3}$$

where,

- T_{wt} refers to trust weight assigned based on the trust level of user U_N. (For e.g., VHGT has T_{wt} of 1 and MT has T_{wt} of 0.6 and VLWT has T_{wt} of 0.2)
- tr_{U_T} represents the trust value of the Target User U_T presented in fuzzy number format as shown in Table 6.
- I_j represents items (books) which are not given any review
- NB represents the number of neighbors chosen

Consider the randomly chosen reviewer say reviewer 72 (R_{72}) requesting for the recommendation of k users (Let k = 10). The similarity (S) between the requester and the rest of the users is calculated. Then it is boosted using Eq. (2). The Table 7 shows the similarity and boosted similarity (S_b) score of the top-k reviewer. The reviewers are sorted based on similarity from highest to lowest. Though both similarities show the highest score for the top reviewers, the trust level differs. The trust level of highly matched reviewer with R_{72} is 'HT'. Therefore, the top-10 reviewers are expected to have the trust level of 'HT'. But, in case of without boosting, top 4th, 6th and 10th reviewers have other trust level ('MT') instead of 'HT'. Similarly, in case of boosting, the top 10th reviewer has different trust level.

Table 7. Similarity and Boosted similarity score of top-k reviewer.

Without boost			With boost			Top-k reviewer
Reviewer number	S	Trust level	Reviewer number	S_b	Trust level	
722	1	HT	28	1	HT	1
600	1	HT	29	1	HT	2
335	1	HT	722	1	HT	3
251	1	MT	600	1	HT	4
216	1	HT	335	1	HT	5
212	1	MT	165	1	HT	6
194	1	HT	177	1	HT	7
177	1	HT	194	1	HT	8
165	1	HT	216	1	HT	9
633	0.998	MT	158	0.998	MT	10

Therefore, the prediction error is more in without boosting and lesser in boosted method. Boosting the similarity appropriately ranks the top-k reviewers. This way the proposed system gets a reduced MAE and RMSE.

4 Experiments and Result Analysis

To evaluate the performance of gain weighted trust boosted recommender system, experiments are conducted on the popular book based social network Goodreads data set. The aim of these experiments is to present a comparative study of proposed recommendation strategy in fuzzy trust concept. Also proposed trust boosted model is evaluated against other weight strategies. The performance of the proposed recommendation strategy is measured with respect to quality of predictions and quality of recommendations. The quality of prediction is done by measuring Mean Absolute Error (MAE) and Root Mean Squared Error (RMSE). Similarly the quality of recommendation is done by measuring precision@k and recall@k and Average Precision (AP). Leave-one-out method is used to evaluate recommendation systems [14]. This technique involves withholding one rating and trying to predict it with remaining ratings. Then the predicted rating can be compared with the actual rating and the difference will be considered as the prediction error.

4.1 Evaluation of Different Weight Strategies

The different weight strategies considered for evaluation are expected weight method; preference based method [20] and proposed trust boosted method. The MAE, RMSE and AP measures are evaluated for the above mentioned weight strategies. The Fig. 4 shows the MAE value obtained for the cosine similarity method. From this figure, it is observed that the proposed trust boosted method shows the less prediction error (MAE) than the other two methods. Similarly, Fig. 5 shows the RMSE value obtained for the cosine similarity measure. From the figure it is observed that the proposed trust boosted method shows the less prediction error (RMSE) when compared with the expected weight method. The preference based method shows more error rate than the other two methods.

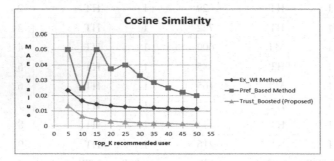

Fig. 4. MAE for cosine similarity measure

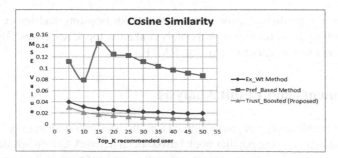

Fig. 5. RMSE for cosine similarity measure

The AP value is shown in Fig. 6 for the above similarity measure. The precision value for the proposed method is higher than the other two methods. The AP is almost same for top-5 and top-10 users. Up to top 20 users precision value is greater than or cqual to 0.90. After that the precision value is start decreasing and for top-50 user, the precision value is very less in preference based method.

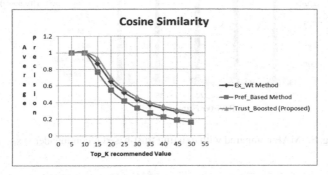

Fig. 6. Average precision for cosine similarity measure

4.2 Comparison with Other Trust-Based Recommender System

The proposed recommender system is compared with other trust-based recommender system. The evaluation is done on MAE and RMSE. First, the proposed method (boosted similarity) is compared against without boosting the similarity. The MAE of this is shown in Fig. 7. The experiment is carried out with Jaccard, Dice and Cosine similarity measure. All these three measures show the lesser prediction error while boosting the similarity than without boosting the similarity. In Jaccard repetition of a word does not reduce the similarity but Cosine measure reduce the similarity.

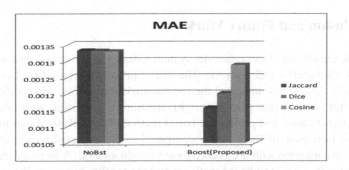

Fig. 7. MAE for with and without boost

The Fig. 8 and 9 shows the MAE and RMSE values of the proposed approach with existing trust-based recommender system respectively. The reviewers are chosen through random sampling The existing approaches considered are Tidal trust, Mole trust, Fuzzy Trust Filtering, Ensemble and Hybrid. The MAE value is checked for few randomly selected reviewers. The graph shows the reduced prediction error in the proposed method.

Similar to MAE, the RMSE value is checked with few randomly selected reviewers. The graph shows the reduced prediction error in the proposed method.

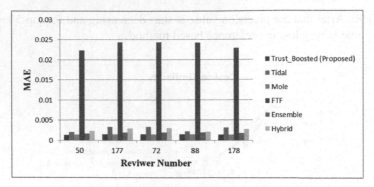

Fig. 8. MAE compared with existing trust-based recommender system

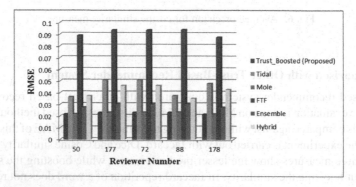

Fig. 9. RMSE compared with existing trust-based recommender system

5   Conclusion and Future Work

In this work, trust-boosted recommender system is designed to recommend top-k reviewers of the book based social network. The use of provenance based trust computation from multiple aspects has improved the recommendation quality. Also performance of proposed trust boosted (the gain as weight) measure is compared with other weights such as expected value, and preference based method. The analysis shows that the precision@k is increased 10.166% when compared to the expected weight method and 2.186% when compared with preference based weight method. Also proposed approach is compared with other trust based methods and the results shows that the prediction is achieved with minimum MAE and RMSE. The future work is to recommend the top-k reviewers to a group of users. That is to develop a group recommender system

Acknowledgements. This Publication is an outcome of the R&D work undertaken in the project under the Visvesvaraya PhD Scheme (Unique Awardee Number: VISPHD-MEITY-2959) of Ministry of Electronics & Information Technology, Government of India, being implemented by Digital India Corporation (formerly Media Lab Asia).

References

1. Abdul-Rahman, A., Hailes, S.: Supporting trust in virtual communities. In: Proceedings of the 33rd Hawaii International Conference on System Sciences, Maui, HI, USA, pp. 1769–1777. IEEE (2000)
2. Andersen, R., et al.: Trust-based recommendation systems: an axiomatic approach. In: Proceeding of the 17th International Conference on World Wide Web, pp. 199–208. ACM, New York (2008)
3. Avesani, P., Massa, P., Tiella, R.: Moleskiing.it a trust-aware recommender system for ski mountaineering. Int. J. Infonomics, 1–10 (2005)
4. Smyth, B., O'Donovan, J.: Trust in recommender systems. In: 10th International Conference on intelligent user Interfaces, San Diego, California, USA, pp. 167–174. ACM (2005)
5. Ziegler, C.N., Lausen, G.: Propagation models for trust and distrust in social networks. Inf. Syst. Front. **7**(4–5), 337–358 (2005)
6. Olaru, C., Wehenkel, L.: A complete fuzzy decision tree technique. J. Fuzzy Set Syst. **138**, 221–254 (2003)
7. Teekaraman, D., Sendhilkumar, S, Mahalakshmi, G.S.: Semantic provenance based trustworthy user classification on book-based social network using fuzzy decision tree. Int. J. Uncertain. Fuzziness Knowl. Based Syst. **28**(1), 47–77 (2020)
8. Falcone, R., Pezzulo, G., Castelfranchi, C.: A fuzzy approach to a belief-based trust computation. In: Falcone, R., Barber, S., Korba, L., Singh, M. (eds.) TRUST 2002. LNCS, vol. 2631, pp. 73–86. Springer, Heidelberg (2003). https://doi.org/10.1007/3-540-36609-1_7
9. Golbeck, J: Computing and applying trust in web-based social networks, PhD Dissertation, University of Maryland at College Park College Park, MD, USA (2005)
10. Hang, W.C., Singh, M.P.: Trust based recommendation based on graph similarities (2010). http://www.csc.ncsu.edu/faculty/mpsingh/papers/mas/aamas-trust-10-graph.pdf
11. Tang, J., Gao, H., Liu, H.: mTrust: discerning multi-faceted trust in a connected world. In: Proceedings of the Fifth ACM International Conference on Web Search and Data Mining, Seattle, Washington, USA, pp. 93–102. ACM (2012)
12. Josang, A.: A logic for uncertain probabilities. Int. J. Uncertain. Fuzziness Knowl. Based Syst. **9**, 279–311 (2001)
13. Janowicz, K.: Trust and provenance you can't have one without the other. Technical report, Institute for Geoinformatics, University of Muenster (2009)
14. Massa, P., Avesani, P.: Trust-aware recommender systems. In: Proceedings of the First ACM Conference on Recommender Systems, Minneapolis, USA, pp. 17–24. ACM (2007)
15. Nicholas, J.J.: Smith: Vagueness and Degrees of Truth. Oxford Scholarship (2009)
16. Victor, P., Cornelis, C., De Cock, M., Pinheiro da Silva, P.: Gradual trust and distrust in recommender systems. Fuzzy Sets Syst. **160**(10), 1367–1382 (2009)
17. Victor, P., Cornellis, C., Cock, M.D., Teredesai, A.M.: Trust and distrust-based recommendations for controversial reviews. IEEE Intell. Syst. **26**, 48–55 (2011)
18. Guha, R., Kumar, R., Raghavan, P., Tomkins, A.: Propagation of trust and distrust. In: Proceedings of the 13th International Conference on World Wide Web, pp. 403–412. ACM, New York (2004)
19. Mitra, S., Konwar, K.M., Pal, S.K.: Fuzzy decision tree, linguistic rules and fuzzy knowledge-based network generation and evaluation. IEEE Trans. Syst. Man Cybern. Part C (Appl. Rev.) **32**(4), 328–339 (2002)
20. Boulkrinat, S., Hadjali, A., Mokhtari, A.: Towards recommender systems based on a fuzzy preference aggregation. In: 8th Conference of the European Society for Fuzzy Logic and Technology, Atlantis Press, University of Milano-Bicocca, Milan, Italy (2013)

21. Sinha, R.R., Swearingen, K.: Comparing recommendations made by online systems and friends. In: DELOS Workshop, Personalization and Recommender Systems in Digital Libraries, Dublin, Ireland (2001)
22. Ram, S., Liu, J.: Understanding the semantics of data provenance to support active conceptual modeling. In: Chen, Peter P., Wong, Leah Y. (eds.) ACM-L 2006. LNCS, vol. 4512, pp. 17–29. Springer, Heidelberg (2007). https://doi.org/10.1007/978-3-540-77503-4_3

Sensitive Keyword Extraction Based on Cyber Keywords and LDA in Twitter to Avoid Regrets

R. Geetha[(⊠)] [iD] and S. Karthika

Department of Information Technology, Sri Sivasubramaniya Nadar College of Engineering,
Chennai, India
geethkaajal@gmail.com, skarthika@ssn.edu.in

Abstract. Twitter is the most popular social platform where common people reflect their personal, political and business views that obliquely build an active online repository. The data presented by users on social networking sites are usually composed of sensitive or private data that is highly potential for cyber threats. The most frequently presented sensitive private data is analyzed by collecting real-time tweets based on benchmarked cyber-keywords under personal, professional and health categories. This research work aims to generate a Topic Keyword Extractor by adapting the Automatic Acronym - Abbreviation Replacer which is specially developed for social media short texts. The feature space is modeled using the Latent Dirichlet Allocation technique to discover topics for each cyber-keyword. The user's context and intentions are preserved by replacing the internet jargon and abbreviations. The originality of this research work lies in identifying sensitive keywords that reveal Tweeter's Personally Identifiable Information through the novel Topic Keyword Extractor. The potential sensitive topics in which the social media users frequently exhibit personal information and unintended information disclosures are discovered for the benchmarked cyber-keywords by adapting the proposed qualitative topic-wise keyword distribution approach. This experiment analyzed cyber-keywords and the identified sensitive topic keywords as bi-grams to predict the most common sensitive information leaks happening in Twitter. The results showed that the most frequently discussed sensitive topic was 'weight loss' with the cyber-keyword 'weight' of the health tweet category.

Keywords: Twitter · Cyber-keywords · Privacy leaks · Regrets · Social media

1 Introduction

One of the most popular social networking media, Twitter, is now emerging as a platform to share what users feel, experience and comment both on personal and nation's current affairs with 280 characters per tweet. Almost all tweets contain some personal value when tweeted or re-tweeted by the users. Though sharing personal information in social media gives pleasure, the extent to which the message travels have no bounds or could not be limited as the user believes. Also, the users feel free to publish their opinion and

A. Chandrabose et al. (Eds.): ICCIDS 2020, IFIP AICT 578, pp. 59–70, 2020.
https://doi.org/10.1007/978-3-030-63467-4_5

thoughts online than sharing it in-person because this networking platform provides a wide range of audience which is both undefined and unlimited.

Gathering private information in this digital era has become much easier because of the advancements in modern communication technologies. Tommasel and Godoy (2016) stated that the access to social media posts are also widely open to every user irrespective of their age, location, expertise and professional community boundaries which leads to leaks in Personally Identifiable Information (PII) either knowingly or unknowingly. The publicly available personal information is at high risk of potentially being misused by cyber criminals eventually resulting in diverse effects like identity theft, child trafficking, business loss and professional black marks. A lot of issues related to privacy are arising in social media and plenty of privacy protection technologies are designed for users which tend to be independent and application specific. Such privacy protection mechanisms usually aim at protecting the information and control information disclosure to unintended audiences.

In-spite of all, the protection mechanisms available for both online and offline, there is a drastic increase in cyber-crimes. Therefore, some kind of prevention strategies should be adopted in the user's side to withstand and overcome such cyber-crimes. One such strategy is proposed in this research work by highlighting the importance of text that users post on social media. Users should be ethically aware of what content is presented to whom, who all are accessible to what all contents and what contents should be excluded in the post. The authors define the sensitivity of a user tweet based on many existing researches and user experiences.

Sensitivity

In context with this research work, sensitivity in a social media post, say Twitter, should be broadly categorized into three domains namely personal, professional and health-oriented information shared publicly in a social media platform. The sensitivity in a user message can be referred to *"a private identity or incriminating information disclosed by either the data owner itself or others without the consent or knowledge of the data owner"*. Here, the data owner can signify to a person or organization or community who can be identified or traced through the sensitive content posted by the social media user.

Hence the task of sensitive text classification is integrated for identifying the tweet sets, text categories and topics, taking text analytics to the next level as described by Hu et al. (2012). The objective of the research work is to formulate better tweet pre-processing rules, impose exact words for internet jargons and abbreviations in tweets, identify the sensitive keywords for personal, professional and health related tweets by modeling a framework called Sensitive Topic Keyword Extractor.

The major contributions of this research work are listed as follows:

1. To build an Automatic Acronym-Abbreviation Replacer for twitter community
2. To determine what is sensitive in personal, professional and health tweet domains
3. To generate unique topic models using LDA for personal, professional and health tweet domains
4. To investigate sensitive keywords that co-occur with the cyber-keywords in personal, professional and health tweet domains

2 Related Work

2.1 Twitter and Privacy Concerns

Lu et al. (2015) defined two privacy attributes namely confidentiality and universality with a set of 53 keywords. The privacy and security levels of Internet of Things (IoT) and Cyber Physical Systems (CPS) are classified by the Privacy Information Security Classification (PISC Model). Various challenges that are related to detecting privacy related information from unstructured texts are discussed by Sleeper et al. (2013) using natural language processing, ontology and machine learning approaches. It also deals with the problem of identifying the user's perception, domain specificity, context dependence, privacy sensitivity classification, data linkages in the social media messages. Wang et al. (2011) conducted a survey for capturing regrets that were caused by misunderstanding, misuse, misjudging of online posts and discovered the mechanisms preferred by the users to overcome the regrets such as apology, decline or delay the request, etc.

2.2 Twitter Topic Identification

The research work carried out by Aphinyanaphongs et al. (2014) presented a feasibility study of identifying and inferring the usage patterns and most frequent places where users posts alcohol related tweets using various classification techniques such as support vector machines, Naive Bayes, random forest and Bayesian logistic regression. Since natural language processing tools cannot be directly applied to social media texts, it is necessary to modify the extracted data so that it resembles standard texts. Baldwin et al. (2013) used data from various sources like Twitter, YouTube and Facebook comments, forums, blogs, Wikipedia and British National Corpus (BNC) to build corpus and performed lexical analysis, grammatical checks, language modeling and corpus similarity. A novel, unsupervised approach for detecting topic on twitter by using Formal Concept Analysis (FCA) is presented by Cigarrán et al. (2016). To handle the adverse effects of feature sparsity encountered by Tesfay et al. (2016) and curse of dimensionality in short text message classification, Lei et al. (2012) introduced the concept of boosting called LDABoost designed with the Naïve Bayes algorithm as a weak classifier. To improve the topic learning in social media messages, Mehrotra et al. (2013) proposed a novel approach of pooling and aggregating tweets in the tweet pre-processing step carried out for LDA. Whereas to learn the topic and behavior interests of the users, Qiu et al. (2013) introduced Behavior LDA (B-LDA) model using Twitter messages.

3 Methodology

The proposed Sensitive Topic Keyword Extractor analyses large volume of user generated tweets collected using the Twitter Streaming API for 68 cyber-keywords grouped under three categories namely personal, professional and health domains as derived from the findings of Lu et al. (2015) and Mao et al. (2011). The raw tweets are cleaned and prepared with pre-processing rules resulting with tweets suitable for text classification and analysis. The data annotation process for identifying sensitive tweets was performed using the most popular and reliable crowd sourcing platform called Amazon Mechanical Turk (AMT).

The LDA, being the most powerful probabilistic model for text classification is opted for identifying the topic keywords for each tweet category. The objective of the Sensitive Topic Keyword Extractor is to identify the next level of sensitive keywords that would cause potential vulnerability when used with the benchmarked cyber-keywords. The sensitive keywords that occur with the cyber-keywords are extracted using the Variational Expectation Maximization (VEM), VEM-Fixed and Gibbs model by filtering the non-cyber keywords present in keyword extraction process for various topics. Identifying such sensitive keywords will help users in realizing the potential threat caused by posting information related to those keywords. A detailed architecture of the proposed system, Sensitive Topic Keyword Extractor is represented in Fig. 1.

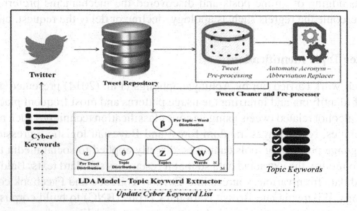

Fig. 1. Architecture of the proposed system – Sensitive Topic Keyword Extractor

3.1 Tweet Cleaning and Pre-processing

The tweets are short text messages posted by users without any formal template or sentence structure. The tweets tend to have short forms, internet jargons, abbreviations, symbols, numbers, date and time, email, URLs and other non-dictionary words as mentioned by Sproat et al. (2001). Therefore, it is necessary to wisely handle such inappropriate content and perform data transformation operations as listed in Table 1. During the process of cleanup, it is necessary to replace acronyms with corresponding abbreviations. Hence a dedicated list of acronyms and abbreviations is prepared with respect to twitter slang and internet jargons which will be used to handle the automatic replacement of short forms in the tweet.

3.2 Topic Keyword Detector Using LDA Model

The pre-processed tweets are taken to construct the term-document matrix which maps the words to its importance in the tweet and the overall tweets set under each category say

Table 1. A list of pre-processing rules used by Sensitive Topic Keyword Detector.

Pre-Processing Rule
Convert t to lowercase.
Remove"@" in user mentions and "#" in hashtags.
Replace URLs → 'url', date → "ddmmyy", time → "hhmm", phone numbers → "ph_no".
Remove stopword and perform stemming for each word.
Remove symbols, punctuations, numbers, unwanted whitespaces and new lines.
Replace acronyms with corresponding abbreviations and internet jargons.

personal, professional and health. There are basically three features considered for text classification namely Feature Frequency (FF), Term Frequency – Inverse Document Frequency (TF-IDF), Feature Presence (FP). The LDA model, an unsupervised probabilistic model generally combines topic, context and word through probability by focusing only words. The topics and keywords are extracted based on two parameters namely per document distribution and per topic word distribution. The LDA model generally has three steps in detecting topics. Firstly, the model identifies the term distribution β for each topic specified. Secondly, the proportion of each topic is computed for each tweet. Thirdly, for every word w_i in tweet t, a multinomial function is selected for each topic associated with that word; a word w_i is selected from that multinomial function based on the probability distribution for each topic. The domain-wise sensitive keywords are extracted by setting a threshold value $\emptyset_{tc}$ for each topic in a tweet category, the keywords are selected based on the conditional probability values. The results showed a set of well associated domain-wise sensitive keywords for all three models taken into consideration namely VEM, VEM_Fixed and Gibbs model used for topic learning and detection. The VEM is similar to Expectation Maximization technique with the difference of iteratively performing the expectation step until the likelihood becomes computationally intractable.

4 Results and Discussions

4.1 Dataset

The benchmarked cyber-keywords pre-defined by Lu et al. (2015) comprised of 23 keywords for personal topics, 22 keywords for professional topic and 7 for health topics. Since there are only 7 cyber-keywords related to health, an additional set of 16 health-related cyber-keywords were added by filtering out the most frequently used health keywords in internet as mentioned by Geetha et al. (2019). Finally, a list of 68 cyber-keywords are defined and used for tweet collection. The authors avoided manual inclusion of keywords in the cyber-keyword list due to the factor that it might lead in compromising the quality and relevance of data collected for sensitive data analysis. A large volume of real time tweets were collected that counted to 800900 tweets. This collection contributed about 36% of personal tweets, 32% of professional tweets and 32% of health-related tweets. As depicted in Fig. 2, on building the corpus data for analysis, the authors chose to consider only the original tweets by eliminating the retweets from the data repository. The tweet corpus built comprised of 35%, 33% and 32% of the total tweets collected for personal, professional and health related tweets.

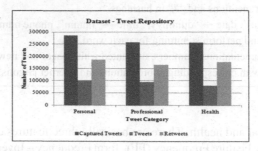

Fig. 2. The statistical representation of tweets collected for three tweet categories.

Dataset Annotation

The original tweets are further considered for annotation process as a Human Intelligence Task (HIT) that is performed by the AMT domain expert workers. The authors presented a detailed documentation composed of the following three itineraries to the AMT domain expert workers.

(i)   What is sensitivity in a social media text?
(ii) A ruleset for identifying the presence of sensitive private data or PII.
(iii) A set of gold data with sensitive and insensitive tweets in three tweet domains.

The set of gold data that were submitted to the AMT workers was the driving factor to arrive at a clear understanding and converge the rules and assumptions on sensitivity with the given tweets to identify the presence of PII. A sample gold data submitted to AMT domain expert workers is presented in Table 2.

On submission of the dataset, the authors received annotation results from 670 AMT works for 54176 tweets. The authors selected 153 out of 670 AMT domain experts by considering their knowledge in social media, turnaround time and consistency in annotating 50 tweets per session. The annotation process was completely assisted by the authors in case of any queries from the AMT domain expert workers. The annotation results submitted by the AMT domain expert workers were evaluated by applying the approach of multiple grading and gold standard evaluation. The tweets after the evaluation process is preprocessed with a set of standard data cleaning steps from Table 1 and stored in the tweet corpus as described in the Table 3.

4.2 Sensitive Topic Keyword Extractor Using LDA Model

The pre-processed tweets are taken for corpus building which in-turn is given for generating the bag-of-words by forming term document matrix. For text categorization,

Table 2. A sample of gold data submitted to the AMT domian expert workers.

Tweet domain	Gold set data	Annotation
Personal tweet	I got into a car accident last night and I think I'm still in shock	Sensitive
	It's getting easy for criminals to get anyone's mobile call record illegally. @hydcitypolice @CPHydCity @KTRTRS @amjedmbt @asadowaisi	Insensitive
Professional tweet	@AmericanAir Hi. I purchased a ticket online, the money was taken from my bank account, but I have no record locator or conf. email. Help!!	Sensitive
	These cards are submitted like insurance in the pharmacy. So the company now has your name, address, DOB, phone number, and med history.	Insensitive
Health tweet	Having RA has been an absolute test for me but I'm learning to live life w/this disease & it's getting more manageable day by day ??	Sensitive
	Diabetes-related eye disease often has no symptoms until it reaches an advanced stage #WorldSightDayAU	Insensitive

Table 3. Transformation of a sample tweet in data cleaning and pre-processing stage.

Sample tweet	FYI: For the next 7 days I will not have access to SMS or mobile phone. Please use email, WhatsApp, Viber, ... https://t.co/IDmYVyJnGy
Pre-processed tweet	For your information for the next 7 days i will not have access to short message service or mobile phone please use email whatsapp viber url
Tweet in corpus	Information next day access short message service mobile phone email whatsapp viber

feature selection is based on the selection of appropriate words by using dictionaries and NLP standards. Thus, this paper uses optimal text replacement strategies for building better feature space by using the proposed Automatic Acronym - Abbreviation Replacer (AAAR) specially designed for twitter. The AAAR was built with 1200 internet jargons and twitter slangs that are most popularly used by users and frequently encountered in our dataset. The AAAR is subject to change and update in accordance with the evolving practices and trends of online social networks.

The generated matrix is applied to three different models of LDA namely VEM, VEM-Fixed and Gibbs model to find topic keywords provided the k number of keywords to be found under n topics. The three category of tweet sets are taken individually and applied for LDA with k = 10; n = 5 and k = 20 and n = 10. Therefore, it generates 5 sets of 10 topic keywords for 5 topics and 10 sets of 20 topic keywords for 10 topics for all three LDA models. The sensitive keywords are identified by filtering out the cyber-keywords from the generated topic keywords.

Table 4 projects the overview of the identified sensitive keywords under each tweet category with cyber-keywords. The α value, parameter for finding the per topic distribution in tweet category denoted the confidence of the topic assignment. It can be inferred that, higher the value of α, more even the distribution of topics over the tweets. The lower the value of α, higher the percentage of tweets been assigned to one particular topic.

Table 4. A list of bench-marked cyber-keywords and identified sensitive keywords by the proposed strategy for personal, professional and health related tweets.

Tweet category	No. of cyber-keywords	Sample cyber-keywords	Identified sensitive keywords
Personal	23	car, chat, family, hobby, marriage, phone number. religion, spouse	friend, today, message, happy, time, god, live, stay, nice, great, work, read, application, hope
Professional	22	company address, insurance, investment, bank account, password, credit score, passport	twitter, real, money, order, private, trump, pay, state, incoming, market, debt. business, tax, team
Health	23	birth, blood type, height, disease, vaccine, therapy	food, body, world, home, mom, sinus, cancer, care, women, water, flu, sex

Therefore, Fig. 3 shows the topic distribution values achieved for all three tweet categories with three LDA models. It can be inferred from the topic modeling results that as the number of topic increases, the α value reduces which denotes the uneven distribution of topic keywords. The VEM model performs better than other two models with better variations in topic distribution for all three tweet categories as depicted in Fig. 4. The topic distribution for personal tweets with 20 keywords under 10 topics was 19.28085 whereas for 10 keywords with 5 topics was 24.58494. This clearly shows that as the number of topics increases, the keywords are more widely spread. The same behavior can be observed in all three tweet categories. This effect is due to the high sparsity that exists in social media messages which should be handled by adapting feature selection mechanisms.

The optimal model for sensitive keyword extraction was identified to be LDA with VEM model using TF-IDF for feature extraction with higher F1 score as shown in Fig. 4. The topic-wise keyword distribution achieved for each tweet category and a threshold $\emptyset_{tc} = 0.3$ determined by evaluating the model with varying iterations $nIter = \{10, 20, 30, 40, 50\}$ derived a set of 93 sensitive keywords.

The authors combined all cyber-keywords and identified sensitive keywords into bigrams from which the most frequently discussed sensitive topic would be identified as shown in Fig. 5. It can be inferred from the bi-gram analysis that in personal tweet category, the cyber-keyword 'phone' and sensitive keyword 'love' were frequently detected

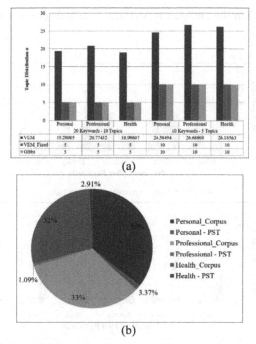

(a)

(b)

Fig. 3. a. The topic distribution parameter - α value achieved for three tweet category for VEM, VEM_Fixed and Gibbs Model to extract 20 keywords under 10 topics and 10 keywords under 5 topics. **b.** The proportion of Potentially Sensitive Tweets (PST) in the tweet corpus for three categories of tweet.

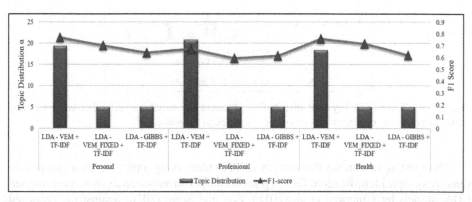

Fig. 4. The topic-wise keyword distribution (10 topics – 20 keywords) and F1 score achieved through various sampling models of LDA for three tweet categories.

to be occurring together. Similarly, the professional tweet category has 'account' and 'money' and health tweet category has 'weight' and 'loss' as the most frequently occurring cyber-keyword and sensitive keyword respectively. These co-occurring keywords are subject to vary when the analysis is performed with future user tweets in the same tweet domains.

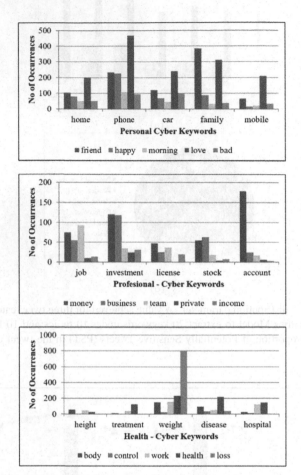

Fig. 5. The Topic-wise keyword distribution obtained for benchmarked cyber-keywords and sensitive keywords of three tweet categories modeled as bi-grams.

The existing researches that employ LDA in identifying topics has many scopes for sensitivity topic identification. Table 5 presents the performance analysis of most relevant LDA models by Gunawan et al. (2018), Zou and Song (2016) against the proposed method using the F1 measure thereby highlighting the suitability and uniqueness of the proposed approach for sensitive data identification in Tweets. Therefore, from the observed research results, the authors concluded that users should be cautious when using these cyber and sensitive keywords together because they are believed to have high potential in revealing personally identifiable information.

Table 5. Comparison of results for the sensitive data for various LDA models.

Dataset		
8,00,900 Tweets in three tweet domains		
Personal	2,86,249	
Professional	2,58,493	
Health	2,56,158	
Performance Comparison - F1 measure		
Twitter LDA – GIBBS (Gunawan et al. 2018)		
Personal	Professional	Health
0.642	0.59	0.667
LDA + TF-IDF (Zou et al. 2016)		
Personal	Professional	Health
0.695	0.653	0.638
Proposed Method		
LDA – VEM + TF-IDF		
Personal	*Professional*	*Health*
0.7581	*0.6682*	*0.7558*
LDA – VEM-Fixed + TF-IDF		
Personal	Professional	Health
0.6998	0.5891	0.7145

5 Conclusion

This research work discovers sensitive keywords in addition to the benchmarked cyber-keywords that are more frequently used along the cyber-keywords in user tweets. An enhanced data cleaning mechanism is implemented in replacing the abbreviations and internet jargons by building an Automated Acronym - Abbreviation Replacer module. The major contribution of this research work is building a framework called Sensitive Topic Keyword Extractor which the identified of 93 sensitive keywords that are potentially sensitive which when used with the benchmarked 68 cyber-keywords. Therefore, the sensitive keywords are also considered as critical or sensitive keywords and will be used in tweet collection and tweet repository building for the future works which will result in having a set of 161 query keywords for tweet extraction.

References

Tommasel, A., Godoy, D.: Short-text feature construction and selection in social media data: a survey. Artif. Intell. Rev. **49**(3), 301–338 (2016). https://doi.org/10.1007/s10462-016-9528-0

Hu, X., Liu, H.: Text analytics in social media. Min. Text Data, 385–414 (2012)

Lu, X., Qu, Z., Li, Q., Hui, P.: Privacy information security classification for internet of things based on internet data. Int. J. Distrib. Sens. Netw. **11**(8), 932–941 (2015)

Sleeper, M., et al.: I read my Twitter the next morning and was astonished: a conversational perspective on Twitter regrets. In: Proceedings of the SIGCHI Conference on Human Factors in Computing Systems, pp. 3277–3286 (2013)

Wang, Y., Norcie, G., Komanduri, S., Acquisti, A., Leon, P.G., Cranor, L.F.: I regretted the minute I pressed share: a qualitative study of regrets on Facebook. In: Proceedings of the Seventh Symposium on Usable Privacy and Security, p. 10. ACM (2011)

Aphinyanaphongs, Y., Ray, B., Statnikov, A., Krebs, P.: Text classification for automatic detection of alcohol use-related tweets: a feasibility study. In: 2014 IEEE 15th International Conference on Information Reuse and Integration (IRI), pp. 93–97 (2014)

Baldwin, T., Cook, P., Lui, M., MacKinlay, A., Wang, L.: How noisy social media text, how diffrnt social media sources? In: IJCNLP, pp. 356–364 (2013)

Cigarrán, J., Castellanos, Á., García-Serrano, A.: A step forward for topic detection in Twitter: an FCA-based approach. Expert Syst. Appl. **57**, 21–36 (2016)

Tesfay, W.B., Serna, J., Pape, S.: Challenges in detecting privacy revealing information in unstructured text. In: PrivOn@ ISWC (2016)

Lei, L., Qiao, G., Qimin, C., Qitao, L.: LDA boost classification: boosting by topics. EURASIP J. Adv. Signal Process., 233 (2012)

Mehrotra, R., Sanner, S., Buntine, W., Xie, L.: Improving LDA topic models for microblogs via tweet pooling and automatic labeling. In: Proceedings of the 36th International ACM SIGIR Conference on Research and Development in Information Retrieval, pp. 889–892 (2013)

Qiu, M., Zhu, F., Jiang, J.: It is not just what we say, but how we say them: LDA-based behavior-topic model. In: Proceedings of the 2013 SIAM International Conference on Data Mining, pp. 794–802. Society for Industrial and Applied Mathematics (2013)

Mao, H., Shuai, X., Kapadia, A.: Loose tweets: an analysis of privacy leaks on Twitter. In: Proceedings of the 10th Annual ACM Workshop on Privacy in the Electronic Society, 1–12 (2011)

Sproat, R., Black, A.W., Chen, S., Kumar, S., Ostendorf, M., Richards, C.: Normalization of non-standard words. Comput. Speech Lang. **15**(3), 287–333 (2001)

Gunawan, D., Rahmat, R.F., Putra, A., Pasha, M.F.: Filtering spam text messages by using Twitter-LDA algorithm. In: 2018 IEEE International Conference on Communication, Networks and Satellite (Comnetsat), pp. 1–6. IEEE, November 2018

Zou, L., Song, W.W.: LDA-TM: a two-step approach to twitter topic data clustering. In: 2016 IEEE International Conference on Cloud Computing and Big Data Analysis (ICCCBDA), pp. 342–347. IEEE, July 2016

Geetha, R., Karthika, S., Pavithra, N., Preethi, V.: Tweedle: sensitivity check in health-related social short texts based on regret theory. Procedia Comput. Sci. **165**, 663–675 (2019)

Sentiment Analysis of Bengali Tweets Using Deep Learning

Kamal Sarkar(✉) (iD)

Department of Computer Science and Engineering, Jadavpur University, Kolkata, India
jukamal2001@yahoo.com

Abstract. Sentiment analysis is the research area that deals with analysis of sentiments expressed in the social media texts written by the internet users. Sentiments
of the users are expressed in various forms such as feelings, emotions and opinions. Tweet sentiment polarity detection is an important sentiment analysis task
which is to classify an input tweet as one of three classes: positive, negative and
neutral. In this study, we compare various deep learning methods that use LSTM,
BILSTM and CNN for the sentiment polarity classification of Bengali tweets.
We also present in this paper a comparative study on the Bengali tweet sentiment
polarity classification performances of the traditional machine learning methods
and the deep learning methods.

Keywords: Sentiment polarity classification · Machine learning · Deep
learning · Bengali tweets · LSTM · BILSTM · CNN · Naïve Bayes · Support
Vector Machines

1 Introduction

There has been a tremendous surge of data on the internet after the social media boom.
Since the advent of the Social Web, the amount of textual content on the Internet has
exponentially grown and provides a great deal of potential for analysis. Since millions
of tweets are generated every day, machine learning can be more useful for efficient
and fast data analysis to better understand the activities of the social media users as
well as the users' feedback on various social and political issues. The online merchants
allows the customer to give feedback on various products. The opinions and comments
on the other issues related to crickets, sports, movies, music etc. are also available on
the Internet.

After collecting a vast amount of users' feedback or opinions/comments, the next
task is to analyze them to derive knowledge that can help the online merchants in improving the customer services and launching new products to satisfy customer needs. The
opinions and comments related to various political and social events can be useful in
policy making as well as enriching the various academic fields such as politics, political
science, sociology, psychology. Sentiment of a text refers to the expression of people's
thought, judgment or attitude prompted by feelings [1] in the text. Sentiment polarity

© IFIP International Federation for Information Processing 2020
Published by Springer Nature Switzerland AG 2020
A. Chandrabose et al. (Eds.): ICCIDS 2020, IFIP AICT 578, pp. 71–84, 2020.
https://doi.org/10.1007/978-3-030-63467-4_6

classification is one of the important sentiment analysis tasks. In sentiment polarity classification task, sentiment expressed in an input tweet is classified into one of three categories - positive, negative or neutral.

Since manual processing of the vast amount of social media texts for knowledge extraction is a difficult task, there is a need for an automatic system which can determine the polarity of the sentiments of social media texts coming from multiple heterogeneous sources.

Since the Bengali language is the 7th most spoken language in the world, with approximately 215 million speakers all over the world. Bengali is a primary language in Bangladesh, and in India - West Bengal, Tripura and Assam and recently Bengali tweets are becoming available on the various social media, there is a need for a tool for sentiment classification of Bengali tweets. But the most existing sentiment classification systems have been developed for processing English texts and a limited numbers of research works on Bengali sentiment classification are reported in the literature.

In this paper, we focus on developing sentiment classification system for Bengali tweets and report the results obtained for Bengali sentiment classification using the several deep learning algorithms such as LSTM (Long Short Term Memory), BILSTM (Bidirectional Long Short Term Memory) and CNN (Convolutional Neural Networks).

In the rest of the paper, Sect. 2 discusses the existing works related to our work. In Sect. 3, we present descriptions of the model. In Sect. 4, we discuss the experimental setup, evaluation and results. We conclude in Sect. 5.

2 Related Work

The earliest works on sentiment polarity classification used lexicon based sentiment polarity detection methods [2] that used information retrieved from sentiment lexicon to determine sentiment polarity (positive class or negative class) of the text. The simpler version of sentiment lexicon is a mere collection of positive and negative terms [3]. Since manual construction of such lexicon is a laborious task, automatic construction of sentiment lexicon has also been proposed [4, 5]. Some researchers have studied on automatic prediction of semantic orientation of adjectives and adjectival phrases [6, 7]. The approach proposed in [8] determines polarity of words. SentiWordNet [9] is a more informative sentiment lexicon which is created by analyzing the glosses (associated with synsets) retrieved from WordNet [10]. The main criticism of the lexicon based sentiment analysis is that maintaining a universal sentiment lexicon is difficult.

Compared to sentiment lexicon based approach, machine learning based approach has been more popular due to the fact that the machine learning based approach does not require in-depth domain knowledge for crafting rules from data and it automatically learns the relationship between input and the output from the data provided to the system. So, the most existing works on sentiment polarity detection considers this task as a classification problem and use a number of machine learning algorithms for course grained (document level) sentiment polarity classification [11–18]. The more fine-grained analysis has also been done using the linguistic features combined with machine learning methods [19–22]. Though sentence level sentiment polarity classification [23] is not uncommon today, the substantial portion of previous research works focused on sentiment classification of product and movie reviews [24–27].

In supervised learning method for sentiment polarity detection, a labeled corpus converted to a collection of labeled feature vectors is provided to the learning algorithms. This labeled data is used to train a machine learning (ML) algorithm and then the trained model is applied to classify new unseen texts. Since ML algorithm requires each instance of text to be transformed into feature vector, the common features used for feature representation are individual words, n-grams, surrounding words, punctuations and the popular supervised learning algorithms used for this task are SVM (Support Vector Machines), K-NN (K-nearest neighbor), decision tree and Naïve Bayes [28–32] and ANN (Artificial Neural Networks) [33].

Some approaches to sentiment polarity classification are the hybrid approaches [34] that combine both supervised and unsupervised learning for analyzing reviews. Sentiment analysis of micro-blog posts and tweets have been presented in [35, 36].

Within the last several years, deep learning based approaches to sentiment analysis have been applied and many research papers have reported efficacy of deep learning based approaches applied to sentiment polarity detection tasks [8, 37, 39].

Most of the existing works on sentiment polarity detection discussed above has dealt with detecting polarity of non-Indian language texts. Since many Indians are now actively participating in writing their opinions and comments on Twitter or other micro-blog sites using Indian languages, opinionated texts in Indian languages are now available on the Internet. But there is a limited number of research works on sentiment analysis of tweets or comments or opinions written in Indian languages. So, there is a need for a system that can classify a large volume of Indian social media texts.

For sentiment analysis of tweets in Indian languages, a few attempts have been made so far. A contest on sentiment analysis in Indian languages (SAIL) tweets @ MIKE 2015 was held in 2015 at IIIT Hyderabad, India [40].

Some studies on sentiment analysis for Hindi language have been presented in [41–43].

Some baseline results on aspect-based sentiment analysis has been reported in [44]. Deep learning based sentiment analysis system that uses LSTM and CNN for sentiment analysis of Bengali tweets has been presented in [45] and [38].

Unlike the prior work on tweet sentiment analysis for Bengali tweets, which applied Naïve Bayes and SVM [35, 46], we have used in this present work the bidirectional recurrent neural networks named LSTM (Long Short Term Memory) for sentiment classification. We have also compared the sentiment classification performances of LSTM, BILSTM and CNN.

3 Model Development

Our proposed system for sentiment polarity detection uses a special kind of deep learning models. Our proposed system has several components - (1) preprocessing (2) development of deep learning models

3.1 Preprocessing

Since tweet data is noisy, we preprocess our data to clean unimportant characters. From tweets in our corpus, the characters [,-, _, =, :, +, $,@,~,!, ,,;,/,^,),(,],{,},<,> are removed.

At this stage, we use SentiWordNet to augment the tweet terms by searching terms in SentiWordNet and retrieving the corresponding sentiment tags. We have used for our present work, the SentiWordNet [47], downloaded from http://amitavadas.com/sen tiwordnet.php. This is basically a collection of three types of polarity words-positive, negative and neutral words. How the terms of a tweet are augmented is illustrated as follows. For example, the following two tweets:

Tweet1: *অতি অসাধারণ <Positive> ! ক্যামেরার কাহিনী দেখে পুরাই চমৎকার হইয়া গেলাম!(Awesome! I was totally amazed by the camera's story!)*

Tweet2: *জঘন্য মুভি, আমার কাছে মনে হয়েছে পুরো ৩ টা ঘন্টা নষ্ট করেছি (Damn movie, I feel like I've wasted an entire hour)*

They are augmented as follows:

Tweet1: *অতি অসাধারণ <Positive> ! ক্যামেরার কাহিনী দেখে পুরাই চমৎকার <positive> হইয়া গেলাম!*

Tweet2: *জঘন্য <negative> মুভি, আমার কাছে মনে হয়েছে পুরো ৩ টা ঘন্টা নষ্ট <negative> করেছি*

Since the words "অসাধারণ (Awesome)" and "চমৎকার (Excellent)" of tweet1 are found in SentiWordNet, they are augmented with <positive>). Since the words "জঘন্য (Damn)" and "নষ্ট (wasted)" of tweet2 are found in SentiWordNet, they are augmented with <negative>).

This augmentation is done because we observe that the external knowledge of word level sentiments can help in classifying overall polarity of tweets. After processing all the training tweets in the above mentioned way, they are submitted to the deep learning algorithms for model development.

3.2 Deep Learning Models

For implementing our proposed sentiment classification system, we have considered three deep learning algorithms - (1) Recurrent Neural Networks called LSTM, (2) Bidirectional Recurrent Neural Networks called BILSTM and (3) Convolutional Neural Networks (CNN). Unlike the traditional sentiment analysis model that uses multinomial Naïve Bayes and SVM along with bag-of-words tweet representation [35, 46], the deep learning based models can take into account the entire sequence of tokens contained in a tweet while detecting sentiment polarity of the tweet. Since the deep learning models with word embedding layer transforms the input words into the dense word vectors, the data sparseness problem arising out from the bag-of-words tweet representation used by the traditional machine learning based models is overcome to some extent by the deep learning models.

3.2.1 LSTM and BILSTM Based Models

The important feature of LSTM based deep learning based model is that it can process sequence of tokens in the input tweet to encode it to a contextual embedding that passes through the final *softmax* layer that classifies the tweet into one of the three sentiment classes - positive, negative and neutral.

Thus the LSTM based model can make use of the appropriate context words in the sequence of tweet words for sentiment polarity prediction.

A LSTM (Long Short Term Memory) is a special type of recurrent neural networks, which can process relatively long sequence of data in better way because it does not suffer from so-called gradient vanishing problem [48]. The details of LSTM with its working principle can be found in [49].

The outline of the broad architecture that we have used for developing our LSTM based sentiment polarity detection model is given in Fig. 1. The output of LSTM unit goes to the softmax layer which predicts the class of the input tweet.

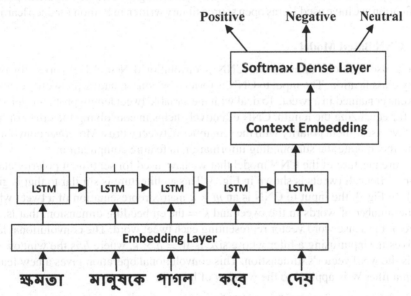

Fig. 1. Unrolled LSTM over a sample input tweet **"ক্ষমতা মানুষকে পাগল করে দেয়"** (Power makes a man mad)

While LSTM can process the input in only forward direction (from the left to right of the input word sequence), the bidirectional LSTM (BILSTM) can capture the better contextual information by processing the input word sequence in both the forward and backward directions.

In practice, architecture involves two recurrent layers side-by-side, then providing the input sequence as input to the first layer and providing a reversed copy of the input sequence to the second. The outputs of two parallel LSTM layers are finally merged for improving the performance of Recurrent Neural Networks (RNNs). Figure 2 shows how bidirectional LSTM works.

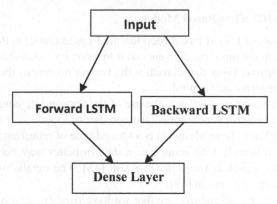

Fig. 2. Bidirectional LSTM architecture

For implementation of our LSTM based deep learning models for sentiment polarity classification, we have used Keras open source library written in Python for deep learning.

3.2.2 CNN Based Model

We have also used one dimensional CNN (Convolutional Neural Networks) for tweet polarity classification. The input to CNN is formed by concatenating the word vectors of the tokens contained in a tweet. To deal with the variable tweet length problem, zeros are added for equalizing the length. CNN effectively helps in convolving the effect of local (word level) sentiment polarity with the semantics of tweet words. Moreover convolution helps to incorporate the surrounding information in feature computation.

The architecture of the CNN model that we have used for sentiment polarity classification of Bengali tweets is shown in Fig. 3. This architecture is similar to that is given in [38]. In Fig. 3, the input to CNN is an $m \times k$ matrix representation of a tweet where m = the number of words in the tweet and k = the embedding dimension, that is, k is the size of the dense word vector representing each tweet word. The convolutional layer convolves the input using a filter whose size is $W = h \times k$, where h is the window size and k is the word vector's dimenstion. This convolutional operation gives a new feature f_i when a filter W is applied to the window of h words:

$$f_i = g\big(W.X_{(i:i+h-1)} + b\big)$$

Where $X_{(i:i+h-1)}$ is the vector formed by concatenating the word vectors corresponding to the words falling in the window that spans from the ith word to the $(i + h - 1)^{th}$ word in the tweet, b is the bias weight and g is the non-linear transformation functions such as hyperbolic tangent or rectilinear.

The filter is slided right with stride equal to 1 and each possible window of word vectors are convolved to a new feature. This gives a feature map:

$$f = [f_1 f_2, \ldots, f_{(m-h+1)}].$$

Then max pooling [35] is applied on all the obtained feature maps (the number of filters determine number of feature maps) to select the most salient feature per feature map f. Using pooling the problems related to the variable tweet length are also overcome.

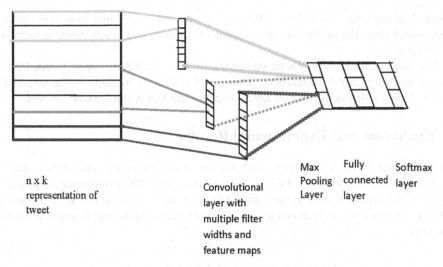

n x k
representation of
tweet

Convolutional
layer with
multiple filter
widths and
feature maps

Max   Fully
Pooling   connected   Softmax
Layer   layer   layer

Fig. 3. CNN architecture

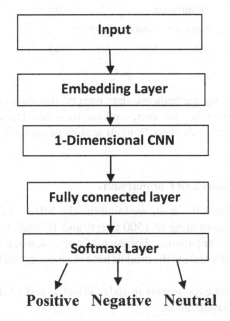

Fig. 4. Steps of our implementation of CNN model

Thus each filter can give one convoluted feature and we obtain a feature vector of length N if N filters are used (For the different filters, the different window sizes can be used).

The output of the convolution layer is then connected to a fully connected (FC) layer which receives as input a vector of dimension N. Here the convolutional layer acts as feature learning layer. The output of this fully connected layer is then passed to *softmax* layer which gives the probability distribution over three class labels-positive, negative and neutral.

For implementing the CNN based model, we have used Keras open source library written in Python for deep learning. The model is developed using Keras sequential model. Figure 4 shows the various layers used for implementing the CNN model.

4   Evaluation and Experimental Results

We have used Bengali sentiment analysis datasets used in the shared task named SAIL@MIKE 2015, held at IIIT Hyderabad, India [40]. The distribution of positive, negative and neutral tweets in the training set and the size of the test set is shown in Table 1. The training set contains a total of 1000 tweets and the test contains a total of 500 tweets.

Table 1. Description of SAIL 2015 Bengali dataset

Training set			Test set
Negative	Positive	Neutral	
356	368	276	500

Since the deep learning methods are data hungry, this dataset is not sufficient for training deep learning models. However, we have used this dataset because this is the only dataset on which many existing Bengali sentiment analysis techniques had been tested.

4.1   Experimental Results and Comparisons

Since training set is relatively small, we combine the MIKE training set and test set to form a single dataset consisting of 1500 tweets and 10-fold cross validation is done for model performance comparisons. For each of our developed models, we compute average accuracy over 10 folds and the models have been compared based on the obtained average accuracy.

We conducted several experiments to judge effectiveness of deep learning models for sentiment polarity classification.

Experiment 1: In this experiment, we have developed a simple LSTM based model which has been discussed in Sect. 3.

Experiment 2: In this experiment, we have also developed an bidirectional LSTM model as discussed in Sect. 3.

Experiment 3: In this experiment, we have developed a CNN based model as discussed in Sect. 3.

4.1.1 Model Configuration

The deep learning models have several tunable hyperparametes which need to be properly tuned to obtain the best possible results. We have tuned the hyperparameters in sequential manner. For example, initially hyperparameters are set to default values as per Keras implementation of LSTM based model, then batch size is tuned by varying its value from low values to high values and the batch size that gives the best results is noted. Now by setting the batch size to the obtained optimal value, other hyperparametes except embedding size are set to the default values and embedding size is tuned. Thus the best configuration of each model is found.

We have obtained the best results for our LSTM based model with the following configurations:

embedding_size = 100
SpatialDropout = 0.4
Dropout = 0.2 and recurrent_dropout = 0.2
dimension of the output space = 100
Loss function = "categorical_crossentropy"
optimizer = "Adam"
epochs = 25
batch_size = 100

For the bidirectional LSTM model, we obtained the best results with the following settings of the model hyperparameters:

embedding_size = 32
Dropout = 0.2
dimension of the output space = 32
Loss function = "categorical_crossentropy"
optimizer = "RmsProp"
epochs = 20
batch_size = 64

For the CNN base model, the best results are obtained with the following model configuration:

embedding_size = 32
Activation of CNN layer = "relu"
Number of filters = 32, kernel_size = 3, strides = 1
Max pooling size = 3
Size of the first dense layer = 32
Activations of the dense layer = "tanh"
Dropout = 0.2
Loss function = "categorical_crossentropy"
optimizer = "Adam"
epochs = 20
batch_size = 64

The 10-fold cross validated results of our developed deep learning models with the above mentioned configurations are shown in Table 2.

Table 2. Performances comparisons of our proposed deep learning based models for Bengali sentiment classification

Models	Accuracy (%)
The BILSTM based model	**55.73**
The LSTM based model	55.27
The CNN based model	51.93

As we can see from Table 2, BILSTM performs the best among the three deep learning models we have implemented.

4.2 Model Comparisons and Discussion

By comparing the results of all the deep learning based models, we can observe that, for Bengali sentiment classification task, the BILSTM based model can perform slightly better than the other deep learning models.

To judge the effectiveness of the deep learning models, we have compared them with the traditional machine learning algorithms. Since some previous works suggest that two machine learning algorithms- Multinomial Naive Bayes [35] and SVM with the linear kernel [46] are effective in Bengali sentiment polarity classification, we have chosen these two algorithms for comparisons with our developed deep learning based models. Multinomial Naive Bayes and SVM based models have been developed on WEKA[1] platform with the same feature set and parameter settings as mentioned in [35] and [46] respectively. The comparisons of the results are shown in Table 3.

Table 3. Performances comparisons of our proposed deep learning based model and two existing machine learning based models for Bengali sentiment classification

Models	Accuracy (%)
The BILSTM based model	**55.73**
The LSTM based model	55.27
The CNN based model	51.93
SVM with unigram features [46]	53.73
Multinomial Naïve Bayes with unigram and bigram features[35]	53.07

It is evident from Table 3 that performances of LSTM and BILSTM based models are slightly better than the traditional machine learning algorithms-Multinomial Naive

---

[1] http://www.cs.waikato.ac.nz/ml/weka/.

Bayes and SVM though the CNN based model performs like the traditional machine learning algorithms. BILSTM performs the best because of the fact that it analyzes the tweet from both sides: left-to-right and right-to-left, whereas the LSTM processes the tweet from left-to-right only.

We can also see that deep learning models outperform the traditional machine learning models by small margin. It is also unclear to us why CNN gives relatively poor performance. One of the possible reasons for such results of deep learning models may be the nature of SAIL 2015 dataset and the size of the dataset. Since deep learning model is highly data hungry and SAIL 2015 dataset is insufficient for training the deep learning models, the deep learning models could not outperform the traditional machine learning models by large margin. Though SAIL 2015 dataset is small, we have applied deep learning on this dataset for several reasons - (1) to the best of our knowledge, this is only benchmark dataset for Bengali tweet sentiment analysis, (2) *No free lunch* theorem states that no learning algorithm can outperform another when evaluated over all possible classification problems and (3) to investigate whether deep learning can be the automatic choice for Bengali tweet sentiment analysis task or not.

5   Conclusion

In this work, we have presented several deep learning based models for sentiment polarity classification of Bengali tweets. SentiWordNet has been incorporated in each model to utilize external knowledge.

We feel that we need more training data for improving accuracy of the deep learning models for sentiment polarity classification. With the availability of relatively large training data and the hybrid deep learning models such as BILSTM with additional attention layers can be used for developing more enhanced deep learning model. For improving the performance of the CNN based model, transfer learning mechanism can be integrated with the model in future. With slight modifications, our proposed deep learning models can easily be applied to the tweet data sets for other Indian languages.

Acknowledgment. This research work has received support from the project entitled "Indian Social Media Sensor: an Indian Social Media Text Mining System for Topic Detection, Topic Sentiment Analysis and Opinion Summarization" funded by the Department of Science and Technology, Government of India under the SERB scheme.

References

1. Bowker, J.: The Oxford Dictionary of World Religions. Oxford University Press, Oxford (1997)
2. Melville, P., Gryc, W., Lawrence, R.D.: Sentiment analysis of blogs by combining lexical knowledge with text classification. In: Proceedings of the 15th ACM SIGKDD International Conference on Knowledge Discovery and Data Mining, pp. 1275–1284. ACM (2009)
3. Ramakrishnan, G., Jadhav, A., Joshi, A., Chakrabarti, S., Bhattacharyya, P.: Question answering via Bayesian inference on lexical relations. In: Proceedings of the ACL 2003 Workshop on Multilingual Summarization and Question Answering, vol. 12, pp. 1–10. Association for Computational Linguistics (2003)

4. Jiao, J., Zhou, Y.: Sentiment polarity analysis based multi-dictionary. Phys. Procedia **22**, 590–596 (2011)
5. Macdonald, C., Ounis, I.: The TREC Blogs06 collection: creating and analysing a blog test collection. Department of Computer Science, University of Glasgow Tech Report TR-2006-224, 1, 3-1 (2006)
6. Hatzivassiloglou, V., McKeown, K.R.: Predicting the semantic orientation of adjectives. In: Proceedings of the 35th Annual Meeting of the Association for Computational Linguistics and Eighth Conference of the European Chapter of the Association for Computational Linguistics, pp. 174–181. Association for Computational Linguistics, July 1997
7. Wiebe, J.: Learning subjective adjectives from corpora. In: AAAI/IAAI, pp. 735–740, July 2000
8. Riloff, E., Wiebe, J.: Learning extraction patterns for subjective expressions. In: Proceedings of the 2003 Conference on Empirical Methods in Natural Language Processing, pp. 105–112. Association for Computational Linguistics, July 2003
9. Esuli, A., Sebastiani, F.: SENTİWORDNET: a publicly available lexical resource for opinion mining. In: Proceedings of LREC, vol. 6, pp. 417–422, May 2006
10. Fellbaum: "WordNet". Blackwell Publishing Ltd. (1999)
11. Zhao, J., Liu, K., Wang, G.: Adding redundant features for CRFs-based sentence sentiment classification. In: Proceedings of the Conference on Empirical Methods in Natural Language Processing. Association for Computational Linguistics, pp. 117–126 (2008)
12. Joachims, T.: Making Large Scale SVM Learning Practical. Published in book-Advances in Kernel Methods, pp. 169–184 (1999)
13. Pang, B., Lee, L., Vaithyanathan, S.: Thumbs up?: sentiment classification using machine learning techniques. In: Proceedings of the ACL-02 Conference on Empirical Methods in Natural Language Processing, vol. 10, pp. 79–86. Association for Computational Linguistics (2002)
14. Dave, K., Lawrence, S., Pennock, D.M.: Mining the peanut gallery: opinion extraction and semantic classification of product reviews. In: Proceedings of the 12th İnternational Conference on World Wide Web, pp. 519–528. ACM (2003)
15. Mullen, T., Collier, N.: Sentiment analysis using support vector machines with diverse information sources. In: EMNLP, vol. 4, pp. 412–418 (2004)
16. Pang, B., Lee, L.: Opinion mining and sentiment analysis. Found. Trends Inf. Retr. **2**(1–2), 1–135 (2008)
17. Goldberg, B., Zhu, X.: Seeing stars when there aren't many stars: graph-based semi-supervised learning for sentiment categorization. In: Proceedings of the First Workshop on Graph Based Methods for Natural Language Processing, Association for Computational Linguistics, pp. 45–52 (2006)
18. Miao, Q., Li, Q., Zeng, D.: Fine grained opinion mining by integrating multiple review sources. J. Am. Soc. Inf. Sci. Technol. **61**(11), 2288–2299 (2010)
19. Riloff, E., Wiebe, J.: Learning extraction patterns for subjective expressions. In: Proceedings of the 2003 Conference on Empirical Methods in Natural Language Processing, pp. 105–112. Association for Computational Linguistics (2003)
20. Prabowo, R., Thelwall, M.: Sentiment analysis: a combined approach. J. Inf. **3**(2), 143–157 (2009)
21. Narayanan, R., Liu, B., Choudhary, A.: Sentiment analysis of conditional sentences. In: Proceedings of the 2009 Conference on Empirical Methods in Natural Language Processing, vol. 1, pp. 180–189. Association for Computational Linguistics (2009)
22. Wiegand, M., Balahur, A., Roth, B., Klakow, D., Montoyo, A.: A survey on the role of negation in sentiment analysis. In: Proceedings of the Workshop on Negation and Speculation in Natural Language Processing, pp. 60–68. Association for Computational Linguistics (2010)

23. Yu, H., Hatzivassiloglou, V.: Towards answering opinion questions: separating facts from opinions and identifying the polarity of opinion sentences. In: Proceedings of the 2003 Conference on Empirical Methods in Natural Language Processing, pp. 129–136. Association for Computational Linguistics, July 2003

24. Ku, L., Liang, Y., Chen, H.: Opinion extraction, summarization and tracking in news and blog corpora. In: AAAI Spring Symposium: Computational Approaches to Analyzing Weblogs (2006)

25. Kim, J., Chern, G., Feng, D., Shaw, E., Hovy, E.: Mining and assessing discussions on the web through speech act analysis. In: Proceedings of the Workshop on Web Content Mining with Human Language Technologies at the 5th International Semantic Web Conference (2006)

26. Pang, B., Lee, L.: A sentimental education: sentiment analysis using subjectivity summarization based on minimum cuts. In: Proceedings of the 42nd Annual Meeting on Association for Computational Linguistics, p. 271 (2004)

27. Zhu, F., Zhang, X.: Impact of online consumer reviews on sales: the moderating role of product and consumer characteristics. J. Mark. **74**(2), 133–148 (2010)

28. Pang, B, Lee, L.: A sentimental education: sentiment analysis using subjectivity summarization based on minimum cuts. In: Proceedings of ACL (2004)

29. Chen, C., Tseng, Y.D.: Quality evaluation of product reviews using an information quality framework. Decis. Support Syst. **50**(4), 755–768 (2011)

30. Kang, H., Yoo, S.J., Han, D.: Senti-lexicon and improved Naïve Bayes algorithms for sentiment analysis of restaurant reviews. Expert Syst. Appl. **39**(5), 6000–6010 (2012)

31. Clarke, D., Lane, P., Hender, P.: Developing robust models for favourability analysis. In: Proceedings of the 2nd Workshop on Computational Approaches to Subjectivity and Sentiment Analysis, pp. 44–52. Association for Computational Linguistics (2011)

32. Reyes, A., Rosso, P.: Making objective decisions from subjective data: detecting irony in customer reviews. Decis. Support Syst. **53**(4), 754–760 (2012)

33. Moraes, R., Valiati, J.F., Neto, W.P.G.: Document-level sentiment classification: an empirical comparison between SVM and ANN. Expert Syst. Appl. **40**(2), 621–633 (2013)

34. Martín-Valdivia, M.T., Martínez-Cámara, E., Perea-Ortega, J.M., Ureña-López, L.A.: Sentiment polarity detection in Spanish reviews combining supervised and unsupervised approaches. Expert Syst. Appl. **40**(10), 3934–3942 (2013)

35. Sarkar, K., Chakraborty, S.: A sentiment analysis system for Indian language tweets. In: Prasath, R., Vuppala, A.K., Kathirvalavakumar, T. (eds.) MIKE 2015. LNCS (LNAI), vol. 9468, pp. 694–702. Springer, Cham (2015). https://doi.org/10.1007/978-3-319-26832-3_66

36. Li, Y.M., Li, T.Y.: Deriving market intelligence from microblogs. Decis. Support Syst. **55**(1), 206–217 (2013)

37. Ouyang, X., Zhou, P., Li, C.H., Liu, L.: Sentiment analysis using convolutional neural network. In: Proceedings of IEEE International Conference on Computer and Information Technology, Ubiquitous Computing and Communications, Dependable, Autonomic and Secure Computing, Pervasive Intelligence and Computing (CIT/IUCC/DASC/PICOM), pp. 2359–2364 (2015)

38. Sarkar, K.: Sentiment polarity detection in Bengali tweets using deep convolutional neural networks. J. Intell. Syst. **28**(3), 377–386 (2019)

39. Wang, X., Jiang, W., Luo, Z.: Combination of convolutional and recurrent neural network for sentiment analysis of short texts. In: Proceedings of COLING 2016, the 26th International Conference on Computational Linguistics: Technical Papers, pp. 2428–2437 (2016)

40. Patra, B.G., Das, D., Das, A., Prasath, R.: Shared task on sentiment analysis in Indian languages (SAIL) tweets - an overview. In: Prasath, R., Vuppala, A.K., Kathirvalavakumar, T. (eds.) MIKE 2015. LNCS (LNAI), vol. 9468, pp. 650–655. Springer, Cham (2015). https://doi.org/10.1007/978-3-319-26832-3_61

41. Arora, P.: Sentiment analysis for Hindi language. MS by Research in Computer Science (2013)
42. Joshi, A., Balamurali, A.R., Bhattacharyya, P.: A fall-back strategy for sentiment analysis in Hindi: a case study. In: Proceedings of the 8th ICON (2010)
43. Sharma, Y., Mangat, V., Kaur, M.: A practical approach to sentiment analysis of Hindi tweets. In: 2015 1st International Conference on Next Generation Computing Technologies (NGCT), pp. 677–680, September 2015
44. Rahman, M., Kumar Dey, E.: Datasets for aspect-based sentiment analysis in bangla and its baseline evaluation. Data 3(2), 15 (2018)
45. Sarkar, K.: Sentiment polarity detection in Bengali tweets using LSTM recurrent neural networks. In: 2019 Second International Conference on Advanced Computational and Communication Paradigms (ICACCP), Gangtok, India 2019, pp. 1–6 (2019). https://doi.org/10.1109/icaccp.2019.8883010
46. Sarkar, K., Bhowmik, M.: Sentiment polarity detection in Bengali tweets using multinomial Naïve Bayes and support vector machines. In: CALCON 2017, Kolkata (2017)
47. Das, A., Bandyopadhyay, S.: SentiWordNet for Indian languages. In: Proceedings of 8th Workshop on Asian Language Resources (COLING 2010), Beijing, China, pp. 56–63 (2010)
48. Bengio, Y., Simard, P., Frasconi, P.: Learning long-term dependencies with gradient descent is difficult. IEEE Trans. Neural Netw. 5(2), 157–166 (1994)
49. Hochreiter, S., Schmidhuber, J.: Long short-term memory. Neural Comput. 9(8), 1735–1780 (1997). https://doi.org/10.1162/neco.1997.9.8.1735

Social Media Veracity Detection System Using Calibrate Classifier

P. SuthanthiraDevi$^{(\boxtimes)}$ ⓘ and S. Karthika

Department of Information Technology, SSN College of Engineering, Kalavakkam 603110,
Tamil Nadu, India
devisaran2004@gmail.com, skarthika@ssn.edu.in

Abstract. In the last decade, social media has grown extremely fast and captured tens of millions of users are online at any time. Social media is a powerful tool to share information in the form of articles, images, URLs and, videos online. Concurrently it also spreads the rumors. To fight against the rumors, media users need a verification tool to verify the fake post on Twitter. The main motivation of this research work is to find out which classification model helps to detecting the rumor messages. The proposed system adopts three feature extraction techniques namely Term Frequency-Inverse Document Frequency, Count-Vectorizer and Hashing-Vectorizer. The authors proposed a Calibrate Classifier model to detect the rumor messages in twitter and this model has been tested on real-time event#gaja tweets. The proposed calibrate model shows better results for rumor detection than the other ensemble models.

Keywords: Rumor · Count vectorizer · TF-IDF · Hashing vectorizer · Ensemble learning

1 Introduction

On the emergence of online social networking services, many researchers have been interested to analyze the veracity of social media data. Nowadays social media sites like Twitter, Facebook has more popularity than other micro blogging services. It requires minimum time and cost to share information and the level of usage increases the volume and velocity of the data. People spend time on social media is increased gradually [1]. At the same time, this social media platform is speeding up the dis-closure of data and broadcasting the incorrect information. A huge amount of rumor messages spread over this media during crisis time. The definition of a rumor is "An item of circulating information whose veracity status is yet to be verified at the time of posting" [2]. Rumors can affect the society as well as individuals in the following ways (i) It can disturb the authenticity of the news media. (ii) This rumor information influences the media users to accept biased stories. Some of the people and companies disseminate the rumor news for their political and financial gain [3]. For example, during the US presidential election 2016, most of the posts on social media were fake [4]. In India, during the national

© IFIP International Federation for Information Processing 2020
Published by Springer Nature Switzerland AG 2020
A. Chandrabose et al. (Eds.): ICCIDS 2020, IFIP AICT 578, pp. 85–98, 2020.
https://doi.org/10.1007/978-3-030-63467-4_7

election 2019, various Whatapp users were created to spread the rumor message against the current ruling party [5]. During times of crisis like cyclone Gaja, a huge amount of tweets are generated by people and institutions who report various news and information related to the cyclone. A total of around 90,867 tweets are collected using various keywords and hashtags such as #CycloneGaja, #SaveDelta, and @TNSDMA, which includes images, videos, and texts. Sample of fake images and twitter posts which are generated by various users are shown in Fig. 1.

At present detecting rumor on social media is the biggest challenge for government officials. It is important to deal with the issue of rumor message dissemination on twitter during the crisis time. The main objective of this work is to identify the tweet posts as rumor or not by using the Ensemble Classifiers such as Bagging, Boosting and Calibrate classifiers. These classifiers were trained and tested with the aid of the cyclone event #gaja dataset. The authors propose a Calibrate classifier to identify the rumor tweets with significant accuracy as compared to the state-of-art machine learning algorithms. The models have been evaluated with three feature extraction techniques namely Term Frequency-Inverse Document Frequency (TF-IDF), Count-Vectorizer (CV) and Hashing-Vectorizer (HV). A brief outline of the related work in the field of rumor detection is discussed in Sect. 2. Static analysis for the dataset has been explained in Sect. 3. The proposed methodology for rumor detection is explained in Sect. 4. The performance of the classifier results are discussed in Sect. 5. Section 6 concludes this research work.

Fig. 1. Sample rumor tweet and image for #gaja dataset

2   Related Work

Jing Ma et al. proposed a kernel learning method for detecting rumors in microblog posts. This method learns discriminate clues for detecting rumors and measure similarity among the propagation trees. It overcomes the drawbacks of the feature-based method and allows further information discriminations. Two twitter dataset was tested on PTK model and achieves 75% accuracy in one dataset and 73% accuracy in another dataset [6].

Kwon et al. analyze the difference between the rumors and non-rumors based on the network, temporal, linguistic and user features. The time window algorithm examines rumor characteristics over short and long time windows. The authors compare the prediction level over different time windows and it was observed that, during the initial period the user features were effective for predict rumors. Linguistic features were stable and powerful predictors of rumors over a time. The Network features were used to predict information spreading on a network over a longer time period [7].

ZheZhoo et al. designed a rumor detection approach by clustering the tweets and each cluster contains enquiry patterns. The clusters are ranked based on statistical features and then compared the properties of the whole cluster into a signal tweet. By this method, the rumors in an early stage can be detected effectively and in order to improve the detection method first, they improve the filtering mechanism and correction signal [8].

Michal lukasik et al. suggests an approach for classifying judgments of rumors in both supervised and unsupervised domain adaptation. The multi task learning approach was performed effectively when compared to single-task learning [9].

Zilong et al. tracks both fake news and real news from the Twitter message in Japan and Weibo in China. Both media has spread fake news distinctively from multiple broadcasters. The real news has spread using dominant sources [10]. The authors analyze the predictability feature of this difference of the propagation networks to detect fake news in an early stage. They demonstrate filtering out fake news from the beginning of their propagation using collective structural signals [7] (Tables 1 and 2).

Table 1. Summary of machine learning techniques and evaluation metrics

References	Proposed approach	Features	Evaluation metrics
Zilong Zhao et al.	Collective Structural Model	Topological features	Heterogeneity measure
Kwon et al.	SpikeM Model	Structural, Linguistic, Temporal features	F1-Score
Arkaitz Zubiaga et al.	General Methodology	User and Twitter feature	Nil
AditiGupta et al.	Semi-supervised SVM Rank	TF-IDF	ROC_AUC curve
Sandeep Soniet et al. [15]	Predictive accuracy, cue words and cue groups	Cue word, Cue word group	Precision, Recall

Table 2. Taxonomy of machine learning algorithms for credibility analysis in Twitter

References	Types of event		Detection method	Data set	Inference
	Specified	Unspecified	Supervised		
Zilong Zhao et al. [11]	✓		✓	Twitter data from Japan and Weibo in China	Identified the fake news at an early stage using collective structural signals
Kwon et al. [12]	✓	✓	✓	KoreanFanDeath, LadyGaga, Montauk Data	Compare the prediction level over different time windows, during the initial period the user features are effective for predict rumors, linguistic features are stable and powerful predictors of rumors over time
ArkaitzZubiaga et al. [13]	✓		✓	Ferguson unrest, Ottawa shooting, Sydney siege, CharlieHebdoshooting, German wings plane crash	True rumors tend to be resolved faster than false rumors. Rumors in their unverified stage produce distinctive bursts in the number of retweets within the first few minutes, substantially more than rumors proven true or false
Aditi Gupta et al. [14]	✓		✓	The Boston Marathon blasts in the US, Typhoon Haiyan/Yolanda in the Philippines	A real-time web-based tool to check the information credibility based on the score

3   Dataset

Due to the cyclonic storm 'Gaja' over the Bay of Bengal rainfall started between Cuddalore and Bamban on 15-November,2018 at 5.30 P.M. There were many rumor news are disseminated in twitter during this event. The authors have collected 90,867 tweets from 24,534 unique users with the aid of hashtags namely #cyclonegaja, #savedelta, and @TNSDMA. The #gaja corpus consists of source tweets, retweets and replies tweets.

The distribution of length of a tweet in terms of word counts are to be analyzed. Figure 2 shows the length of the tweet Vs total length of the tweets related to this event. It has been analyzed that there are 2,500 unique words to be identified in the dataset. The length of the each tweets are varies between 5 to 35 word counts. The collected tweets are used to model the classifier.

The authors validate the data annotation work through Fleiss kappa coefficient. It is used to measure the Inter Rated Reliability with three annotators for classifying the

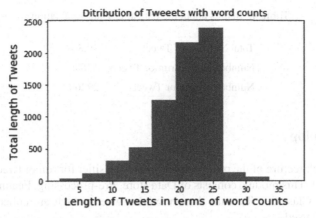

Fig. 2. Distribution of Tweets with word counts for the #gaja dataset

rumor or non-rumor tweets. This is derived by

$$k = \frac{\hat{p} - \widehat{p_e}}{1 - \widehat{p_e}} \tag{1}$$

Where $1 - \widehat{p_e}$ is the degree of agreement that is achievable than the chance and $\hat{p} - \widehat{p_e}$ is the degree of agreement actually achieved above chance. The following Table 3 shows that the sample annotation process of rumor tweets.

Table 3. Sample tweets for Data Annotation

Tweet	A1	A2	A3
Gale wind speed reaching 70–80 kmph gusting to 85 kmph prevails over East central and adjoining West central & Southeast Bay of Bengal	NR	NR	NR
Schools and colleges in Tamil Nadu remained shut today and offices and business establishments were asked to relief?	R	NR	R
Kind people of twitter #help	NR	NR	NR
#GAJA is very likely to further intensify during the next few days and become a severe cyclonic storm over the Bay of Ben	NR	NR	R
Cyclone Gaja To Turn Into? Severe Cyclonic Storm? In Next 24 h, Tamil Nadu And Andhra Pradesh To Witness Heavy Rainfall?	NR	R	R

The annotator1 (A1) validate nearly 70% of the tweets as non-rumor and annotator2 (A2), annotator3 (A3) validate 60% of the tweets as non-rumor. The excepted probability of the overall annotation is calculated using k. The k value for this annotation process is 0.58.It has been observed that our agreement is moderate. The statics of the data annotation is shown in Table 4.

Table 4. Statics about the data after Data Annotation

Total Number of Tweets	90,867
Number of Non-rumor Tweets	63606
Number of Rumor Tweets	27260

4   Methodology

The overall architecture of the proposed Calibrate classifier for rumor tweet detection is shown in Fig. 3. This module consists of Data store, Pre-processing, Feature generation, and Ensemble Classifier. To train these models, tweets which are collected from the #gaja event are used.

4.1   Data Store

Twitter allows us to mine the data of any user using Twitter API or Tweepy. The Streaming API works by making a request for a specific type of data filtered by keyword, user, geographic area. Tweets were collected using various hash-tags like #CycloneGaja, @TNSDMA, #SaveDelta, etc. The tweets collected in a streaming fashion represent the tweets that were posted in that particular time duration.

4.2   Data Pre-processing

To remove stop words such as pronouns, conjunctions, and prepositions there were eight preprocessing rules are applied. This noise reduction in the text helps to improve the performance of the classifier and remove the textual content not related to the event. The preprocessing rules are Convert to lowercase, RT removal, Replacement of User-mentions, URL Replacement, Hash Character Removal, Removal of Punctuations and Symbols, Lemmatization and Stop word Removal. The preprocessed data is then fed into feature generation module.

4.3   Feature Generation

In this module, the features are extracted from the pre-processed data. In order to convert the collected text document into an integer or floating-point values are known as feature vectorization.

Count Vectorizer (CV)
Count Vectorizer converts the word into the matrix of token counts. A CV is based on count of the word occurrences in the document. CV is selected for feature extraction because it has performed both the tokenization and counting the occurrence of the word in the data. It is observed that CV converts each word in the document as a vector value (integer). Each vector consists of the feature name and its corresponding word occurrences. Each column in the matrix contains the terms like cyclonegaja, gate, and speed

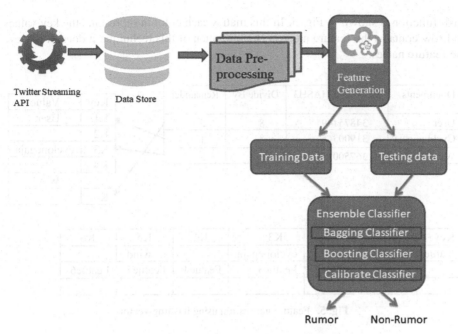

Fig. 3. The architecture of the proposed rumor detection system

as feature names. The rows (doc1, doc2) represents the frequency of words retrieved from the vector and their corresponding word count. The following Fig. 4 shows that the sparse matrix for sample #gaja dataset. In this matrix doc0 and doc1 represent the number of words retrieved from the dataset. The limitation of this technique is less frequency terms have more influence than the high frequency words.

	cyclonegaja	gale	kmph	reaching	speed	user	warning	wind
Doc0	1	1	0	0	0	1	1	1
Doc1	0	1	1	1	1	0	0	1

Fig. 4. Feature generation using count vectorizer

Hashing Vectorizer (HV)

Hashing Vectorizer tokens are encoded as a numerical index. It requires only limited amount of memory for feature generation. This HV is chosen for simplifying the implementation of the bag-of-words and improves the scalability. It has generated the hash value for a given dataset. The most popular MURMURHASH3 hashing algorithm is applied to the hashed words to generate a random number. These values are divided by the length of the data and find the corresponding remainder value.

Based on the remainder values every word is stored into the corresponding key-value pairs. The hash value of the textual data is estimated using MURMURHASH3 hash function is shown in Fig. 5. In this matrix each column represents the key values and row contains the feature names. The limitation of this technique it doesn't retrieve the feature names.

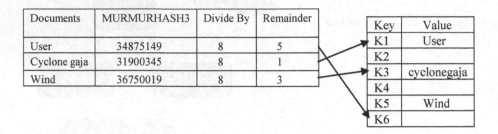

Documents	MURMURHASH3	Divide By	Remainder
User	34875149	8	5
Cyclone gaja	31900345	8	1
Wind	36750019	8	3

Key	Value
K1	User
K2	
K3	cyclonegaja
K4	
K5	Wind
K6	

Key	K1	K2	K3	K4	K5	K6
Value	user		cyclonegaja		wind	
	Featue1	Featue2	Featue3	Featue4	Featue5	Featue6

Fig. 5. Feature generation using hashing vector

Term Frequency-Inverse Document Frequency (TF-IDF)

TF-IDF is used to generate a weighted matrix for the important words in the dataset. The following Fig. 6 shows that the TF-IDF weighted matrix for preprocessed data. It is observed that TF-IDF are tokenized the documents, learn the word and assign weights for each word. In this matrix the columns are represented by the token and the row (doc0, doc1) represent weighted value for number of words retrieved from the dataset. Beel et al. [17] showed that 83% of text based categorizations are done by using the tf-idf vectorization technique.

	cyclonegaja	gale	kmph	reaching	speed	user	warning	wind
Doc0	0.499221	0.3552	0.000000	0.000000	0.000000	0.499221	0.499221	0.3552
Doc1	0.000000	0.3552	0.499221	0.499221	0.499221	0.000000	0.000000	0.3552

Fig. 6. Feature generation using TF-IDF

4.4 Ensemble Classifier

The feature generated dataset is fed into the classifier for the detection of the rumor tweet. Ensemble methods are to build a learning algorithm in a statically and computational way. It is used to deal imbalanced data efficiently. Rumor detection system is implemented by using ensemble methods. In this research work the ensemble methods such as Bagging, Boosting and Calibrate Classifiers are used to detect the rumors. The bagging classifier extracts a subset of the training dataset from multiple models. Boosting classifiers learns to fix the prediction errors of a prior model chain. Calibrate classifiers are used to combine the predictions of multiple models. In this experiment, the authors have used non-linear classifier to detect the rumors. But these classifiers are generally predicting uncalibrated results. Calibrate classifier is used to turn the output of the model into well-calibrated continues probabilities of the models. Rumor and Non-rumor tweets are classified with the aid of the ensemble learning classifiers.

5   Results and Discussions

5.1   Pre-processing

The tweets repositories of 90,689 tweets were annotated by manually as either Rumor or Non-rumor class with respect to the ground truth obtained by official user of @TNS-DMA. The tweets are pre-processed by applying rules as discussed in data pre-processing section. The following Table 5 shows the original and preprocessing tweets.

Table 5.   Pre-processing of sample tweet for #gaja event

Original tweet	RT @ndmaindia: #CycloneGaja #Wind Warning: Gale wind speed reaching 70–80 kmph gusting to 85 kmph prevails over East central and adjoining W?
Preprocessed tweet	user cyclonegaja wind warning gale wind speed reaching kmph gusting to kmph prevails over east central and adjoining

5.2   Feature Generation

The preprocessed tweet messages are applied for feature generation. Both content and contextual features are computed using three methods such as TF-IDF, Count Vectorizer and Hashing Vectorizer techniques. The following Fig. 7 shows the vector generation for the above-preprocessed tweet. The conversion of textual content into the numerical vector is implemented using TF-IDF, CV and HV techniques. The vector graph shows that the learned vocabulary and number of document frequencies. In this graph where x-axis represents number of token values, whereas y-axis represents the size of the sample data. Fig. 7(a), 7(b) and 7(c) are visualizing the encoded vector values for the preprocessed text. The normalization value for TF-IDF is (0, 1), CV (1,2) and for HV (−1,1). In Fig. 7(a) show the result of vector using TF-IDF, most frequently used words in the documents are shadowed between 0.20 to 0.25 and less frequent values are showed near (0.30,0.45)

Figure 7(b) show the results of CV, the frequent occurrence of the words is shadowed near 1 and least is shadowed near 2. In Fig. 7(c) show the hash values for most frequent and less frequent values between (−1,1). The generated feature values are fed into classifier to detect the rumor tweet in #gaja dataset.

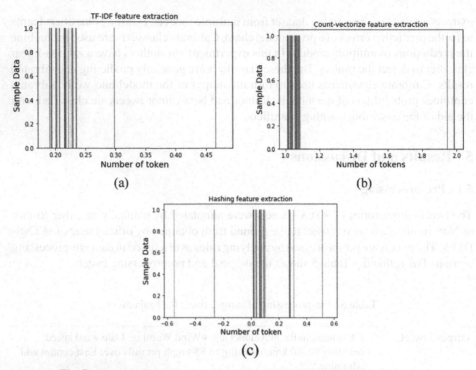

Fig. 7. Numerical vector generation using TF-IDF, CV and HV techniques

5.3 Calibrate Classifier for Rumor Detection System

In this experiment, the classifiers are trained using #gaja dataset. This dataset is divided into 80% for training and 20% for testing. Three vectorizer techniques are applied on the following classifiers.

Bagging Classifier
Bagging or Bootstrap aggregation takes multiple samples from the original dataset and train the classifier. This classifier can be trained on models $Xm = \{x_1, x_2, \ldots x_n\}$ using the original dataset $Ym = \{y_1, y_2, \ldots y_m\}$, then the average model is derived by

$$x = \frac{1}{N} \sum_{N=1}^{N} x_n \qquad (2)$$

where 'n' represents the total number of data and x_n represents the number of models. The Bagging is implemented using the Decision Tree classifier with hundred numbers of trees.

Boosting Classifier
This classifier trains weak models using training data. It has computed the error of the model and gives more importance to the mistake models. Retrain the model by using

weighed training samples. The probability of the selected classifier is derived by

$$P = \frac{1 + m_i^n}{\sum_{j=0}^{N} 1 + m_j^n} \qquad (3)$$

where n represents the total number of data and m_i^n represents the number of models. The boosting is implemented using the AdaBoost classifier with seventy number of trees.

Calibrate Classifier

Calibrate classifier combines the predictions from multiple machine learning classifiers such as Logistic Regression, Decision Tree and Support Vector Machine. All predicted values are added and averaged to form the probability vector is derived by

$$\hat{Y} = \arg \frac{1}{N_c} \sum_c (p1, p2 \ldots pn) \qquad (4)$$

where $p1, p2 \ldots pn$ represents the number of predictions and N is the number of classifiers.

The Bagging, Boosting and Calibrate classifiers are implemented with the aid of vectorizer values for detecting the rumors. The detection of rumor instances using TF-IDF technique performs well for #gaja events. It can achieve accuracy of 95.7% for bagging, boosting and 96.4% for calibrate classifier. It can learn the vocabulary from the data and apply inverse frequency weights to encode the data. CV only counts the word occurrence and does not assign any weighted values. HV doesn't return feature names for further analysis. The authors have inferred that TF-IDF vectorizer Vs classifier perform well on a high volume of data. Since, TF_IDF assigns accurate weighted vector for each word in the dataset. The rumor classification results for the #gaja dataset with respect to the accuracy are shown in Fig. 8.

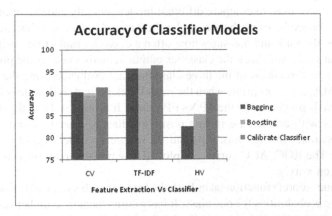

Fig. 8. Performance analysis using accuracy metric for #gaja data

The Calibrate Classifier with TF_IDF performs better due to the fact it combines multiple classifiers for predicting the rumor tweet. The results of the calibrate classifier

Table 6. Comparison of existing models with a proposed system

Author	Proposed approach	Feature generation method	Accuracy
Sawinder et al.	Classify news article as fake or real using various machine learning classifiers	TF_IDF	93.8
		CV	89.3
		HV	86.3
Proposed calibrate classifier	Rumor message detection system is proposed to achieve high accuracy	TF_IDF	96.42
		CV	91.45
		HV	89.91

are compared to the existing classifier on false news detection as shown in Table 6. Based on the comparison of existing model, the authors conclude that the performance of the Calibrate Classifier is better than other ensemble classifiers.

6 Evaluation Metrics

The classifiers of ensemble learning algorithms are evaluated based on the ROC-AUC curves. ROC curve is used to predict the probability of binary classification. This curve takes the true positive and false positive values. The True Positive and False Positive Rate are derived by

$$TP\ Rate\ =\ TP/(TP + FN) \tag{5}$$

$$FP\ Rate\ =\ FP/(FP + TN) \tag{6}$$

The ROC curve used to compare different models directly and the summary of the classifier is evaluated by using the area under the curve. All the three classifiers are evaluated based on the rank and measures how often a classifier ranks true positives higher than true negative. It can check the classifier output actually matches the probability of the event. The performances of the three classifiers are evaluated using the ROC-AUC curve. ROC-AUC are appropriate when the model with perfect skill value between (0,1). The ROC curve is plotted using the TP Vs FP values. It is observed that the raise in false positive values with an increase in true positive values by varying a threshold of the ensemble classifiers. In Fig. 9(a), 9(b) and 9(c) represents the TP and FP values for all classifiers. The ROC_AUC curves are plotted true positive rates on y-axis and false positive rates on x-axis.

The roc_auc_score() function takes both the true values between (01) from the training and testing probabilities for one class. It has returns the auc score between (0.0,0.1) for non-rumor and rumor values respectively. Calibrate Classifier outperforms the other classifiers with the true-positive rate of 0.894.

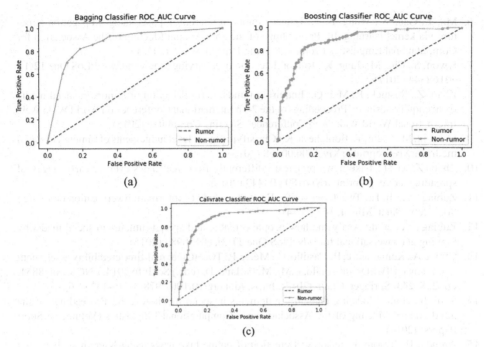

Fig. 9. True positive and false positive rate for the rumor detection system

7 Conclusion

The main goal of this research work is to detect the rumor messages in social media and analyze the best classifier to prevent the rumor message dissemination. The authors are experiments with the #gaja dataset using three vectorizer techniques TF-IDF, CV, and HV. Ensemble classifiers are used to classify the rumor and non-rumor messages. The results show that the Calibrate Classifier outperforms than the bagging and boosting classifiers. The experimental results are evaluated with the aid of the ROC-AUC metrics and it proves the calibrate classifier results are accurate.In future, the rumor detection system to classify the rumor data based on the retweet count.

References

1. https://www.socialmediatoday.com/marketing/how-muchtime-do-people-spend-social-media-infographic/
2. Zubiaga, A., Aker, A., Bontcheva, K., Liakata, M., Procter, R.: Detection and resolution of rumours in social media: a survey. ACM Comput. Surv. (CSUR) **51**(2), 32 (2018)
3. Shu, K., Sliva, A., Wang, S., Tang, J., Liu, H.: Fake news detection on social media: A data mining perspective. ACM SIGKDD Explor. Newsl. **19**(1), 22–36 (2017)
4. https://www.independent.co.uk/life-style/gadgets-andtech/news/tumblr-russian-hacking-us-presidential-election-fakenews-internet-research-agency-propaganda-bots-a8274321.html. arXiv:1809.01286v1
5. Fake news on whatsapp. http://bit.ly/2miuv9j. Accessed 27 Aug 2019

6. Ma, J., Gao, W., Wong, K.-F.: Detect rumors in microblog posts using propagation structure via kernel learning. In: Proceedings of the 55th Annual Meeting of the Association for Computational Linguistics (Volume 1: Long Papers), vol. 1 (2017)

7. Kwon, S., Cha, M., Jung, K.: Rumor detection over varying time windows. PLoS One 12(1), e0168344 (2017)

8. Zhao, Z., Resnick, P., Mei, Q.: Enquiring minds: Early detection of rumors in social media from enquiry posts. In: Proceedings of the 24th International Conference on World Wide Web. International World Wide Web Conferences Steering Committee (2015)

9. Lukasik, M., Cohn, T., Bontcheva, K.: Classifying tweet level judgements of rumours in social media. arXiv preprint arXiv:1506.00468 (2015)

10. Zhao, Z., et al.: Fake news propagate differently from real news even at early stages of spreading. arXiv preprint arXiv:1803.03443 (2018)

11. Zubiaga, A., Ji, H.: Tweet, but verify: Epistemic study of information verification on twitter. Soc. Netw. Anal. Min. 4, 163 (2014)

12. Zubiaga, A., et al.: Analysing how people orient to and spread rumours in social media by looking at conversational threads. PloS One 11(3), e0150989 (2016)

13. Gupta, A., Kumaraguru, P., Castillo, C., Meier, P.: TweetCred: real-time credibility assessment of content on Twitter. In: Aiello, L.M., McFarland, D. (eds.) SocInfo 2014. LNCS, vol. 8851, pp. 228–243. Springer, Cham (2014). https://doi.org/10.1007/978-3-319-13734-6_16

14. Soni, S., et al.: Modeling factuality judgments in social media text. In: Proceedings of the 52nd Annual Meeting of the Association for Computational Linguistics (Volume 2: Short Papers) (2014)

15. Ahmed, H., Traore, I., Saad, S.: Detection of online fake news using N-gram analysis and machine learning techniques. In: Traore, I., Woungang, I., Awad, A. (eds.) ISDDC 2017. LNCS, vol. 10618, pp. 127–138. Springer, Cham (2017). https://doi.org/10.1007/978-3-319-69155-8_9

16. Kaur, S., Kumar, P., Kumaraguru, P.: Automating fake news detection system using multi-level voting model. Soft. Comput. 24(12), 9049–9069 (2019). https://doi.org/10.1007/s00500-019-04436-y

17. Beel, J., et al.: Paper recommender systems: a literature survey. Int. J. Digit. Librar. 17(4), 305–338 (2016)

A Thesaurus Based Semantic Relation Extraction for Agricultural Corpora

R. Srinivasan$^{(\boxtimes)}$ and C. N. Subalalitha

SRM Institute of Science and Technology, Kattankulathur 603203, India
srinirvs89@gmail.com, subalalitha@gmail.com

Abstract. Semantic relations exist two concepts present in the text. Semantic relation extraction becomes an essential part of building an efficient Natural Language Processing (NLP) applications such as Question Answering (QA) and Information Retrieval (IR) system. Automatic semantic relation extraction from text increases the efficiency of these systems by aiding in retrieving more accurate information to the user query. In this research work, we have proposed a framework that extracts agricultural entities and finds the semantic relation exist between entities. Entity extraction is done using a Parts Of Speech (POS) tagger, Word Suffixes and Thesaurus without using any of the external domain-specific knowledge bases, such as Ontology and WordNet. Semantic relation exists between entities are done by using Multinomial Naïve Bayes (MNB) classifier. This paper extracts two entities, namely disease and treatment and focuses on two semantic relations namely "Cure" and "Prevent". The "Cure" semantic relation expresses the remedial measure for the diseases that prevail in the crops, and the "Prevent" semantic relation shows the precautionary measures that could prevent the crop from being affected. The proposed approach has been trained with 2281 sentences and tested against 553 sentences and then evaluated using standard metrics.

Keywords: Agricultural entity · Semantic relation · Multinomial Naïve Bayes · Feature extraction

1 Introduction

Agriculture plays a vital role in the world economy [1]. Several organisations, such as Food and Agriculture Organizations (FAO), International Fund for Agricultural Development (IFAD), National Farmers' Union (NFU), International Federation of Agricultural Producers (IFAP) and many others have been working in the agriculture domain for several years to increase the food production. As a result, there has been a vast increase in agricultural data in an unstructured manner on the World Wide Web (WWW) [2]. This aids in the computational

Supported by SRM Institute of Science and Technology.

analysis of the data which can bring out fruitful solutions by the computer science researchers, thereby making a significant contribution in solving the problems involved in food production.

Extraction of semantic relations between entities is an intuitive and vital role in Natural Language Processing (NLP) applications. NLP is a sub-domain of Artificial Intelligence which automates the text analysis to build novel applications. Several techniques are used to identify the semantic relations between entities, such as Rule-based approaches, Knowledge-based approaches, Link-based approaches, Machine Learning based approaches and Deep Learning based approaches. The semantic relations between the agricultural entities have not been explored to our knowledge, which is done by the proposed method. In the healthcare sector, the extraction of agricultural entities and semantic relation between entities is an important research topic in the field of agriculture.

The goal of the proposed work is focused on two tasks: The first task is to identify the agricultural entities namely, "disease" and "treatment" using various techniques such as POS tagger, Word Suffixes and Thesaurus. The thesaurus can be extracted from the National Agricultural Library (NAL) [3]. The second task is to identify the semantic relations namely, "Cure" and "Prevent". This proposed work can be a pointer to build agricultural domain-specific Information Retrieval (IR) systems. These types of IR systems can be a significant help for the farmers, agricultural and NLP researchers.

The main contributions of this paper are:

1) Extracting Entities: By using POS tagger, Word suffixes and Thesaurus to identify the "disease" and "treatment" entities and then the sentences are classified into positive or negative sentences. A positive sentence contains both "disease" and "treatment" entities within a sentence, else treated as a negative sentence.
2) Extraction of Semantic relation using Multinomial Naive Bayes (MNB) classifier: Once the sentences are classified, the next task is to identify the semantic relation between the sentences. The semantic relation that is addressed here, cure, prevent and irrelevant entities. The paper has organized as follows: Sect. 2 discusses literature survey, Sect. 3 defines the proposed approach to detect and classify the meaningful sentences, Sect. 4 examines the evaluation results obtained, and Sect. 5 describes conclusions and future work.

2   Literature Survey

The author proposed a platform for extracting medical entities and identifying relationship from texts. Genia tagger has been used to identify the entities and determine semantic types using MetaMap [4]. The linguistic patterns had been used along with the help of domain knowledge to extract 16 types of medical entities. A precision of 75.72% and a recall of 60.46% is obtained using linguistic patterns.

Xin has presented a novel approach for semantic relation extraction, which combines both the pairwise relation and the link-based relation within words

[5]. The pairwise and Link-based approach is used to measure the relationship between the phrases. The author combines the approach mentioned above to generate a document clustering model. The author proposed a two-phase method which includes entity identification and relationship integration [6]. In the entity identification phase, the ML algorithms, namely Support Vector Machine (SVM), and Decision Tree (DT) are combined using statistical features. Relationship identification between the entities has been made by clustering the semantic entities. Min-Ling Zhang et al. (2009) have proposed a feature selection mechanism which incorporates Multi-Label Naïve Bayes (MNB) to improve its performance [7]. Principal Component Analysis (PCA) has been used to remove irrelevant and repeated features. Naive Bayes classifiers had been widely used for various natural language processing tasks [8].

Koichi Takeuchi et al. (2005) have used Support Vector Machine (SVM) to extract biological entities like scientific names, protein names, genes and viruses [9]. The entities are obtained from various combinations of features, such as orthographic features, context window and head noun features. Ensemble technique combined with fuzzy logic to extract the disease entities with the help of orthographic features. It provides the promising result of 94.66%, 89.12%, 84.10%, and 76.71% of F-measure for various corpora [10].

Oana Frunze et al. (2011) has used various ML methodologies that are suitable for identifying the health-care information [11]. It extracts sentences from published papers that have the mention of diseases and treatments and identifies the semantic relations between the entities. Three semantic relations, namely Cure, Prevent and Side-effect have been identified. The main idea of this work is carried out the semantic relationships in biomedical text and identifying the best Machine Learning algorithm for their dataset.

Archana Chaudhary et al. (2016) have presented a hybrid ensemble technique that combines more than one machine learning algorithm for diagnosing the oil-seed disease [12]. The oil-seed dataset is developed from different sources, which include 24 nominal attributes. It provides humidity, soil moisture, temperature and symptoms to diagnose the oil-seed disease.

As far as NLP is concerned, agricultural semantic extraction from texts has never been attempted to the best of our knowledge. In Machine Learning point of view also, agricultural domain-specific text data remain unexplored. The proposed system differs from the state of the art by attempting the semantic relation classification using MNB classifier on agricultural domain-specific documents. This proposed system might be a future pointer to develop many other useful applications for agricultural domain-specific texts using both advanced NLP and ML techniques. The next section describes the proposed methodology.

3   Proposed Methodology

3.1   System Architecture

The architecture of the proposed method is shown in Fig. 1. The two steps are performed in this work provide the interface for information extraction framework that is accomplished to identify and extract agricultural information. The first task determines the meaningful sentences on diseases and treatments topics, while the second one performs a classification of sentences according to the semantic relations that exist between diseases and treatments.

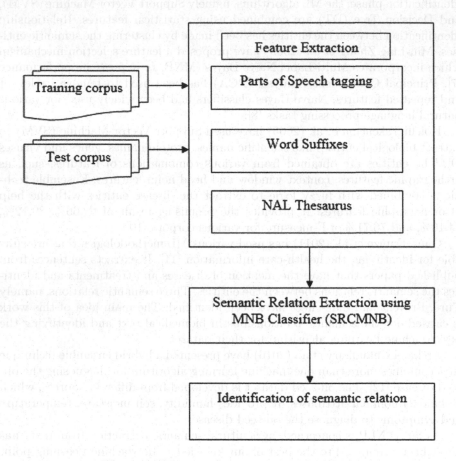

Fig. 1. Architecture for proposed methodology.

In the first step (Feature Extraction and Sentence Selection) agriculture sentences are collected from various online resources which include web pages, research articles and abstracts that discuss diseases and treatments. It extracts meaningful information that contains both disease and treatment information. The second step (Semantic Relation using MNB classifier) identifies semantic relations in the sentences from the meaningful information (e.g., the output of the first step). This research work focuses on two semantic relations, namely "cure" and "prevent".

3.2 Dataset Details

We have created our in-house agricultural data set by web scraping, and the same is available at [13] named as agriculture. Table 1 details the in-house agricultural data and the same has been used for our research. The numbers in parentheses which describe the count of training and testing data. For example, prevent relation consists of 173 sentences, 140 is used or training set and 33 is used for testing set.

Table 1. Agriculture dataset

Total sentences	Semantic relation	Example	Disease identified	Treatment identified
Cure (701,160)	Treatment cures disease	Wiping plants with soapy water is to heal aphids	Aphids	Soapy water
Prevent (140,33)	Treatment avoid Disease	Soluble Boron (0.01%) can be done at monthly intervals to tackle ringspot in papaya	Ringspot	Soluble Boron
Irrelevant (1458,360)	No disease and treatment relation	Agriculture is the process of producing food	No disease name	No treatment name

Since it is an in-house dataset, it should be tested using inter-rater reliability tool. The inter-rater reliability is test validity tool to measure the score given by the human experts. Human experts will classify the sentences based on their meaning. In our dataset, human experts to check the sentences fall into cure category or prevent category. It is not necessary to check the irrelevant category. Irrelevant category may not contain about the agricultural entities. Cure and Prevent category contains the agricultural entities that are used to classify the sentences. The dataset consist of 2834 sentences, 1034 sentence fall into either cure category or prevent category.

Table 2 describe the percentage agreement for the cure and prevent sentences. In Table 2, 0 represents cure category and 1 represents prevent category. For example, the sentence 1 all the annotator classified the sentence as 0, whereas 0 represents cure category and agreement is 100%. Similarly, inter-rater reliability is calculated for all the cure and prevent category sentences.

Table 2. Percentage agreement across multiple annotators

Sentence	Annotator 1	Annotator 2	Annotator 3	% agreement
Sentence 1	0	0	0	100
Sentence 2	1	1	1	100
Sentence 3	0	1	1	66.66
.....	.....	.....	.....	.....
.....	.....	.....	.....	.....
.....	.....	.....	.....	.....
Sentence 1034	1	1	1	100
Inter-rater reliability				89.23

3.3 Feature Extraction

Parts of Speech. Stanford POS tagger is used in our model to extract the entities. A Part-Of-Speech Tagger (POS Tagger) is a software that reads a text and assigns parts of speech to each word such as noun, adjective, adverb, verb, etc., although generally computational applications use more fine-grained POS tags like 'noun-plural'. At a first step, a sentence is passed into POS tagger and finds the parts of speech. Noun always refers to the entities and identifies the pattern for the sentence.

Example 1 (Training Data)

- Soluble Boron (0.01%) <treatment name> can be done at monthly intervals to tackle ringspot <disease name> in papaya
- POS of Treatment Name: Soluble <Adjective> Boron <Noun>
- POS of Disease Name: ringspot <Noun>
- Pattern identified for extracting treatment name: Adjective, Noun(feature)

In Example 1, we identified the feature for extracting treatment name. The feature is Adjective followed by a Noun is probably a treatment name. This step is used to extract the feature of treatment entities. Patterns are considered a feature by applying the method as mentioned above. Once the features are extracted, stored in two different feature set namely Disease feature set and Treatment feature set. Most frequent feature are listed in Table 3.

Table 3. Features of disease and treatment entities using POS

Entity type	Example	Feature
Disease	Aphids	Noun
Disease	Late Blight	Noun, Noun
Treatment	Liquid copper	Adjective, Noun
Treatment	Wiping plants with soapy water	Verb, Noun, Preposition, Adjective, Noun

Word Suffixes. Word suffixes are a group of characters added to the ending or in between the texts. In our paper, word suffixes play an important role to extract disease or treatment entities. The first step of feature extraction is done by POS tagger. It extracts the patterns from the tagged entity that helps to identify the entity in the test data. But not all the patterns which obtain the proper disease and treatment names. The output of the first step is passed into the second step. The second step verifies the entities is disease name or treatment name. Word suffix identifies the minimal amount of disease and treatment entities. Some of the disease names follow the common word ending suffixes and are listed in Table 4.

Table 4. Features of disease and treatment entities using word suffixes

Entity type	Example	Suffix	Feature
Disease	Leaf spot	Spot	Spot
Disease	Late blight	Blight	Blight
Treatment	Psychodynamic therapy	Therapy	Therapy
Treatment	Overdiagnosis	Diagnosis	Diagnosis

NAL Thesaurus. The next step is done by analysing the definitions given in the NAL thesaurus that contains the seed words and meanings. The analysis done by NAL is a collection of descriptions of agricultural entities developed in conjunction with the formation of the National Agricultural Library (NAL) Thesaurus. NAL thesaurus Contains over 255,390 terms, including 148,282 descriptors in English and Spanish. The 2017 edition contains 5,223 definitions, and the NAL Thesaurus Staff composes it. Features are extracted from POS Tagging and Word Suffixes. Entities are passed through the NAL Thesaurus. NAL Thesaurus verifies that the entity is present or not.

Let $S = w1, w2, \ldots, wn$ where S is a sentence and w represents the words which may contain the definition of a disease or treatment name is a context word in NAL thesaurus. If a sentence carries both the entities, the sentence is labelled as a positive sentence. If a sentence not containing any of the information or non-relevant information, labelled as a negative sentence.

3.4 Semantic Relation Classifier Based on Multinomial Naïve Bayes (MNB)

MNB is a supervised Machine Learning model widely used as a probabilistic classifier model for classifying the text. MNB classifiers are widely used for text classification process [14]. MNB classifier is based on Bayes' theorem with strong independent attribute assumption [15]. MNB classifier assumes not only the independent attributes but also finds the frequency of the all attribute belong to the same class. Frequency count of an attribute is important that decides the probability of an attribute falls into class [16]. Once it classifies the positive, the next step is to identify the semantic relation occurs in a sentence by using MNB. Rosario and Hearst described the nine semantic relations in bio-science text [17]. The proposed work describes the two semantic relations such as "cure" and "prevent" in the agricultural texts. By applying MNB classifier to extract the semantic relations from the identified positive sentence.

function TRAINMNB(D,C)
For each class c belongs to C
Calculate prior[c]
Ns = total number of positive sentences in D
Nc = total number of positive sentences from D belongs to class c

$$prior[c] = \frac{N_c}{N_s} \tag{1}$$

Calculate conditionalprobabilty[w,c]
W ← All the words of positive sentences in D
tot[c] ← append(S) for S belongs to D with class c
u ← number of unique words of D
For each word w in W
count(w,c) ← no of occurrences of w in tot[c]

$$conditionalprobabilty[w,c] = \frac{count(w,c)+1}{\sum_{w^i in W} count(w^i,c)+u} \tag{2}$$

return W, prior, conditionalprobability

function TESTMNB(testdata, W, prior, conditionalprobability, C)
for each class c ∈ C
sum[c] ← prior[c]
for each position of j in testdata
word ← testdata[j]
if word ∈ W
sum[c] ← sum[c] * conditionalprobabilty[word,c]
return argmax_c sum[c]

4   Results and Discussion

Accuracy, Recall, Precision and F-score metrics have been used to evaluate the proposed system [18]. A confusion matrix is developed before evaluating the classification model [14]. Table 5 describes the confusion matrix used in the proposed system.

In Table 5, TP denotes no of sentences that are correctly predicted as positive. FP means no of sentences misclassified as positive. FN denotes no of sentences misclassified as negative. TN denotes no of sentences that are correctly predicted

Table 5. Confusion matrix

True Positive (TP)	False Positive (FP)
False Negative (FN)	True Negative (TN)

as negative. Accuracy is the ratio of the number of correctly predicted sentences to the total sentences. Accuracy is calculated by using the following equation

$$Accuracy = \frac{TP + TN}{TP + FP + TN + FN} \tag{3}$$

Recall defined the ratio of the correctly predicted relevant sentences to the overall observation in the relevant sentences. The following equation is used to calculate the recall.

$$Recall = \frac{TP}{TP + FN} \tag{4}$$

Precision is defined as the ratio of correctly predicted relevant sentences to the total predicted relevant sentences and can be calculated as follows.

$$Precision = \frac{TP}{TP + FP} \tag{5}$$

F-score is the weighted harmonic mean between precision and recall.

$$F - score = \frac{2 * recall * precision}{recall + precision} \tag{6}$$

4.1 Evaluation and Performance Analysis of the Sentence Selection Module

Table 6 describes the dataset used for sentence selection. Sentences which contain information about both "disease" and "treatment" are labelled as positive sentence. As mentioned above in Sect. 3.2, sentences which contain "disease" or "treatment" information or sentences containing neither of this information are considered as negative sentence.

Table 6. Dataset used for sentence selection

Sentence	Positive	Negative
Training	1141	1458
Testing	193	360

POS tagger extracts 139 features from the training sentences. Word suffixes extract 30 features from the training data. By combining the features extracted from the various techniques and mapped with the NAL Thesaurus. POS taggers are not sufficient for extracting disease and treatment entities. Table 4 "Aphids" is a word refers to a noun. But all the nouns are not disease or treatment entities. Precision value is low by applying POS tagger alone. Entities retrieved by the system are more, but actual disease and treatment entities are less. The next step is to combine POS tagger with word suffixes. Word suffixes are to verify the entities are disease or treatment. All the disease entities are not terminated with word suffixes. For example, an accelerometer is word occurs in test data. By applying POS tagger, an accelerometer is a noun and checks the word suffixes. Word suffixes having "cele" are considered as a feature. "cele" means swelling that considered as a feature. An accelerometer falls into this positive category, but it does not contain a disease or treatment entities.

The next step is to apply NAL Thesaurus to verify the entity is correct or not. By applying POS tagger, word suffixes and thesaurus gives good accuracy. Table 7 describes the evaluation metrics for sentence selection. After applying the POS, Word Suffixes, and Thesaurus to the test data, 144 sentences are predicted as positive sentences, 266 sentences are identified as negative sentences, and the remaining sentences are misclassified. Accuracy is less due to entities are not repetitive in nature. Most of the entities are unique in our dataset. Due to this reason, accuracy falls down to 74.14%. By applying Eq. 3 through 6, the values of Recall, Precision, F-score and Accuracy are calculated.

Table 7. Evaluation metrics for sentence selection

Method	Recall	Precision	F-score	Accuracy
POS + Word Suffixes + Thesaurus	60.50	74.61	66.73	74.14

4.2 Output for Semantic Relation

The second step is to identify the sentence that contains a cure, prevent or irrelevant semantic relation. The second step is a three way classification technique to identify the semantic relation from the first step. In Table 8, which describes Recall, Precision, F-Score value for each classification algorithm. The output of the Table 8 depends on the output of the previous table. In Table 7, the overall

Table 8. Performance comparison for semantic relation (first step followed by second step)

Classification algorithm	Disease	Treatment	Other
	R/P/F-score	R/P/F-score	R/P/F-score
MNB	83.01/96.95/89.44	92.31/80.00/85.72	96.97.97/89.96/93.33
SVM	90.83/83.9/87.23	100/61.54/76.19	90.53/97.00/93.65
Random Forest	85.09/82.20/83.62	93.33/42.42/58.33	89.32/96.91/92.96

accuracy is obtained as 71.80%. 410 sentences are correctly classified as true positive and true negative in the first step.

Table 8, describes the Recall, Precision, F-score value of all category. For example, "immune" is the name of the disease as well as name of the treatment. The word "immune" followed by "disorder", then it is a disease entity. The word "immune" followed by "therapy" or "inhibitors", then it is a treatment entity. Based on the meaning or, it is considered to be a disease entity or treatment entity. After applying Support Vector Machine (SVM), Random Forest, Multinomial Naive Bayes (MNB), MNB outperforms the best result. It calculates the probability value of each and every word and find out the overall probability of the sentences. Due to word count of each word MNB obtains the best F-score value in the semantic relation.

Table 9. Performance comparison for semantic relation (second step independent of first step)

Classification algorithm	Disease	Treatment	Other
	R/P/F-score	R/P/F-score	R/P/F-score
MNB	81.31/89.94/85.41	76.92/66.67/71.43	95.97/91.99/93.94
SVM	88.13/81.50/84.69	84.62/55/66.67	89.35/96.09/92.6
Random Forest	89.71/62.24/73.49	90.48/52.78/66.67	79.23/96.76/87.12

Table 9 describes the Recall, precision, F-score value of second step which is independent of the first step. Naïve Bayes algorithm outperforms the best result for this dataset in all aspects. The first task is followed by the second task which produces the F-score value of 93.94%. In other case, second task is independent of the first task and produces the F-score value of 93.33%.

5   Conclusion

This paper presented a novel approach to classify agricultural documents based on Multinomial Naïve Bayes classifier. The first task identifies the disease and treatment relation with the help of POS Tagger, Word Suffixes and Thesaurus. The second task is to identify the semantic relationships between text present in a sentence with the help of MNB method. The study focused on two semantic relations between disease treatment relation in the agricultural text. The accuracy of 71.9% and F-score of 71.43% in the semantic relation are achieved. The performance of an MNB approach shows a better accuracy for the classification process. In future, we are researching how to identify the other types of agricultural relations that relate the agricultural entities in text. Future work can overcome ambiguous entities using different techniques and tries to improve accuracy. This kind of semantic relation extraction will lead to building useful IR Systems and QA systems like agricultural chatbots. Also, various ML algorithms can be tried to achieve better accuracy of the dataset.

A   Appendix

The term NER was introduced in 1996 at the Message Understanding Conference to refer the entities [19]. NER is defined as to identify and classify the information elements called Named Entities [20]. Biomedical Named Entity Recognition (BNER) is used to determine the biological entities such as protein names, genes, disease name in biomedical texts [9,21]. Please note that the first paragraph of a section or subsection is not indented. The first paragraph that follows a table, figure, equation etc. does not need an indent, either.

References

1. Alston, J., Pardey, P.: Agriculture in the global economy. J. Econ. Perspect. **28**, 121–46 (2002)
2. Janssen, S.J.C., et al.: Towards a new generation of agricultural system data, models and knowledge products: information and communication technology. Agric. Syst. **155**, 200–212 (2017)
3. National Agricultural Library - Thesaurus. https://agclass.nal.usda.gov/download. shtml. Accessed 4 Feb 2019
4. Ben Abacha, A., Zweigenbaum, P.: Automatic extraction of semantic relations between medical entities: a rule based approach. J. Biomed. Semant. **2**(5), S4 (2011)
5. Cheng, X., Miao, D., Wang, C.: A link-based approach to semantic relation analysis. Neurocomputing **154**, 127–138 (2015)
6. Wang, D., Liu, X., Luo, H., Fan, J.: A novel framework for semantic entity identification and relationship integration in large scale text data. Future Gener. Comput. Syst. **64**, 198–210 (2016)
7. Zhang, M.L., Peña, J.M., Robles, V.: Feature selection for multi-label naive Bayes classification. Inf. Sci. **179**(19), 3218–3229 (2009)
8. Altheneyan, A.S., Menai, M.E.B.: NaïVe Bayes classifiers for authorship attribution of arabic texts. J. King Saud. Univ. Comput. Inf. Sci. **26**(4), 473–484 (2014)
9. Takeuchi, K., Collier, N.: Bio-medical entity extraction using support vector machines. Artif. Intell. Med. **33**(2), 125–137 (2005)
10. Bhasuran, B., Murugesan, G., Abdulkadhar, S., Natarajan, J.: Stacked ensemble combined with fuzzy matching for biomedical named entity recognition of diseases. J. Biomed. Inf. **64**, 1–9 (2016)
11. Frunza, O., Inkpen, D., Tran, T.: A machine learning approach for identifying disease-treatment relations in short texts. IEEE Trans. Knowl. Data Eng. **23**(6), 801–814 (2011)
12. Chaudhary, A., Kolhe, S., Kamal, R.: A hybrid ensemble for classification in multi-class datasets: an application to oilseed disease dataset. Comput. Electron. Agric. **124**, 65–72 (2016)
13. Agricultural dataset. https://drive.google.com/file/d/1b1TfA25dqXFxdH6U9eW2 MP2S12ae8IaI/view?usp=sharing. Accessed 4 Feb 2019
14. Zhang, W., Gao, F.: An improvement to naive Bayes for text classification. Procedia Eng. **15**, 2160–2164 (2011)
15. John, G.H., Langley, P.: Estimating continuous distributions in Bayesian classifiers. In: Proceedings of the Eleventh Conference on Uncertainty in Artificial Intelligence, UAI 1995, pp. 338–345 (1995)

16. Bermejo, P., Gámez, J., Puerta, J.: Improving the performance of Naive Bayes multinomial in e-mail foldering by introducing distribution-based balance of datasets. Expert Syst. Appl. **38**, 2072–2080 (2011)
17. Rosario, B., Hearst, M.: Classifying semantic relations in bioscience texts. In: Proceedings of the 42nd Annual Meeting of the Association for Computational Linguistics (ACL-2004), Barcelona, Spain, pp. 430–437 (2004)
18. Powers, D.: Evaluation: from precision, recall and F measure to ROC, informedness, markedness and correlation. J. Mach. Learn. Tech. **2**, 37–63 (2007)
19. Chinchor, N.A.: Overview of MUC-7. In: Seventh Message Understanding Conference (MUC-7): Proceedings of a Conference Held in Fairfax, Virginia (1998)
20. Marrero, M., Urbano, J., Sánchez-Cuadrado, S., Morato, J., Gómez-Berbís, J.M.: Named entity recognition: fallacies, challenges and opportunities. Computer Stan. Interfaces **35**(5), 482–489 (2013)
21. Gridach, M.: Character-level neural network for biomedical named entity recognition. J. Biomed. Inf. **70**, 85–91 (2017)

Irony Detection in Bengali Tweets: A New Dataset, Experimentation and Results

Adhiraj Ghosh[1] and Kamal Sarkar[2(✉)]

[1] Manipal Institute of Technology, Manipal, India
adhiraj.ghosh@learner.manipal.edu
[2] Jadavpur University, Kolkata, India
jukamal2001@yahoo.com

Abstract. Irony detection is a difficult task because the intended meaning of a sentence differs from the literal meaning or sentiment of that sentence. Most existing work on this subject has focused on irony detection in the English language. Since no public dataset is available for this task in the Bengali domain, we have created a Bengali irony detection dataset that contains a total of 1500 labeled Bengali tweets. This paper presents the description of the Bengali irony detection dataset developed by us and reports some results obtained on our Bengali irony dataset using several widely used machine learning algorithms such as Naïve Bayes, Support Vector Machine, K-Nearest Neighbor and Random Forest.

Keywords: Irony detection · Machine learning · Bengali · Tweets · Naïve Bayes · Support Vector Machine · K-Nearest Neighbor · Random Forest

1 Introduction

There has been a tremendous surge of data on the internet after the social media boom. Since the advent of the Social Web in 2006, the amount of textual content on the internet has exponentially grown and provides a great deal of potential for analysis. Since millions of tweets are generated every day, machine learning provides the necessary platform for adequate data analytics to better understand the activities of the average social media user as well as the users' feedback on various social and political issues. So, to understand the users' activities and comments on social media, understanding of varied features of language is needed. When the automatic system is used to manipulate and process a large amount of social media texts, the system should have the ability to understand the intricate features of language. Ironic texts are more difficult to understand than the general non-ironic texts.

Irony is popularly defined as a literary device wherein the literal meaning of a sentence differs from the figurative meaning the sentence is trying to portray [1]. This paper focuses on comprehending the most common forms of irony by enlisting certain characteristic features of each type of irony to develop a system to classify tweets

A. Chandrabose et al. (Eds.): ICCIDS 2020, IFIP AICT 578, pp. 112–127, 2020.
https://doi.org/10.1007/978-3-030-63467-4_9

correctly. Automatic irony detection provides a more fine-grained understanding of sarcasm detection [2] and sentiment analysis [3] and the multiclass analysis streamlines the Bengali user's propensity for the usage of the language. Irony detection and analysis is important for the development of sociolinguistics and analysis the uses of language constructs by specific communities. Automatic irony detection also has many other applications, for example, online harassment detection, ironic speech understanding and more importantly, sentiment analysis. It was observed that sentiment analysis system shows relatively poor performance on ironic texts compared to non-ironic texts. Hence, an added analysis of irony in the tweets propels such tasks towards more effectiveness.

The most common approaches to Irony detection in English Twitter use machine learning algorithms [4, 5]. But a huge stride in automatic irony detection in English language was made by Hee et al. [6–8] as a part of a shared task, SemEval-2018 Task 3 on Irony detection in English tweets [9]. Some of the most notable results of the irony detection tasks for English tweets [10–14] have been obtained using various techniques that are based on machine learning algorithms.

The Bengali language is the 7th most spoken language in the world, with approximately 215 million speakers all over the world. Bengali is a primary language in Bangladesh, and in India - West Bengal, Tripura and Assam. In the past few years, South Asian nations like India and Bangladesh have witnessed a data revolution, which was prompted by greatly diminished costs of mobile data in both countries, resulting in a massive boom in the number of active internet users in the countries. The latest reports on the number of active internet users in India 604.21 million[1] and that of Bangladesh being 91.421 million[2]. These staggering numbers provide a parameter to apprehend the activity of most of the Bengali speaking population in the countries.

Twitter has been the platform for posting opinions and comments on various social and political issues. With the evolution of a script-based keyboard and the advent of the smartphone, Twitter sees plenty of users' opinions written in several languages and scripts the people from various corners of the world are using this platform and posting opinions and comments in their own languages.

The focus of this work is on developing a corpus for Bengali irony detection and developing baseline systems for irony detection in Bengali. Section 2 describes the Bengali irony dataset created by us. In Sect. 3, we have provided a detailed description of the various types of irony we have considered for our experiments and the distinction factors among the four types of irony considered in implementing our classification model. Section 4 is where we have discussed the feature extraction process. The four widely used baseline classification algorithms have been used to implement our models-Naïve Bayes, Support Vector Machine, K-Nearest Neighbors and Random Forest which have been described in Sect. 5. Section 6 provides the results of our models on our dataset and finally, in conclusion section, we describe future scope and the importance of study in the field of Indian languages.

---

[1] https://main.trai.gov.in/sites/default/files/PIR_04042019_0.pdf.

[2] http://www.btrc.gov.bd/content/internet-subscribers-bangladesh-january-2019.

2 Corpus Development

It is to be noted that there is a difference between sentiment, satire and irony when it comes to a linguistic definition and scope. While there is research in sentiment analysis in Bengali in [3], research is lacking in a fine-grained model for irony in Bengali. In that regard, a major obstacle faced in the development of irony analysis system for Bengali tweets was the lack of any publicly available dataset for Bengali Irony Detection. To that end, a total of 1500 tweets were collected from Twitter and manually annotated by us for corpus development. Due to the low number of tweets available in the Bengali language as compared to English, the tweets have been collected within the time period spanning from 17/10/2010 to 4/1/2019. The tweets that have been collected and annotated contain the bare textual content and does not contain traces of metadata, which include twitter handle, display name, timestamps, user ids, locations, etc. All mentions, tags, URLs and punctuation were removed too as they are not significant for irony classification. Finally, all English letters were converted to lowercase to prevent duplication in the vocabulary list.

The entire corpus was created in a text file with UTF-8 (Unicode Transformation Format) encoding. In this way, the.txt file could support all the Bengali script characters. In our case, the comma separates the tweet from its type of irony label.

For irony detection, there are some stark differences from standard sentiment analysis. Firstly, for irony classification, we did not consider stop word removal as this may tamper the overall meaning of the tweets. Next, every emoji was replaced by a universal code as determined by the Unicode Common Locale Data Repository (CLDR), which supports all Unicode characters. Every emoji also has a CLDR short name[3], which is a suitable substitute for the emoji itself for Natural Language Processing applications. Table 1 shows some examples of the representation of emojis in Unicode.

Table 1. Code and the CLDR short names for common emojis.

Emoji	Code	CLDR short name
😀	U + 1F600	grinning_face
😂	U + 1F602	face_with_tears_of_joy
👻	U + 1F47B	ghost

3 Irony Types

We have annotated the tweets at the two level: (1) coarse level and (2) fine-grained level. At coarse level, a tweet is annotated as ironic and non-ironic whereas at the fine-grained

---

[3] https://unicode.org/emoji/charts/full-emoji-list.html.

level, ironic tweets are further classified into four types of irony. We have implemented document classification as a method of manual annotation, where based on a given set of rules that are used to define a class or label and a defined set of parameters to cause a distinction among the four classes, we have managed to label each input tweet based on the given task (coarse-grained annotation or fine-grained annotation).

Four types of irony were considered for our multi-class irony classification- (1) irony-clash, (2) verbal irony, (3) situational irony and (4) non-irony. The distribution of each type of label in our developed irony corpus is shown in Table 2.

Table 2. Distribution of irony types in the corpus.

Irony type	Number of tweets
Verbal_Irony	281
Irony_Clash	258
Situational_Irony	254
No_Irony	707

Our developed Bengali irony dataset contains 707 non-ironic tweets and 793 ironic tweets. Since our research primarily focuses with fine-grained irony analysis, the dataset was limited to 1500 tweets due to the lack of adequate number of ironic tweets present at the aforementioned timeframe and to prevent classification bias for the non-ironic tweets.

Since irony analysis is extremely subjective, the annotation process has been made as objective as possible. Based on objective criteria, at first, a fundamental distinction was made between an ironic statement and an ironic narrative. Then, the ironic statement is divided into three types-Verbal_Irony and Irony_Clash and Situational_Irony. The detailed descriptions of the different types of irony with sample examples are given in the subsequent subsections.

3.1 Verbal Irony

Verbal irony is widely used and well-defined form of irony [15]. According to the core premise of irony, verbal irony describes a statement that has an implicit sentiment of irony. Burgers [16] describes the four aspects of verbal irony: implicit, evaluative, differentiable from a non-ironic statement, and the comprehension of an opposing sentiment. Keeping these aspects in mind, a statement that contains verbal irony should:

- have a detectable sentiment in the literal composition of the tweet
- have the opposite sentiment in the intended meaning of the tweet
- have only one sentiment in the words used in the tweet and the opposite in its intention. For example, the words of a tweet should have a positive sentiment literally and a negative sentiment intentionally and vice versa
- not have neutral sentiment attached to either the literal composition or the intended meaning of the tweet

- have emojis of the same sentiment as that of the overall literal sentiment of the tweet, if present in the tweet.

A few examples of verbal irony are given below:

1. "আপনার হৃদয় শিরিষ-কাগজ এর মতন মসৃণ" (Your heart is as smooth as sandpaper). Here, the overall sentiment of the tweet is positive with the use of the word শিরিষ- (smooth). But the intended meaning uses the কাগজ (sandpaper) and infers a contradiction, since sandpaper is rough, thereby providing an opposite, negative sentiment to the meaning behind the tweet. The tweet does not have a neutral sentiment. Therefore, since the tweet fulfils 4 out of 5 criteria for verbal irony, we annotate the above tweet as Verbal_Irony.

2. খেলা দেখে অনেক মজা পাইতাছি । খেলা তো নয় মনে হয় কমেডি হইতাছে।।।: face_with_tears_of_joy: #LOL #BANvAFG" (Watching the game, I am having a lot of fun. The match feels like a comedy show, not a cricket game: face with tears of joy: #LOL #BANvAFG). This tweet has positive sentiment, including emojis. The tweet has only one sentiment and an opposing intention and hence fulfils almost all the criteria for verbal irony.

3.2 Irony by Clash

This category describes an expressive statement where there exists a literal and opposite intended meaning of the tweet and where the positive polarity and the negative polarity that defines the inversion between the literal and intended meanings are present in the tweet itself, as opposed to verbal irony, where the intended meaning has to be inferred by the reader. Therefore, tweets with the Irony_Clash label have some distinctive features. These kinds of tweets should:

- have both sentiments of positive and negative polarities in the literal meaning of the tweet, providing a literal distinction from verbal irony
- have either of the two sentiments in the intended meaning of the tweet
- not have an overall neutral sentiment just because words of opposite polarities are clashing
- may or may not depend on emojis to provide the opposing sentiment.

A few examples of irony-clash are given below.

1. "৬০,০০০ টাকার মোবাইল ফোন   Bluetooth   নাই!! তাহলে সেই আপেল হাতে না রেখে খাওয়াটাই ভালো!" (A 60,000 rupees phone doesn't even have Bluetooth in it!! Then, it's better if I ate that apple rather than keep it in my hand.) In this tweet, the sentiment clash occurs between 'doesn't' and 'better'. The tweet is ironic due to the similarities made between a costly phone and an everyday fruit and the intended meaning of the tweet has a negative sentiment as well, without having a neutral overall sentiment. Hence, it is classified as Irony_Clash.

3.3 Situational Irony

While verbal irony and irony by clash described irony existent in statements, situational irony is evident in the narratives of situations in text. The irony is evident when the outcome of the narrative defies the standard expectation of the same. In literary texts, situational irony arises when the readers were aware of the outcome of the actions of the characters when the characters themselves did not. According to Shelley [17], situational irony is derived from the schema-recognition system of human cognizance where we script a narrative for a situation, failing to comply with which is deemed ironic. The situational ironic tweets have the following characteristics. The tweets should:

- be in the form of a narrative of situation, with an identifiable flow of events
- exist in a particular time frame, from start to finish
- refer to a subject of the narrative, living or non-living, whose actions are deemed as ironic
- must operate on a distinction between what is being claimed in the text and what is being inferred by the reader
- invoke a sense of higher understanding that is not implicit to the tweet, which is the burden of the reader to comprehend
- may or may not depend on emojis to add to the narrative.

An example of Situational irony is given below.

1. "একটি অ্যাম্বুলেন্স রাস্তায় একটি লোককে চাপা দিল ।" (The ambulance ran over a man on the street). This tweet depicts the narrative of an ambulance on the street, which had run over a man, where we see the beginning and end of the narrative. The irony is denoted by the contrast between ambulance, the vehicles which are associated with hospitals and, by extension, life and "লোক কে চাপা দিল" (ran over the man) which signifies death. Hence, the tweet can be classified as Situational_Irony.

3.4 No Irony

The easiest to understand, a tweet is non-ironic if the statement or narrative does not have an inversion between its literal meaning and its intended meaning. Such tweets require no added analysis and are much easier to create and annotate. Most of the tweets on Twitter are non-ironic, in all languages. For example:

1. "মানুষের জীবনে শৈশব হল সবচেয়ে গুরুত্বপূর্ণ ।" (Childhood is the most important part of human life.) This tweet has no implicit meaning, is not a narrative or situation, does not have a clash in the statement and does not have an inverted intended sentiment. Hence the tweet is non-ironic.

4 Feature Extraction

For any machine learning algorithm to be applied on irony detection, it is important to design features that can discriminate among tweets. Since it is preferred not to use text as input for normal machine learning models. Hence it is necessary to implement word embedding, which is a methodology where words, after tokenization, are defined by real-valued vectors in a vector space. For our classification task, we have used a Term Frequency-Inverse Document Frequency (TFIDF) based model, which is a frequency-based word embedding technique, that involves representing each tweet in a vector space where a feature corresponds to a distinct term of the corpus. This was implemented after the tokenisation of the entire corpus and the subsequent development of a vocabulary list for the same. Here the feature is a distinct term and feature value is the TFIDF weight of the term calculated as the product of term frequency (number of times a term occurs in a tweet) and inverse document frequency calculated based on corpus statistics [18].

The TFIDF model is a mechanism for information retrieval that highlights the importance of a word in a corpus [19]. By using this model, we give less importance to terms that are present throughout the corpus since they have poor discriminating power. Using TFIDF model, the entire corpus is converted to tweet-term matrix wherein each row of the matrix corresponds to the vector representation of a tweet. The formula which calculates the TFIDF of a term in a tweet d is:

$$\text{tfidf}(d, t) = \text{tf}(d, t) * \text{idf}(d, t) \tag{1}$$

where tf(d, t) indicates the frequency of the term t in the tweet d.

The inverse document frequency, for n documents in the corpus is defined as:

$$\text{idf}(d, t) = \log\left[\frac{n}{\text{df}(d, t) + 1}\right] + 1 \tag{2}$$

where the document frequency function df(d, t) returns the value of how many tweets of the corpus contain the term t at least once.

There may exist many terms (for example, articles and preposition), in an extreme case, which exist in all the documents in a corpus. Mathematically, the idf of these terms should be 0. To prevent the TFIDF to be 0, there is a 1 added to the end of the formula. A smoothing factor has been incorporated into our system to prevent zero divisions. This is done by adding 1 to the numerator and denominator of the idf formula, which is now defined as follows:

$$\text{idf}(d, t) = \log\left[\frac{n + 1}{\text{df}(d, t) + 1}\right] + 1 \tag{3}$$

We have considered the above-mentioned smoothing factor while computing the tweet-term matrix.

Finally, we obtain the tweet-term matrix where each row corresponds to a tweet and each column corresponds to each distinct term in the corpus and each value in the row is the TFIDF weight of the corresponding column term if the term is present in the corresponding tweet. This value is set to 0 if the term is not present in the tweet. Each row is labeled with the label of corresponding tweet.

Using the TFIDF weight criterion in the irony classification task provides valuable insight into how effective the irony classification corpus is to identify patterns and distribution of features across the corpus, which effectively helps us determine how these features are more likely to express which type of irony.

5 Model Development

After developing the tweet-term matrix for our entire corpus, we have divided the dataset into training and testing set. Each model is developed using the machine learning algorithm trained with the training set. For the irony classification, we have considered several machine learning algorithms, such as Naïve Bayes, Support Vector Machine, K-Nearest Neighbor and Random Forest. Each of the four algorithms and the rationale behind using them to test our corpus have been described below.

5.1 Naïve Bayes

Naïve Bayes (NB) Classifier finds probability of a tweet t being in the class C based on the following equation:

$$P(t|C) = P(w_1 w_2 .. w_k|C) = P(C) \prod_{i=1}^{k} P(w_i|C) \tag{4}$$

where:

- Tweet t is represented as vector, $[w_1 \ w_2 — w_k]$ and w_i is the TFIDF weight of the term corresponding to the i-th column of tweet-term matrix described in the earlier sections,
- $P(w_i \ |C)$, the probability that i-th feature of the tweet t with value w_i belongs to class C, is calculated based on the assumption that the values of the feature are normally distributed in the class C. So, the observation value w_i of the i-th feature of the tweet, its expected value in class C and the variance of the values of the feature in class C are plugged into the equation of normal (Gaussian) distribution to compute the probability $P(w_i \ |C)$.
- P(t|C) is called posterior distribution [20] and P(C) is called prior probability calculated as ratio of size of C and the sum of sizes of all classes considered in developing the model.

Given a test tweet, the probability of the tweet being in each class is computed and its label is decided by comparing those probabilities.

Smoothing is applied to deal with data sparseness problem. In our model, we have used a smoothing parameter α, for which we have used Laplace smoothing ($\alpha = 1$) [21]. The Naïve Bayes classifier will converge more quickly than discriminative statistical models and algorithms like logistic regression, which is a significant rationale for using this algorithm for testing on the dataset.

5.2 Support Vector Machine

Vapnik introduced the utility of the Support Vector Machine (SVM) for classification in 1995 [22] and why this method of supervised learning is so robust to overfitting. The

SVM algorithm finds the maximum margin hyperplane in the feature space separating one class from another, defined as

$$wx + b = 0 \qquad (5)$$

x being the input vector used for classification and w and b are vectors are learned through the process of implementing the SVM algorithm.

To ensure a globally optimal solution, SVM is used to solve linearly constrained problems like Eq. (6), defined as

$$\min_{w} \frac{1}{2}||w||^2 + C\sum_{i}\xi_i \qquad (6)$$

C being the penalty parameter of error terms or the parameter used to control tolerance of outliers of the feature vector set and ξ is the slack variable used to relax linear separability.

Since we have implemented irony analysis with 4 labels, we have used multi-class SVM that uses one vs. all strategy to develop a group of binary classifiers whose predictions are combined to find the label of the test instance though the alternative method of solving multiclass SVM problems in one step by solving a much larger optimization problem is also used in [23]. In text classification problems, high dimensional spaces and feature vectors are the norm and the SVM algorithm is popular in handling such spaces.

5.3 K-Nearest Neighbor

Fix and Hodges [24] describes the K-Nearest Neighbor (KNN) method as an algorithm that takes an unclassified sample as input and determines the class of the sample by finding the mode of the labels of the K nearest neighbors of the input sample. The K nearest neighbors of the input sample is selected from the training set by computing Euclidean distance between the input sample and the training samples. Based on the value of K, the vector space of inputs is divided accordingly [25]. Euclidian distance between two points x and y in the Euclidian space is calculated as follows:

$$E(x, y) = \sqrt{\sum_{i=1}^{n}(x_i - y_i)^2} \qquad (7)$$

This method employs lazy learning, by not learning feature association with the classes during the training phase, rather makes use of the abstraction of data samples during testing. Since there is no training phase, this classifier parses through the entire training set for each prediction. The KNN algorithm was implemented especially to analyse the model's performance on lazy learning.

5.4 Random Forest

Random forest [26] is a kind of ensemble classifier which combines the predictions of many decision trees using majority voting to determine the class for a test instance. Each decision tree participated in ensembling process is built based on a subset of features chosen randomly from the feature set. The method integrates the idea of "bagging" [27] and the random selection of features.

This algorithm implements multiple decision trees to create subsamples of the data and uses averaging to improve accuracy and stop the occurrence of over-fitting. We use this algorithm for our irony classification task for several reasons - for many datasets, it is proven to be a highly accurate classifier, it runs efficiently on large and high dimensional datasets as the algorithm works by creating subsets of the input data and the process can be split to multiple processors or machines to run parallelly.

6 Experiments and Results

For implementing and testing our models, we have used a manually annotated corpus of 1500 tweets, which resulted in the generation of 1500*20721 feature vector (that is, each tweet is represented as a vector of 20721 dimensions). After splitting the dataset into the training set and the testing set, the machine learning algorithms were trained on the training set to develop the models. Due to the dependence of each term in a document with its neighboring terms, a major feature of the corpus prioritized for the irony classification task was the distinctive manner of expression. This feature was captured and fed into each machine learning model by implementing the word n-gram sequence model the given samples of text used in training.

We have judged the performances of these models for our defined two tasks-(1) irony detection at the coarse-grained level (classification of tweets as ironic or non-ironic) and (2) irony detection at the fine-grained (classification of tweets into one of four classes: Verbal_Irony, Irony_Clash, Situational_Irony and No_Irony).

We have used the standard 10-fold cross-validation [28] which divides the data into equal 10 parts and considers one part consisting of 150 unique tweets as the test set and the remaining 9 parts consisting of 1350 tweets as the training set for each fold. The accuracy on the test set for each fold is recorded and the overall accuracy is computed by averaging the results obtained for 10 folds. To obtain optimal results, we have tuned the parameters of each algorithm to best fit our dataset. The details of parameter tuning have been described in the following section.

For implementation of our models, we have used Google Colaboratory, a Google Cloud Service variant of the Jupyter Notebook, which supports Python 2 and Python 3. The Pandas library was used to simplify and organize the structure of the dataset into dataframes, which is a useful data structure for our text classification task. The library used for the implementation of the machine learning algorithms is scikit-learn. The metric used to assess the quality of the dataset and models for the corpus is accuracy as it is a good metric for the comparison of the four algorithms used and is a relevant metric to check how the model works on unseen data.

6.1 Naïve Bayes

For this model, we did appropriate parameter tuning by setting smoothing parameter, alpha to 1 for Laplace smoothing. The parameter fit_prior [29] was also set. This model achieves the 10-fold cross validated accuracy of 46.53% for fine-grained task and 56.67% for coarse-grained task.

6.2 Support Vector Machine

Due to its inherent ability to deal with high dimensional data, the Support Vector Machine is one of the best models for text classification We have used multi-class SVM for the multiclass irony analysis of Bengali tweets.

We have implemented a Support Vector Classification (SVC) model with several kernels [30] for obtaining the optimized results. We have also varied the value of penalty parameter C for obtaining the best results. We have shown in Table 3 and Table 4 how the performance of Support Vector Classifier on our Bengali irony dataset is affected when the choices of kernel and the cost parameter are varied.

Table 3. Accuracy of Support Vector Machine Classifier for fine-grained irony detection task when the choices of kernel and the cost parameter are varied.

SVC Kernels	C = 1	C = 10	C = 100	C = 1000
Linear	47.20	45.93	45.86	44.27
Polynomial	47.13	46.93	43.40	43.27
RBF	**47.33**	47.13	44.33	43.06

Table 4. Results of Support Vector Classifier algorithm for coarse-grained irony detection task when the cost parameter C is varied.

SVC Kernels	C = 1	C = 10	C = 100	C = 1000
Linear	64.53	65.67	65.07	63.67
Polynomial	**67.47**	66.47	63.87	62.60
RBF	66.67	66.27	63.80	63.20

As we can see from Table 3 and Table 4, the SVM algorithm achieves the best performance for the fine-grained irony classification task using the RBF kernel and C being set to 1 whereas, for coarse-grained irony classification, SVM achieves the best performance when the polynomial kernel is chosen, and C is set to 1.

6.3 K-Nearest Neighbor

Since the value of K affects the performance of K-nearest neighbor algorithm [31], we have varied the values of k from 1 to 99 and the ten-fold cross validation accuracy is calculated for each case. Only odd values of K were considered to avoid the ties among the classes. The optimum value of accuracy was found at k = 79 for fine-grained classification and at k = 7 and 15 for the coarse-grained classification task. Weights of vote cast by each nearest neighbor is set to the inverse of their Euclidian distance.

Since the Euclidian distance method was considered, the power parameter for the Minkowski metric [32] was set as 2. The performances of the KNN model for both the coarse- grained and the fine-grained Bengali irony tweet classification tasks with varying values of K are shown in Table 5 and Table 6.

Table 5. Fine-grained Bengali irony classification performance of K-Nearest Neighbor Algorithm when the value of K is varied.

KNN model	Accuracy (%)
K = 79	**47.93**
K = 85	47.67
K = 81,99	47.60
K = 73	47.53
K = 71, 77, 97	47.47

Table 6. Coarse-grained Bengali irony classification performance of K-Nearest Neighbor Algorithm when the value of K is varied.

KNN model	Accuracy (%)
K = 7, 15	**62.87**
K = 17	62.73
K = 9	62.67
K = 27	62.53
K = 5, 13, 25	62.40

6.4 Random Forest

To obtain the best performance with Random forest, we have varied the number of trees from 2 to 256, at intervals of 2^i, where i is taken from 1 to 8. Values after 256 are not taken as, [33] the model accuracy value stagnates as more trees are used. For the Random Forest Classifier, we have not specified a maximum depth of a tree and kept the minimum value of leaves fixed to 1. The accuracy of the Random Forest model for both the Bengali irony tweet classification tasks with varying number of trees in Table 7 and Table 8. As we can see from Table 7 and Table 8, Random Forest with number of trees set to 128 performs the best for the fine-grained Bengali irony tweet classification task and Random Forest with number of trees set to 256 performs the best for the coarse-grained Bengali irony tweet classification task.

Table 7. Fine-grained Bengali irony classification performance of the Random Forest algorithm when the number of trees is varied.

Random Forest model	Number of trees							
	2	4	8	16	32	64	128	256
Accuracy (%)	36.33	39.80	43.53	45.80	46.67	47.67	**48.13**	47.60

Table 8. Coarse-grained Bengali irony classification performance of the Random Forest algorithm when the number of trees is varied.

Random Forest model	Number of trees							
	2	4	8	16	32	64	128	256
Accuracy (%)	55.27	57.93	59.99	61.73	63.13	64.53	65.93	**66.99**

6.5   Comparisons of Models and Discussion

By comparing the results of all the machine learning algorithms, we can observe that the Random Forest algorithm with 128 trees performed the best for our fine-grained Bengali irony tweet classification task and showed 48.13% accuracy though the KNN algorithm also achieves a very close results with the value of k set to 79, which is surprising considering the fact that K-Nearest Neighbor algorithm is relatively simple and is implemented by lazy learning principle.

But, for coarse-grained Bengali irony tweet classification task, SVM with polynomial kernel and the cost parameter C set to1 performed the best among all of our developed machine learning models. For this task, we also observe that Random Forest model with 256 trees achieved the performance very close to the SVM based model. The SVM model achieves 67.47% accuracy whereas the Random Forest model achieves 66.99% accuracy. Since the coarse-grained Bengali irony tweet classification task is basically a binary classification task, SVM performs the best for this task. Our experimental results reveal that the Random Forest model performs consistently well for both the tasks.

We conclude that the correct classification and detection of irony is linguistically difficult, by the low accuracy results of the fine-grained approach. This may be boiled down to the low number of ironic tweets available right now, which caused a class imbalance problem in the fine-grained corpus. This resulted in the predictions ranging in the 40-50% cross-validated accuracy bracket, while the coarse-grained binary classification for irony yielded better results, ranging in the 60-70% cross-validated accuracy bracket. This is because the corpus was far more balanced for tweets that were classified as simply ironic or not ironic. We discuss solutions for the above problem in Sect. 7.

7   Conclusion

In our research, we created a new dataset for irony classification in the Bengali language. This was done due to the discernible lack of any relevant corpus for our specific research, even though such corpuses are well developed in several other languages. For any new corpus created for natural language processing task, it needs to be tested on several machine learning algorithms to report the baseline results on the dataset. Our paper provides valuable insight into the linguistic analysis required to identify and annotate irony as well as gives an understanding of methods and problems in implementing multi-class irony analysis of Bengali tweets. Though we have used four widely used machine learning techniques for implementing our models, our best models achieve 47.93% accuracy for fine-grained Bengali irony tweet classification task and 67.47% for coarse-grained Bengali irony tweet classification tasks. It shows that classification of Bengali irony tweets is not an easy task.

The major problem we have faced while completing this work is the class imbalance problem. The number of tweets that were not ironic are far greater in number than the rest in the fine-grained dataset. A consequence of the above may be the low separability between the majority class, non-ironic tweets, with the minority classes, which is something which has resulted in the classification accuracy reported in this paper.

The amount of analysis done for this corpus, being the first Bengali corpus for irony classification, should set precedence for other researchers in the field of irony classification of Bengali tweets. On a positive note, the recent surge in activity in the usage of the Bengali script on Twitter will make possible in future to develop the larger corpus and increase it beyond the current 1500 tweet dataset, thus enabling the creation of better feature vectors. We hope that the current corpus will be improved with time with more ironic tweets to handle the class imbalance problem. Further work on irony classification in Bengali will involve word-embedding with Long Short-Term Memory (LSTM) to establish semantic relationships in the corpus and latent semantic analysis for further dimensionality reduction, which may improve the accuracy as well as other performance metrics of the irony classification task.

Acknowledgements. This research work has received partial support from the project entitled "Indian Social Media Sensor: an Indian Social Media Text Mining System for Topic Detection, Topic Sentiment Analysis and Opinion Summarization" funded by the Department of Science and Technology, Government of India under the SERB scheme.

References

1. Sperber, D., Wilson, D.: Irony and the use-mention distinction. Philosophy. **3**, 143–184 (1981)
2. Bouazizi, M., Ohtsuki, T.O.: A pattern-based approach for sarcasm detection on Twitter. IEEE Access. **4**, 5477–5488 (2016)
3. Sarkar, K., Chakraborty, S.: A sentiment analysis system for Indian language Tweets. In: Prasath, R., Vuppala, A.K., Kathirvalavakumar, T. (eds.) MIKE 2015. LNCS (LNAI), vol. 9468, pp. 694–702. Springer, Cham (2015). https://doi.org/10.1007/978-3-319-26832-3_66
4. Barbieri, F., Saggion, H.: Modelling irony in Twitter. In: Proceedings of the Student Research Workshop at the 14th Conference of the European Chapter of the Association for Computational Linguistics, pp. 56–64 (2014)
5. Karoui, J., Zitoune, F.B., Moriceau, V., Aussenac-Gilles, N., Belguith, L.H.: Towards a contextual pragmatic model to detect irony in tweets. In: The 53rd Annual Meeting of the Association for Computational Linguistics and The 7th International Joint Conference of the Asian Federation of Natural Language Processing, pp. 644–650 (2015)

6. Van Hee, C.: Can machines sense irony?: exploring automatic irony detection on social media (Doctoral dissertation, Ghent University) (2017)
7. Van Hee, C., Lefever, E., Hoste, V.: Monday mornings are my fave:)# not exploring the automatic recognition of irony in english tweets. In: Proceedings of COLING 2016, the 26th International Conference on Computational Linguistics: Technical Papers, pp. 2730–2739 (2016)
8. Van Hee, C., Lefever, E., Hoste, V.: Guidelines for annotating irony in social media text. version 2.0. Technical Report 16-01, LT3, Language and Translation Technology Team (2016)
9. Van Hee, C., Lefever, E., Hoste, V.: SemEval-2018 task 3: irony detection in English Tweets. In: Proceedings of the 12th International Workshop on Semantic Evaluation, pp. 39–50 (2018)
10. Ghosh, A., Veale, T.: IronyMagnet at SemEval-2018 task 3: a siamese network for irony detection in social media. In: Proceedings of the 12th International Workshop on Semantic Evaluation, pp. 570–575 (2018)
11. Baziotis, C., et al.: NTUA-SLP at SemEval-2018 task 3: tracking ironic tweets using ensembles of word and character level Attentive RNNs. arXiv preprint arXiv:1804.06659 (2018)
12. Wu, C., Wu, F., Wu, S., Liu, J., Yuan, Z., Huang, Y.: THU_NGN at semeval-2018 task 3: Tweet irony detection with densely connected LSTM and multi-task learning. In: Proceedings of the 12th International Workshop on Semantic Evaluation, pp. 51–56 (2018)
13. Rangwani, H., Kulshreshtha, D., Singh, A.K.: NLPRL-IITBHU at SemEval-2018 Task 3: combining linguistic features and emoji pre-trained CNN for irony detection in tweets. In: Proceedings of the 12th International Workshop on Semantic Evaluation, pp. 638–642 (2018)
14. Rohanian, O., Taslimipoor, S., Evans, R., Mitkov, R.: WLV at SemEval-2018 task 3: dissecting tweets in search of irony. In: Proceedings of the 12th International Workshop on Semantic Evaluation, pp. 553–559 (2018)
15. Wilson, D., Sperber, D.: On verbal irony. Lingua. **87**(1), 53–76 (1992)
16. Burgers, C., van Mulken, M., Schellens, P.: Finding irony: an introduction of the verbal irony procedure (VIP). Metaphor Symbol. **26**, 186–205 (2011)
17. Shelley, C.: The bicoherence theory of situational irony. Cogn. Sci. **25**(5), 775–818 (2001)
18. Ramos, J.: Using TF-IDF to determine word relevance in document queries. In: Proceedings of the first instructional conference on machine learning, vol. 242, pp. 133-142 (2003)
19. Yun-tao, Z., Ling, G., Yong-cheng, W.: An improved TF-IDF approach for text classification. J. Zhejiang Univ. Sci. A **6**(1), 49–55 (2005). https://doi.org/10.1007/BF02842477
20. McCallum, A., Nigam, K.: A comparison of event models for naive bayes text classification. In: AAAI-98 Workshop on Learning for Text Categorization, vol. 752(1), pp. 41–48 (1998)
21. Yuan, Q., Cong, G., Thalmann, N.M.: Enhancing naive bayes with various smoothing methods for short text classification. In: Proceedings of the 21st International Conference on World Wide Web, pp. 645–646. ACM, (2012)
22. Vapnik, V.: The Nature of Statistical Learning Theory. Springer, New York (1995)
23. Hsu, C.W., Lin, C.J.: A comparison of methods for multiclass support vector machines. IEEE Trans. Neural Networks **13**(2), 415–425 (2002)
24. Fix, E., Hodges, J.: Discriminatory analysis: nonparametric discrimination, consistency properties. USAF School of Aviation Medicine, Randolph Field, Texas (1951)
25. D'Agostino, M., Dardanoni, V.: What's so special about Euclidean distance? Soc. Choice Welfare **33**(2), 211–233 (2009)
26. Breiman, L.: Random forests. Mach. Learn. **45**(1), 5–32 (2001)
27. Breiman, L.: Bagging predictors. Mach. Learn. **24**(2), 123–140 (1996)
28. Blum, A., Kalai, A., Langford, J.: Beating the hold-out: bounds for k-fold and progressive cross-validation. In: COLT, vol. 99, pp. 203–208 (1999)
29. Albert, J.: Teaching Inference about Proportions Using Bayes and Discrete Models. J. Stat. Educ. **3**(3) (1995)

30. Cristianini, N., Scholkopf, B.: Support vector machines and kernel methods: the new generation of learning machines. AI Mag. **23**(3), 31 (2002)
31. Yang, Y.: An evaluation of statistical approaches to text categorization. Inf. Retrieval **1**(1–2), 69–90 (1999)
32. Lu, B., Charlton, M., Brunsdon, C., Harris, P.: The Minkowski approach for choosing the distance metric in geographically weighted regression. Int. J. Geogr. Inf. Sci. **30**, 351–368 (2016)
33. Oshiro, T.M., Perez, P.S., Baranauskas, J.A.: How many trees in a random forest? In: Perner, P. (ed.) MLDM 2012. LNCS (LNAI), vol. 7376, pp. 154–168. Springer, Heidelberg (2012). https://doi.org/10.1007/978-3-642-31537-4_13

Computational Intelligence for Image and Video Analysis

Bat Algorithm with CNN Parameter Tuning for Lung Nodule False Positive Reduction

R. R. Rajalaxmi$^{(\boxtimes)}$ (iD), K. Sruthi (iD), and S. Santhoshkumar (iD)

Kongu Engineering College, Perundurai, India
rrrkec.69@gmail.com, ksruthicit@gmail.com, santhosh@kongu.edu

Abstract. Lung cancer, an uncontrolled development of abnormal cells in one or both lungs has been one of the primary causes of cancer related deaths worldwide. Detecting it in the earlier stage is the only solution to reduce lung cancer deaths. The most common tests to look for cancerous cells include X-ray, CT scan, Sputum cytology and biopsy test. CT scan is recognized as one of the effective tools in recognizing it in the earlier stage. Detecting the lung nodules (lesions) sometimes seems to be very difficult in Computer Aided Detection (CAD) systems. Because of the fact that the lung nodules have similar contrast with other structure, there might be a chance in generating numerous false positives. The performance of Convolutional Neural Network (CNN) mainly depends on the hyper parameters selected for a problem. The main motive of the proposed work is to use Bat algorithm to optimize the network hyper parameters such as number of filters in convolution layers, number of neurons and filter size in the CNN to enhance the network performance thereby eliminating the requirement of manual search for optimal hyper parameters. The methodology is validated using important performance validation metrics such as accuracy, sensitivity and specificity. The result shows that CNN in conjunction with Bat algorithm provides better results in the classification of nodules and non-nodules with minimal false positive rate.

Keywords: Deep learning · Lung nodules · Convolutional Neural Network · Bat algorithm · Particle swarm optimization · Hyper parameter

1 Introduction

Deep learning aims in creating artificial neural networks, capable of learning and taking the decisions intelligently with the help of algorithms. Deep learning uses neural networks with several layers of nodes between the input and output layer. Term deep here refers to the number of layers in the network i.e. the more the layers, the deeper the network. Series of layers between input and output perform feature identification. The need for deep learning raised mainly to process huge volume of data, perform complex algorithms, to achieve best performance with large amount of data and for effective feature extraction [10].

Deep learning in health care covers a wider range of problems ranging from cancer screening and disease monitoring to providing personalized treatment suggestions.

A. Chandrabose et al. (Eds.): ICCIDS 2020, IFIP AICT 578, pp. 131–142, 2020.
https://doi.org/10.1007/978-3-030-63467-4_10

Immense amount of data is present from radiological imaging such as X-ray, CT and MRI scans. There is shortage of tools to convert all this data to useful information.

Lung cancer is one of the life-killing dreadful diseases in the developing countries. Computed Tomography (CT), Sputum Cytology, Chest X-ray and Magnetic Resonance Imaging (MRI) are some of the medical imaging techniques employed in the earlier detection of lung cancer. Here, detection means classifying tumour basically into two classes one is noncancerous tumour which is also known as benign and the other one is cancerous tumour which is also referred as malignant. There is very less chances of survival of a lung cancer patient at the advanced stage when compared to the one when diagnosed and treated at the early stage of the cancer.

Deep neural network plays an important role in the recognition of the cancer cells among the normal tissues, thereby providing an effective tool for building an assistive Artificial Intelligence based cancer detection. The cancer treatment will be effective only when the malignant cells are accurately separated from the normal cells. Classification of the tumour cells followed by training of the neural network forms the basis for the deep learning-based cancer detection.

2 Related Works

Andreas Maier and others [9] provided a gentle introduction to deep learning in medical image processing, proceeding from theoretical foundations to applications. The proposed paper discusses about the reasons for the popularity of deep learning, reviews the fundamental basics of perceptron and neural network. It also discusses about medical image processing particularly in image detection and recognition, segmentation, registration and computer aided diagnosis.

A new method of using three-dimensional (3-D) convolutional neural networks (CNNs) for FP reduction in automated pulmonary nodule detection from volumetric CT scans is addressed in the work [2]. The importance and effectiveness of integrating multilevel contextual information into 3-D CNN framework for lung nodule detection in volumetric CT data is discussed using the experimental results.

The authors [3] presented a novel approach, which is Fast and Adaptive Detection of Pulmonary Nodules in Thoracic CT Images Using a Hierarchical Vector Quantization Scheme in which combination of custom rule-based filtering operations, extraction of features, and SVM classifier is employed.

A CAD system that uses deep features extracted from an auto encoder to classify the lung nodules into nodules and non nodules is suggested in the work [5].10 fold cross validation is used for which the results obtained are 75.01% accuracy with a sensitivity rate of 83.35%.

Haeil Lee and others [6] pointed out an approach of using Gaussian weighted average image patches for contextual Convolutional neural networks for lung nodule classification. With the extracted patches, 2D CNN is trained to achieve the classification of lung nodule candidates into positive and negative labels.

A dedicated proposal on survey on deep learning in medical image processing is clearly illustrates about how deep learning algorithms are effectively used for analyzing medical images [7].

The authors [8] used 62,492 slices of nodule-candidates extracted from 1013 CT scans obtained from the LIDC-IDRI database, containing 40,772 nodules and 21,720 non-nodules. The test scheme was designed using two separate strategies. The first was 10-fold cross-validation, and the other was the database division into training data (85.7%) and testing data (14.3%). Five tests were performed with various configurations, hyper parameters, and image sizes and achieved optimal results of 84% of accuracy.

Cascaded CNNs are used [11] to perform as selective classifiers for filtering out apparent non-nodules such as blood vessels or ribs in each cascading stage. To implement such selective classifiers, the CNNs are trained with an inverse imbalanced data set consisting of numerous nodule images and a few non-nodule images. The method was tested on 1348 nodules and 551,062 non-nodules in 888 CT scans obtained from the Lung Nodule Analysis (LNA) database.

Santos and others [12] introduced a methodology for automatic detection of small lung nodules (with sizes between 2 and 10 mm) and performed FP reduction at the end. Tsalli's and Shannon's entropy indexes were used as texture descriptors and SVM to classify suspect regions as either nodules or non-nodules.

A hierarchical learning framework, i.e. Multi-scale Convolutional Neural Networks (MCNN)-is suggested by the authors [13] in order to capture nodule heterogeneity by extracting discriminative features. The methodology was evaluated on CT images from Lung Image Database Consortium and Image Database Resource Initiative (LIDC-IDRI), where both lung nodule screening and nodule annotations are provided.

A new approach based on automated detection of solitary pulmonary nodules using Positron Emission Tomography (PET) and Computed Tomography is presented in the work [14]. An improved false positive (FP) reduction method is identified for the detection of lung nodules in PET/CT images by means of Convolutional Neural Networks (CNNs).

The methodology proposed by Taşci and Ugur [15] used 33 features based on shape and texture information extracted from nodule candidates, and realized tests with four methods of feature selection. The optimum subset is identified by evaluating classifier performance on the top five classifiers among the 10 which is tested based on the area under the ROC curve (AUC).

A new technique of using deep feature fusion from the non-medical training and hand-crafted features is presented by Changmiao and others [16] in order to reduce the false positive results. The experimented result showed 69.3% of sensitivity and 96.2% of specificity.

Semi supervised adversarial model classification (SSAC) of lung nodule on chest CT images has been introduced by Xie and others [17]. This model consists of two networks namely an adversarial autoencoder-based unsupervised reconstruction network and a supervised classification network, and also learnable transition layers. The SSAC model has been extended to the Multi-view Knowledge-based collaborative learning, aiming to employ three SSACs to characterize each nodule's overall appearance, heterogeneity in shape and texture and also to perform such characterization on nine planar views. The model has been evaluated on the widely used benchmark LIDC-IDRI dataset.

Kaur and Sharma [4] proposed a dedicated survey on using nature inspired computing for fatal disease diagnosis. This proposal effectively analyses the efficacy of various nature inspired techniques in diagnosing diverse critical human disorders. The article explains how Genetic Algorithm, Ant Colony Optimization, Particle Swarm Optimization and Artificial Bee Colony optimization have been successfully used in early diagnosis of different diseases. Furthermore, ACO, PSO and ABC are found to be best suited in diagnosing lung, prostate and breast cancer respectively.

Da Silva and others [1] proposed a new model PSO Algorithm with Convolutional Neural Network to evaluate the optimal values of the CNN and to reduce the classification error. This paper presents a methodology for reduction of lung nodule false positive on CT scans.

Nature provides rich models to solve these problems and hence swarm intelligence optimization algorithm has been introduced. The bat algorithm (BA) was proposed by professor Yang based on the swarm intelligence heuristic search algorithm in 2010 and is a kind of effective method to search the global optimal solution. BA has attracted more and more attention because of its simple, less parameters, strong robustness, and the advantage of easy implementation. Therefore, the Bat algorithm is used to optimize a few of the hyper parameters in CNN model, eliminating the requirement of a manual search to identify the optimal hyper parameters for the classification of nodule and non-nodules.

3   Problem Statement

Lung cancer identification is difficult and an extremely complex problem to solve. However, if it is detected in early stage, the patient has a high chance of survivability. The diagnostics data suggests that the highest degree of people who get diagnosed with this type of cancer have already an advanced form of cancer.

4   Proposed Method

The main objective of the proposed work is to classify the CT images into nodules and non-nodules using a CNN in conjunction with the Bat algorithm to optimize the network hyper parameters. With the obtained optimized hyper parameter values, the sensitivity rates can be increased there by efficiently reducing the number of false positives.

The proposed methodology is divided into three steps as described in Fig. 1. To be explained briefly, the first step deals with the preparation of images for lung cancer identification. The second step is the classification into nodules and non-nodules using CNN-based Bat, further followed by result evaluation.

4.1   Dataset Details

The image database used in this research is the pre-processed 50×50 LIDC-IDRI image dataset. The dataset was divided into three sets: training, validation and test set. Table 1 depicts the distribution of the data set.

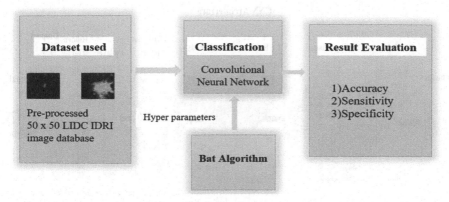

Fig. 1. Proposed methodology

Table 1. Dataset details

Dataset	No. of nodule candidates	No. of non-nodule candidates	Total
Training data set	845	4342	5187
Validation dataset	224	1063	1287
Test data set	182	1318	1500
Total	1251	6723	7974

4.2 Classification Using CNN

The preprocessed CT grey scale images of size 50×50 is submitted to CNN to classify the images into nodule and non-nodules. Figure 2 depicts the CNN architecture. The architecture consists of two convolutional layers with ReLU activation (C1 and C2), two max pooling layers (P1 and P2) after each convolutional layer followed by a dropout layer.

The completely connected layer presents an input layer, a hidden layer followed by ReLU activation and dropout layer and an output layer with sigmoid activation. The total number of filters used in first and second convolutional layers are 32, size of trainable filters are 3 in each convolutional layers C1 and C2, batch size of 32, max pooling is used in the pooling layers with pool size $2 \times 2, 128$ neurons in the hidden layer, the probability of dropout in convolutional layer is 0.35 and probability of dropout used in completely connected layer is 0.04.

4.3 Convolutional Neural Network with Bat Architecture

The performance of plain CNN model purely depends on the network hyperparameters that are assigned manually. In order to obtain optimized hyperparameter value metaheuristic approach can be used to enhance the performance of the network and to eliminate manual search. In the proposed work Bat algorithm, one of the effective metaheuristic methods is used for hyperparameter tuning.

CNN Architecture

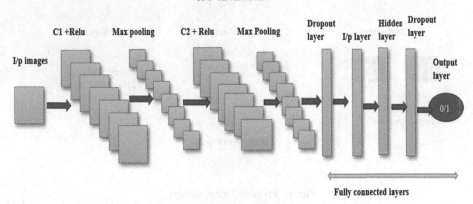

Fig. 2. CNN architecture

In the beginning, the Bat parameters such as number of bat population, maximum number of iterations, upper bound, lower bound, dimension, loudness, pulse rate, maximum and minimum frequency range are initialized as depicted in Table 2. The four parameters in CNN that has to be tuned using Bat algorithm are Number of filters in first and second convolution layer, number of neurons in the hidden layer and filter size. The bat algorithm randomly initializes these hyper parameters based on the upper and lower bound specified in parameter initialization process. The upper and lower bound values for number of filters in first and second convolution layer, number of neurons in the hidden layer are randomly initialized with integer values between 4–100. The filter size is initialized with integer values 3, 5 or 7.

Table 2. Bat-Parameters for initialization

Parameters	Values
Dimension	4
Iterations	3
Bat population	3
Loudness	0.5
Pulse rate	0.5
Minimum and maximum frequency	0.0–0.2

The fitness of each bat is evaluated using the following equation:

$$\text{Fitness} = (2 - 2 * \text{Sensitivity}) + (1 - \text{Specificity}) + (1 - \text{Accuracy}) \quad (1)$$

A higher weight is assigned to the metric sensitivity in order to obtain models with a high capacity to classify nodules correctly. In this way we tend to reduce false positive rates.

Bat Algorithm

Initialize the bat population x_i and v_i (i=1, 2....n)
Initialize frequencies f_i , pulse rates r_i and loudness A_i ---#Min and Max frequency (0.0-0.2)
#Pulse rate and loudness= 0.5
While (t<Max number of iterations)
  Generate new solutions by adjusting frequency
  Update velocities and locations/solutions
  if(rand> r_i)
       Select a solution among the best solutions
       Generate a local solution around the selected best solutions based on
       fitness.
  endif
  Generate a new solution by flying randomly
  if(rand< A_i & f(x_i) < f(x*))
       Accept the new solutions
       Increase r_i and reduce A_i
  end if
  Obtain the fitness for the three bats
  Rank the bats based on minimum fitness value and find the best bat
  End while

The process flow of Bat-CNN is depicted in Fig. 3

Once the terminating criteria is satisfied, the best bat with minimum fitness is obtained by ranking the bats based on minimum fitness as mentioned in Eq. 1. The parameter values for the best bat initialized by bat algorithm are obtained. The best parameter value will be the optimal hyper parameter value for the CNN architecture. The optimized result obtained are 48 filters in first convolutional layer, 90 filters in second convolutional layer, 78 neurons in the hidden layer and filter size of 7.

4.4 Convolutional Neural Network with Particle Swarm Optimization

Since the dataset used is pre-processed lung image datasets a comparison is made with another effective nature inspired algorithm, Particle Swarm Optimization algorithm. PSO is another meta heuristic approach to obtain optimized hyper parameter value to enhance the performance of the network. The same 50 × 50 preprocessed lung CT scan images which is used in CNN based Bat architecture is submitted to CNN based PSO architecture. Initially, the PSO parameters such as swarm size, maximum number of iterations, upper bound, lower bound, dimension, cognitive parameter, social parameter and inertia weight are initialized as depicted in Table 3

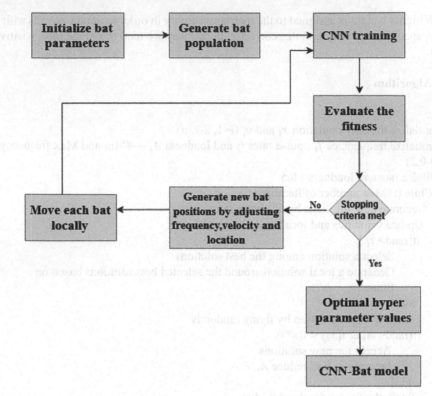

Fig. 3. Process flow of Bat-CNN

Table 3. PSO-Parameters for initialization

Parameters	Values
Dimension	4
Iterations	3
Swarm size	3
Cognitive parameter	2.0
Social parameter	2.0
Inertia weight	0.7

The four parameters in CNN that has to be tuned using PSO algorithm are the number of filters in first and second convolution layer, number of neurons in the hidden layer and filter size. The upper and lower bound values for number of filters in first and second convolution layer, number of neurons in the hidden layer are initialized with integer values between 4–100. The filter size is initialized with integer values 3, 5 or 7.

The fitness of each PSO is evaluated using the following equation:

$$\text{Fitness} = (2 - 2 * \text{Sensitivity}) + (1 - \text{Specificity}) + (1 - \text{Accuracy}) \qquad (2)$$

PSO Algorithm

Initialize the objective function
For each particle
For each dimension
        Initialize position x_i and velocity v_i
While
        For loop over all n particles and d dimensions
        Generate new velocity vi^{t+1}
        Calculate new local solutions $xi^{t+1} = xi^{t+1} + vi^{t+1}$
        Calculate the fitness function value as mentioned in equation 2
        Find the current best value for each particle xi^*
        If the fitness value is better than xi^* in the past history
                Set current fitness value as Xi^*
        End for
        Find the current global best g^* by choosing the value with best fitness value
        Update t=t+1(pseudo time or iteration counter)
End while
Output the final result xi^* and g^* (Obtain local and global best results which is the optimal hyperparameter)

Each particle is attracted towards the position of the current global best g^* and its best position xi^*. When the particle finds a location that is better than any previously found locations, the particle updates the location as the new current best for particle i. In the similar way the current best for all n particles at any time t is obtained. The aim is to find the global best among all the current best solutions until the objective no longer gets improved further or after certain number of iterations. Figure 4 illustrates the process flow of PSO-CNN.

The optimized hyperparameter value obtained using PSO algorithm are 100 filters in first convolutional layer, 44 filters in second convolutional layer, 100 neurons in the hidden layer and filter size of 7.

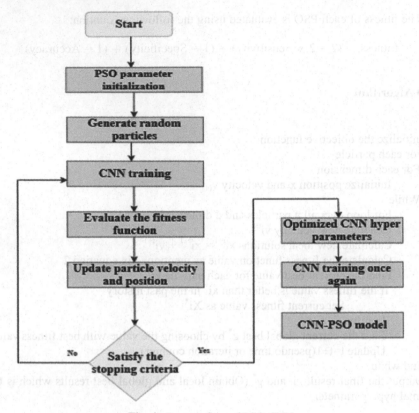

Fig. 4. Process flow of PSO-CNN

5   Experimental Results and Discussion

This section presents and discuss the results obtained with the proposed methodology with reference to lung nodule false positive reduction on CT scans. The entire methodology was implemented in python code using Keras deep learning library and executed on Kaggle GPU and GPU.

Comparison of test accuracy, sensitivity, specificity between plain CNN model, Bat based CNN model, PSO based CNN model is made and the results are depicted in Table 4.

Thus, CNN based Bat algorithm significantly increased the test accuracy, sensitivity and specificity compared to CNN based PSO. With the increased sensitivity rates, the total number of false positives can be reduced. Even though similar works [1] have been done to improve lung nodule classification, Bat algorithm is implemented with reduced datasets. However, the result depicts that the Bat algorithm can provide good results for high dimensional datasets. It is noticed that the feature extracted by the proposed model from the images are indistinguishable between nodule and non-nodule images in some cases. Due to this the model classifies few of the nodule images as non-nodules. The number of hidden layers may be increased to extract the low-level features.

Table 4. Comparative study-plain CNN model, CNN based PSO and CNN based Bat algorithm

Parameters	Plain CNN model	CNN with Bat	CNN with PSO
No. of filters in C1	32	48	100
No. of filters in C2	32	90	44
Number of neurons in hidden layer	128	78	100
Filter size	3	7	7
Sensitivity	0.8120	0.8368	0.8104
Specificity	0.8528	0.9271	0.8612
Accuracy	0.8456	0.9112	0.8504

6   Conclusion

Parameter selection for Convolutional Neural Network is one of the most important problems. Nature inspired algorithm, one of the optimization techniques can provide some efficient results. Bat Algorithm is used for adjusting the CNN parameters to find the optimal CNN hyper parameters. In the proposed work, Bat Algorithm with Convolutional Neural Network is used for controlling the parameter values which gives the better solution by exploring and exploiting. The result shows that the Bat Algorithm with Convolutional Neural Network achieved better classification accuracy, sensitivity and specificity in classifying nodule and non-nodules compared with CNN based PSO and plain CNN models. Thus, better sensitivity results in reduced false positive rates. In future, CNN architecture can be experimented in conjunction with other nature inspired algorithms in order to optimize the parameter values and try to achieve good classification results. The proposed methodology simply classifies CT images into the nodules and non-nodules but does not specify the stages of lung cancer. Future work can be performed on specifying the stages such as initial, moderate and severe cases.

References

1. Da Silva, G.L.F., Valente, T.L.A., Silva, A.C., de Paiva, A.C., Gattass, M.: Convolutional neural network-based PSO for lung nodule false positive reduction on CT images. Comput. Methods Programs Biomed. **162**, 109–118 (2018)
2. Dou, Q., Chen, H., Yu, L., Qin, J., Heng, P.A.: Multilevel contextual 3-DCNNs for false positive reduction in pulmonary nodule detection. IEEE Trans. Biomed. Eng. **64**(7), 1558–1567 (2016)
3. Han, H., Li, L., Han, F., Song, B., Moore, W., Liang, Z.: Fast and adaptive detection of pulmonary nodules in thoracic CT images using a hierarchical vector quantization scheme. IEEE J. Biomed. Health Inf. **19**(2), 648–659 (2014)
4. Kaur, P., Sharma, M.: A survey on using nature inspired computing for fatal disease diagnosis. Int. J. Inf. Syst. Model. Design (IJISMD) **8**(2), 70–91 (2017)
5. Kumar, D., Wong, A., Clausi, D.A.: Lung nodule classification using deep features in CT images. In: 12th Conference on Computer and Robot Vision, pp. 133–138. IEEE, June 2015

6. Lee, H., Lee, H., Park, M., Kim, J.: Contextual convolutional neural networks for lung nodule classification using Gaussian-weighted average image patches. In: Medical Imaging 2017: Computer-Aided Diagnosis, vol. 10134, p. 1013423. International Society for Optics and Photonics, March 2017
7. Litjens, G., et al.: A survey on deep learning in medical image analysis. Med. Image Anal. **42**, 60–88 (2017)
8. Li, W., Cao, P., Zhao, D., Wang, J.: Pulmonary nodule classification with deep convolutional neural networks on computed tomography images. Comput. Math. Methods Med. **2016**, 1–7 (2016)
9. Maier, A., Syben, C., Lasser, T., Riess, C.: A gentle introduction to deep learning in medical image processing. Zeitschrift für Medizinische Physik **29**(2), 86–101 (2019)
10. Pandey, S.K., Janghel, R.R.: Recent deep learning techniques, challenges and its applications for medical healthcare system: a review. Neural Process. Lett. **50**(2), 1907–1935 (2019). https://doi.org/10.1007/s11063-018-09976-2
11. Sakamoto, M., Nakano, H.: cascaded neural networks with selective Classifiers and its evaluation using lung x-ray CT images. arXiv preprint arXiv:1611.07136 (2016)
12. Santos, A.M., de Carvalho Filho, A.O., Silva, A.C., de Paiva, A.C., Nunes, R.A., Gattass, M.: Automatic detection of small lung nodules in 3D CT data using Gaussian mixture models, Tsallis entropy and SVM. Eng. Appl. Artif. Intell. **36**, 27–39 (2014)
13. Shen, W., Zhou, M., Yang, F., Yang, C., Tian, J.: Multi-scale convolutional neural networks for lung nodule classification. In: Ourselin, S., Alexander, Daniel C., Westin, C.-F., Cardoso, M.J. (eds.) IPMI 2015. LNCS, vol. 9123, pp. 588–599. Springer, Cham (2015). https://doi.org/10.1007/978-3-319-19992-4_46
14. Teramoto, A., Fujita, H., Yamamuro, O., Tamaki, T.: Automated detection of pulmonary nodules in PET/CT images: ensemble false-positive reduction using a convolutional neural network technique. Med. Phys. **43**(6Part1), 2821–2827 (2016)
15. Taşcı, E., Uğur, A.: Shape and texture based novel features for automated juxta pleural nodule detection in lung CTs. J. Med. Syst. **39**(5), 46 (2015)
16. Wang, C., Elazab, A., Wu, J., Hu, Q.: Lung nodule classification using deep feature fusion in chest radiography. Comput. Med. Imaging Graph. **57**, 10–18 (2017)
17. Xie, Y., Zhang, J., Xia, Y.: Semi-supervised adversarial model for benign–malignant lung nodule classification on chest CT. Med. Image Anal. **57**, 237–248 (2019)

A Secure Blind Watermarking Scheme Using Wavelets, Arnold Transform and QR Decomposition

Ayesha Shaik$^{(\boxtimes)}$ (iD)

School of Computing Science and Engineering, Vellore Institute of Technology,
Chennai 600127, India
ayeshanoormd@gmail.com

Abstract. In recent years the amount of digitally stored content available as images, videos, documents, etc., has increased exponentially. With the invention of public storages like clouds etc., security and privacy of digital data are of extreme importance. With the availability of powerful editing tools, modification of digital data is no longer a challenging task. Content modification can be done either with positive intentions like image and video enhancement or with malicious intentions like image, video morphing, video piracy, etc. To detect malicious activities, ownership of digital content needs to be established. One possible solution is to embed owner information during the content generation process. So, a secure watermarking (WMG) scheme is proposed using Wavelet transform, Arnold transforms (AT) and QR factorization in this article. The novelty of this technique is the unique way of generating WM (watermark) which makes the WMG secure. The technique is analyzed using the images given in datasets, signal and image processing institute (SIPI), break our watermarking system (BOWS), and Copydays. The experimental results of the proposed scheme are promising.

Keywords: Watermarking · Robustness · Attacks · QR factorization · Arnold transform

1 Introduction

The very fast growth in multimedia communication over the interconnected networks has raised an important issue of security and privacy of the digital data. The data needs to be authenticated as there is a chance of getting attacked or modified by third parties. So, a technique known as digital WMG came into existence for data authentication (DA) and copyright protection (CP) where copyright is inserted into the digital content. The other applications of this scheme are automated control, broadcast monitoring, user identification/authentication and fingerprinting, etc. The digital WMG scheme is designed specifically to domain i.e., spatial and transform domain. Based on the perceptibility of the watermark

© IFIP International Federation for Information Processing 2020
Published by Springer Nature Switzerland AG 2020
A. Chandrabose et al. (Eds.): ICCIDS 2020, IFIP AICT 578, pp. 143–156, 2020.
https://doi.org/10.1007/978-3-030-63467-4_11

(WM), the watermarking technique is classified as VISIBLE and INVISIBLE schemes. Moreover, we can again classify them as blind (B) and non-blind (NB) WMG schemes depending on the requirement of the CI (cover image) during WM extraction. The basic requirements of the WMG scheme are the robustness (Ro) and imperceptibility (Im) of the scheme. Robustness is defined as the tolerance of the WMG scheme towards the attacks. The imperceptibility is defined as the invisibility of the WM i.e., the quality of the watermarked image (WMI). Based on the strength of the WM, the watermarking scheme is classified as robust (R), semi-fragile (SF) and fragile (F). A robust WMG scheme will tolerate a set of attacks, fragile WMG scheme can't tolerate any attack and the WMG scheme that is not R and F can be categorized as a semi-fragile WMG scheme.

In the existing works, the WM is embedded directly on the original data or in the frequency domain. A few works used a combination of discrete wavelet transform (DWT) and the singular value decomposition (SVD) decomposition, where the WM is embedded in the singular values (SVs) of the original values. The disadvantage of these schemes is they are susceptible to the false positive problem of ownership authentication. So, in the proposed work, a combination of DWT and QR decomposition is used followed by AT. Here, AT is used to disorder the pixel values using a secure key such that the third parties will not be able to find out the exact modification done to the data. The combination of these three algorithms and the way they have used in the proposed work both are novel. One more advantage of the proposed work is that a content dependent watermark is generated from the original watermark using QR decomposition which was detailed in detail in the later sections. On the whole, the proposed technique is secure and robust, and it helps to protect the copyrights of the owner.

In this article, a secure WMG scheme using wavelet, Arnold transform and QR factorization is proposed. Section 2 discusses the overview of the WMG schemes existing in the literature. Section 3 presents the preliminaries required to implement the proposed work. Section 4 presents the proposed method in detail. Section 5 gives the results and analysis of the proposed method. Section 6 gives a conclusion followed by the references.

2   Literature Survey

In transform domain, DWT [4], Discrete Cosine Transform (DCT) [9], SVD [5] and Walsh-Hadamard transform (WHT) are existing. In [16] an adaptive WMG scheme is discussed, where the WM is inserted into the most significant part of the CI. A WMG scheme that uses Just-Noticeable Difference (JND) and Fuzzy Inference System (FIS) along with genetic algorithm (GA) is presented in [26]. An SVD WMG technique is discussed in [18] along with Tiny-GA. Fuzzy logic and Tabu search combined digital image WMG scheme is presented in [19]. A color image WMG using QR decomposition is discussed in [25].

A WMG technique using multi-resolution (MR) and complex Hadamard transform is proposed in [10], where the first multi-resolution transform is applied and then insert the WM using Hadamard transform. A watermarking technique which utilizes an optimal transport to map a list of original signals to a list of watermarked signals is discussed in [21]. In [17], the author has proposed a digital image WMG technique using DWT and SVD with least significant bit (LSB)-based techniques to protect copyrights and robust to many attacks.

In [12], the combined 3-level DWT and DCT coefficients are selected to embed a binary WM bit in the cover image (CI) and support vector machine (SVM) is used to retrieve the watermark. Here, the PSNR achieved for 300 images is around 42.45 dB. A review of optical image hiding (IH) and WMG techniques has been discussed in [14]. In this technique, a review of various optical systems and architectures for IH and the summary of processing algorithms related to optical IH are presented. A review of different digital image WMG algorithms in the frequency domain to prove the ownership of the data into the digital image without affecting its Visual Quality (VQ) has been proposed in [15]. Here, the author has found a wide variety of applications and classifications of the same for digital watermarking methods. The author used the Discrete Orthonormal Stockwell Transform (DOST) to achieve improved robustness and imperceptibility of the WMI.

A selected wavelet SVD-based WMG scheme has been presented in [23], in which the author mentioned that the embedding in the RGB and YCbCr color channels achieves high imperceptibility. Here, three different SVD-based image WMG schemes with different wavelet transforms are selected for color image testing and evaluation. A digital image WMG technique based on DWT and encryption has been discussed in [4]. Here, the demonstration of WM inserting and retrieval algorithm using DWT coefficients, distance measures, and encryption has been discussed. Authors of [4] presented that the DWT through multi-resolution analysis provides the much-needed simplicity in WM inserting and retrieval through WM encryption. In [6], the authors discussed the standard WMG system frameworks and listed the needed requirements to design WMG techniques, and reviewed them to find the limitations of state-of-the-art methods.

3   Preliminaries

3.1   QR Factorization

The QR factorization [25] of a matrix will be done using the following Eq. 1.

$$[E_A \ F_A \ G_A] = qr(A) \tag{1}$$

where A is a d $\times$ d matrix that we need to decompose, E_A is an d $\times$ d unitary matrix, G_A is an d $\times$ d permutation matrix and F_A is an d $\times$ d upper triangular matrix. Using Gram-Scmidth orthogonalization technique the columns of E_A are generated from columns of A. If $A = [\ a_1, a_2, \ldots, a_k]$ and $E_A = [e_1, e_2, \ldots, e_k]$, where a_i and e_i are column vectors, then matrix F_A can be calculated as shown below:

$$F_A = \begin{bmatrix} \langle a_1, e_1 \rangle & \langle e_2, q_1 \rangle & \langle e_3, q_1 \rangle & \cdots & \langle e_k, q_1 \rangle \\ \langle a_1, e_2 \rangle & \langle e_2, q_2 \rangle & \langle e_3, q_2 \rangle & \cdots & \langle e_k, q_2 \rangle \\ \vdots & \vdots & \vdots & \ddots & \vdots \\ \langle a_1, e_k \rangle & \langle e_2, q_k \rangle & \langle e_3, q_k \rangle & \cdots & \langle e_k, q_k \rangle \end{bmatrix} \qquad (2)$$

where $\langle \cdot, \cdot \rangle$ denotes an inner product.

3.2 Arnold Transform (AT)

This transform [27] is widely used because of its periodicity. It is usually used for digital encryption and it is the process of realignment of the pixels in the digital data (image). A 2D AT is computed as shown:

$$\begin{pmatrix} k' \\ l' \end{pmatrix} = \begin{pmatrix} 1 & 1 \\ 1 & 2 \end{pmatrix} \begin{pmatrix} k \\ l \end{pmatrix} \bmod S \qquad (3)$$

where k and l are the coordinates of digital image, k' and l' are the coordinates of the scrambled image. S is the size (height or width) of the image (If the image size is $x \times x$ then height will be x). If this operation done repeatedly then the output will be an entirely disordered image compared to the original digital image for a few number of iterations.

3.3 Normalized Cross Correlation (NC)

$$NC(B, \hat{B}) = \frac{\sum_c \sum_d B(c,d) \hat{B}(c,d)}{\sqrt{\sum_c \sum_d B(c,d)^2} \sqrt{\sum_c \sum_d \hat{B}(c,d)^2}} \qquad (4)$$

where B and $\hat{B}$ are original image and the image that is processed.

4   Proposed Watermarking Scheme

The proposed WMG scheme for embedding the WM is given in Fig. 1. The original image I is divided into 4 SBs by applying 2-level DWT. These subbands (SBs) are undergone QR decomposition and Arnold transform to obtain the upper triangular matrices (UTMs) of the original image. The watermark W is converted to hexadecimal form and a predictive watermark is generated as given in Eq. (5). On W, QR decomposition is applied to produce UTMs of the watermark. The UTMs of original data are modified with the UTMs of the WM using the embedding strengths $\alpha_{LL}, \alpha_{LH}, \alpha_{HL}, \alpha_{HH}$ and predicted watermark to obtain the WMI I_W. The WMG procedure and extraction is provided in detail in the algorithms.

The proposed WM embedding scheme is detailed in Algorithm 1. The WM is generated in a unique way in this article. It has all the information related to the

Algorithm 1. Watermark embedding

Input: Original Image I, Watermark W and embedding strengths $\alpha_{LL}, \alpha_{LH}, \alpha_{HL}, \alpha_{HH}$
Output: WMI I_W
1: Apply 2-level DWT on I and W to decompose into sub bands (SBs),
$$(LL_I, \ LH_I, \ HL_I, \ HH_I) = DWT(I)$$
$$(LL_W, \ LH_W, \ HL_W, \ HH_W) = DWT(W)$$

2: Perform QR decomposition on $(LL_I, \ LH_I, \ HL_I, \ HH_I)$
$$QR(LL_I) = Q_{LL_I} \ R_{LL_I} \ P_{LL_I}$$
$$QR(LH_I) = Q_{LH_I} \ R_{LH_I} \ P_{LH_I}$$
$$QR(HL_I) = Q_{HL_I} \ R_{HL_I} \ P_{HL_I}$$
$$QR(HH_I) = Q_{HH_I} \ R_{HH_I} \ P_{HH_I}$$
3: Apply QR on W
$$QR(LL_W) = Q_W \ R_W \ P_W$$
4: Convert W to hexadecimal i.e., W_H
5: Calculate predictive watermark W_B from W as shown.

$$\text{if}(W_i = W_{i+1}) \text{ then } W_{B_i} = 0; \text{ else } W_{B_i} = 1 \qquad (5)$$

6: Apply Arnold transform on $R_{LL_I}, R_{LH_I}, R_{HL_I}$ and R_{HH_I}

$$R_{LL_I}^a = Arnold(R_{LL_I})$$
$$R_{LH_I}^a = Arnold(R_{LH_I})$$
$$R_{HL_I}^a = Arnold R_{HL_I})$$
$$R_{HH_I}^a = Arnold(R_{HH_I})$$

(6)

7: Find

$$R_{LL_I}^{ma} = R_{LL_I}{}^a + \frac{\alpha_{LL} \times R_W \times W_B}{W_H}$$
$$R_{LH_I}^{ma} = R_{LH_I}{}^a + \frac{\alpha_{LH} \times R_W \times W_B}{W_H}$$
$$R_{HL_I}^{ma} = R_{HL_I}{}^a + \frac{\alpha_{HL} \times R_W \times W_B}{W_H}$$
$$R_{HH_I}^{ma} = R_{HH_I}{}^a + \frac{\alpha_{HH} \times R_W \times W_B}{W_H}$$

(7)

8: Multiply the matrices as given below

$$R_{LL_I}^{QR} = Q_{LL_I} \times R_{LL_I}^{ma} \times P_{LL_I}$$
$$R_{LH_I}^{QR} = Q_{LH_I} \times R_{LH_I}^{ma} \times P_{LH_I}$$
$$R_{HL_I}^{QR} = Q_{HL_I} \times R_{HL_I}^{ma} \times P_{HL_I}$$
$$R_{HH_I}^{QR} = Q_{HH_I} \times R_{HH_I}^{ma} \times P_{HH_I}$$

(8)

9: Perform inverse DWT to obtain WMI, I_W

$$I_W = Inverse \ DWT(R_{LL_I}^{QR}, \ R_{LH_I}^{QR}, \ R_{HL_I}^{QR}, \ R_{HH_I}^{QR}) \qquad (9)$$

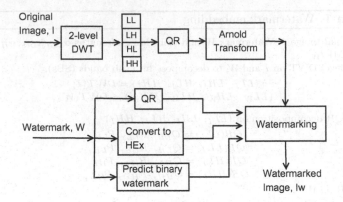

Fig. 1. Block diagram proposed WMG scheme for WM embedding

WM and it is encrypted using the predicted WM and binary WM. According to Algorithm 1, the OI and WM is gone through 2-level DWT to obtain the SBs. The 2-level SBs (LL_I, LH_I, HL_I, HH_I) of the OI will be decomposed into their corresponding unitary matrices and UTMs using QR decomposition. The 2-level sub-band of WM, LL_W will be decomposed into unitary matrix Q_W and UTM R_W. The WM is converted into hexadecimal form W_H, and the WM is predicted to get a binary WM W_B as shown in Eq. (5). The upper triangular matrices (UTMs) $R_{LL_I}, R_{LH_I}, R_{HL_I}, R_{HH_I}$ of the original image SBs will be transformed using AT to obtain $R_{LL_I}^a, R_{LH_I}^a, R_{HL_I}^a, R_{HH_I}^a$. In the transformed UTMs $R_{LL_I}^a, R_{LH_I}^a, R_{HL_I}^a, R_{HH_I}^a$, the unique generated watermark (GW) G_W will be embedded as given in Eq. (6) to obtain $R_{LL_I}^{ma}, R_{LH_I}^{ma}, R_{HL_I}^{ma}, R_{HH_I}^{ma}$. The GW is holding properties of W_H, W_B and the UTM of the WM, $G_W = \frac{\alpha_{SB} \times R_W \times W_B}{W_H}$, where α_{SB} is α_{LL} for LL sub-band, α_{SB} is α_{LH} for LH sub-band, α_{SB} is α_{HL} for HL sub-band and α_{SB} is α_{HH} for HH sub-band respectively. The unitary matrices and permutation matrices are multiplied with the WM embedded UTMs to obtain the product matrices $R_{LL_I}^{QR}, R_{LH_I}^{QR}, R_{HL_I}^{QR}, R_{HH_I}^{QR}$ as given in Eq. (8), and the inverse 2-level DWT is performed on them to obtain WMI I_W.

For retrieval of the watermark, WM extracting algorithm is detailed in Algorithm 2. In this algorithm, the possibly modified WMI, I_W' is undergone 2-level DWT to decompose into sub-bands $(LL_I', LH_I', HL_I', HH_I')$ and perform QR decomposition QR decomposition to obtain the UTMs of all SBs. Inverse AT is applied on those UTMs to produce the $R_{LL_I}^{inv}, R_{LH_I}^{inv}, R_{HL_I}^{inv}, R_{HH_I}^{inv}$ matrices as given in Algorithm 2. Now, extract the WM $W_{ex_{LL}}$ as given in Eq. (11) and calculate W_{LL}^* as given in Eq. (12). The steps are repeated for the other SBs to extract $W_{LH}^*, W_{HL}^*, W_{HH}^*$. After extraction of the WM, the inserted WM is correlated with the extracted WM as given in Eq. (4). If the correlation is high then the ownership identification can be done else it can't be done.

Algorithm 2. Watermark extraction

Input: Possibly   modified   WMI, I'_W,$(\alpha_{LL}, \alpha_{LH}, \alpha_{HL}, \alpha_{HH})$,$W_B$,
  W_H,$(R_{LL_I}, R_{LH_I}, R_{HL_I}, R_{HH_I})$
Output: Extracted encrypted watermark,W^*
1: Apply DWT on I'_W and decompose into sub bands,$(LL'_I, \ LH'_I, \ HL'_I, \ HH'_I)$

2: Perform QR on all the subbands $(LL'_I, \ LH'_I, \ HL'_I, \ HH'_I)$
$$QR(LL'_I) = Q'_{LL_I} \ R'_{LL_I} \ P'_{LL_I}$$
3: Calculate
$$R^{inv}_{LL_I} = invArnold(R'_{LL_I}) \qquad (10)$$

4: Calculate
$$W_{ex_{LL}} = \frac{(R^{inv}_{LL_I} - R_{LL_I}) \times W_H}{W_B \times \alpha_{LL}} \qquad (11)$$

5: Calculate W^*_{LL}
$$W^*_{LL} = Q_W \times W_{ex_{LL}} \times P_W \qquad (12)$$

6: Similarly repeat the steps for all the SBs and extract WMs $W^*_{LH}, W^*_{HL}, W^*_{HH}$
7: Find NC between inserted encrypted and extracted encrypted WM by using Eq. (4).

8: If the NC is greater than the predefined threshold then the ownership is authenticated. Otherwise it is not authenticated.

---

5 Results and Analysis

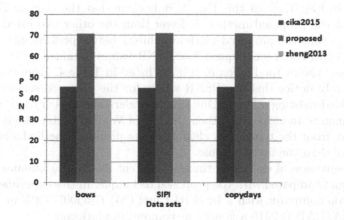

Fig. 2. PSNR values for the images in standard data sets

The proposed WMG method is analyzed using the images dataset [1,2] and [3]. The original and WMIs is shown in Fig. 3, where Fig. 3(a) and Fig. 3(b) are the original flower image and watermarked flower images, Fig. 3(c) and Fig. 3(d) are the original boat image and watermarked boat images, Fig. 3(e) and Fig. 3(f) are the original scenary image and watermarked scenary images, and Fig. 3(g) and Fig. 3(h) are the original mountain image and watermarked mountain images respectively. The average PSNR values for the images in three datasets BOWS, SIPI, and Copydays for the methods presented in [8,28] and for the proposed method are shown in Fig. 2. From this Fig. 2, it is clear that the average PSNR value obtained for the proposed method is higher than the other two listed methods.

The correlation between the inserted and extracted WMs is listed in Table 1 for the proposed method and the method presented in [21]. From this Table 1, one can clearly notice that the correlation values for the proposed scheme is better than the listed existing scheme. One should understand that if the correlation value is closer to zero, the ownership authentication can't be done and if the correlation value is close to 1, then ownership authentication can be done. So, from the table, it is clear that the proposed method shows better performance than the listed scheme. The PSNR values obtained by the proposed technique and the method given in [20] are listed in Table 2. From this Table 2, one can clearly notice that the PSNR values for the proposed scheme is better than the listed existing scheme.

The execution time for the existing techniques and the proposed method is listed in Table 3. The first five rows presents the execution time taken by the existing techniques and the last three rows shows the execution time taken by the proposed method for the three datasets (SIPI, Copydays, and BOWS) respectively. From the table, one can notice that the proposed method consumes lesser time compared to the existing methods. Hence, one can say that the proposed method exhibits better performance compared to the listed existing techniques. The average PSNR values for the images in three datasets BOWS, SIPI, and Copydays for the methods presented in [8,28] and for the proposed method are shown in Fig. 4. From this Fig. 4, it is clear that the average BER value obtained for the proposed method is lower than the other two listed methods, which means that the proposed method exhibits better performance compared to the other two listed techniques. The PSNR values obtained by the proposed technique and the method given in [12] are listed in Table 4. From this Table 4, one can clearly notice that the PSNR values for the proposed scheme is better than the listed existing scheme. One should understand that if the PSNR value is higher (infinite in ideal case), then the VQ of WMI is better else it will be of less VQ. So, from the table, it is clear that the proposed method shows better performance than the listed scheme.

The comparison of execution time of different existing techniques are listed in Table 2 and compared with the proposed technique. In this experimental analysis, a laptop computer with a Intel(R) Core(TM) i7-5500U CPU at 2.40 GHz, Win 10, MATLAB R 2015 a is used as computing platform.

(a) Original flower   (b) Watermarked flower

(c) Original boat   (d) Watermarked boat

(e) Original scenary   (f) Watermarked scenary

(g) Original mountain   (h) Watermarked Mountain

Fig. 3. Original and WMIs for proposed method

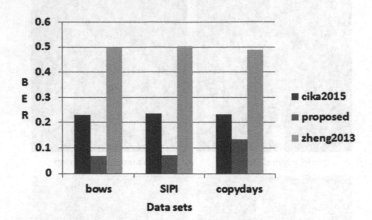

Fig. 4. Bit error rate (BER) values for the images in standard data sets

Table 1. NC correlation values between extracted and original watermarks

Sl. no	Attack	Proposed	TNW [21]
1	Crop	0.896	0.682
2	Gaussian noise	0.9395	0.675
3	Salt and pepper noise	0.93	0.665
4	Histogram equalization	0.9314	0.678
5	Xshear	0.75	0.607
6	Yshear	0.81	0.613

Table 2. PSNR values for the proposed and [20]

Sl. no	Image	Proposed	Majumder [20]
1	Boat	47.77	45.9
2	Monkey	51.4852	45.9
3	Woman	53.9	45.12
4	Peppers	47.6	44.87
5	Nature	55.28	44.8
6	Lena	49.9	44.37
7	Aeroplane	53.4	45.36
8	Correction	54.4	44.86

Table 3. Execution time for the listed techniques

Sl. no	Method	Execution time (in seconds)
1	Chou [7]	2.5122319
2	Golea [11]	2.815017
3	Yashar [22]	1.368692
4	Su [24]	1.855
5	Qingtang [25]	1.11399
6	Proposed (SIPI [3])	0.8525148
7	Proposed (Copydays [1])	0.85324
8	Proposed (BOWS [2])	0.852338

The execution time of the proposed method is very less compared to the listed techniques in the Table 2. In [7] the CI has to be changed to color space for the color quantization and its inverse-transformation is also involved, [22] involves wavelet and QR decomposition but it requires block decomposition which is time consuming and [24] uses schur decomposition which requires about $8N^3/3$ flops, SVD decomposition requires about $11N^3/3$ flops, and the QR decomposition in between [7] which makes it time consuming. The presented method uses 2 level DWT and QR only to a significant SBs which makes it fast compared to the listed techniques.

Evaluation of the digital watermarking techniques is done based on the following categories:

- Theoretical computational complexity: Upper bound of the methodologies used for watermarking
- Practical computational time: System time consumed for watermarking
- Watermarked image quality: peak signal-to-noise ratio, structural similarity index
- Tolerance of the watermarking method against the attacks such as Gaussian noise, Salt & pepper noise, Median filtering, Average filtering, and so on: bit error rate, normalized correlation.

In the Table 3, different watermarking techniques have been compared for the time required for watermarking. This can be done as it is one of the factor to compare these techniques. If we see in detail, the method in [7] have the complexity of $O(n^2)$ and this method is presented in spatial domain, and as the spatial domain techniques are considered to be fragile in nature (not tolerant), this method is not suggested for robust applications. Similarly, the computational complexities of the method given in [11] is $O(n^3)$ as it is using SVD transform, the method in [22] is $O(n^2)$ as it is using DWT transform, the method given in [24] is $O(n^3)$ as it is using Schur decomposition and Schur decomposition uses QR decomposition, and the method given in [25] is $O(n^3)$ as it is using QR decomposition. The proposed method has the computational complexity of $\max(O(n^3),O(n^3))$ which is $O(n^3)$, as it uses DWT and QR decomposition.

Table 4. PSNR values for the proposed and [12]

Sl. no	Image	Proposed	WT-DCT [12]
1	Lena	47.77	42.98
2	Peppers	47.6	43.24
3	Mandril	53.9	41.79
4	Jetplane	48.6	42.92
5	Barbara	55.28	41.49
6	Lake	50.9	42.88
7	X-ray	52.4	43.01
8	Galaxy	51.4	42.45
9	Living room	47.4	42.18
10	Elaine	45.4	41.96

The use of AT encrypts the WM that will make sure that the ordering of the original pixel are not proper and the GW embedded is also a mixture of different variations of the WM which makes the scheme secure. More over the degradation done to the CI is only to the UTM of the SBs which is not significantly perceptual which in turn produces a high quality WMI. The QR decomposition is used to prevent the degradation to the sub-bands, instead the WM is embedded in UTMs of SBs which ensures the less quality degradation to get high quality visual image.

6   Conclusion

A digital image watermarking scheme using 2-level DWT, Arnold transform and QR decomposition is proposed in this article. The cover image is decomposed into the sub-bands and QR is applied on all the subbands, which gives an unitary matrix and an upper triangular matrix. The watermark is undergone discrete wavelet transform and then QR decomposition. The watermark is then transformed to hexadecimal form and transformed to binary based on predictive process. Now the upper triangular matrix of the cover image sub-bands are modified with the upper triangular matrix of the watermark, binary predictive watermark and hex-watermark with the embedding strength. The experimental results of the proposed method are promising. It gives better PSNR and good robustness towards a list of geometric attacks making the scheme robust.

References

1. http://lear.inrialpes.fr/~jegou/data.php#copydays. Accessed May 2014
2. http://bows2.ec-lille.fr. Accessed Aug 2015
3. http://sipi.usc.edu/database/database.php?volume=misc. Accessed July 2015

4. Ambadekar, S.P., Jain, J., Khanapuri, J.: Digital image watermarking through encryption and DWT for copyright protection. In: Bhattacharyya, S., Mukherjee, A., Bhaumik, H., Das, S., Yoshida, K. (eds.) Recent Trends in Signal and Image Processing. AISC, vol. 727, pp. 187–195. Springer, Singapore (2019). https://doi. org/10.1007/978-981-10-8863-6_19

5. Bao, P., Ma, X.: Image adaptive watermarking using wavelet domain singular value decomposition (2005)

6. Begum, M., Uddin, M.S.: Digital image watermarking techniques: a review. Information **11**(2), 110 (2020)

7. Chou, C.-H., Wu, T.-L.: Embedding color watermarks in color images. EURASIP J. Adv. Sig. Process. **2003**(1), 1–9 (2003). https://doi.org/10.1155/ S1110865703211227

8. Cika, P., Skorpil, V.: Robust image watermarking method based on 2D-WHT and SVD. In: Advances in Electrical and Computer Engineering, pp. 134–137 (2015)

9. Cox, I.J., Miller, M.L.: Review of watermarking and the importance of perceptual modeling. In: Electronic Imaging 1997, pp. 92–99 (1997)

10. Falkowski, B., Lim, L.S.: Image watermarking using Hadamard transforms. Electron. Lett. **36**(3), 211–213 (2000)

11. Golea, N.E.H., Seghir, R., Benzid, R.: A bind RGB color image watermarking based on singular value decomposition. In: ACS/IEEE International Conference on Computer Systems and Applications-AICCSA 2010, pp. 1–5 (2010)

12. Islam, M., Kumar, G.R., Shaik, A.S., Laskar, R.H.: WT-DCT domain based image watermarking technique using SVM. TEST Eng. Manage. **82**, 3195–3200 (2020)

13. Jiansheng, M., Sukang, L., Xiaomei, T.: A digital watermarking algorithm based on DCT and DWT. In: International Symposium on Web Information Systems and Applications (WISA 2009), pp. 104–107 (2009)

14. Jiao, S., Zhou, C., Shi, Y., Zou, W., Li, X.: Review on optical image hiding and watermarking techniques. Opt. Laser Tech. **109**, 370–380 (2019)

15. Jibrin, B., Tekanyi, A., Sani, S.: Image watermarking algorithm in frequency domain: a review of technical literature. ATBU J. Sci. Tech. Educ. **7**(1), 257–263 (2019)

16. Joshi, V., Rane, M.: Digital water marking using LSB replacement with secret key insertion technique (2014)

17. Kumar, A.: A review on implementation of digital image watermarking techniques using LSB and DWT. In: Tuba, M., Akashe, S., Joshi, A. (eds.) Information and Communication Technology for Sustainable Development. AISC, vol. 933, pp. 595–602. Springer, Singapore (2020). https://doi.org/10.1007/978-981-13-7166-0_59

18. Lai, C.C.: A digital watermarking scheme based on singular value decomposition and tiny genetic algorithm (2011)

19. Latif, A.: An adaptive digital image watermarking scheme using fuzzy logic and Tabu search. J. Inf. Hiding Multimedia Sig. Process. **4**(4), 250–271 (2013)

20. Majumder, S., Das, T.S., Sarkar, S., Sarkar, S.K.: SVD and lifting wavelet based fragile image watermarking (2010)

21. Mathon, B., Cayre, F., Bas, P., Macq, B.: Optimal transport for secure spread-spectrum watermarking of still images. IEEE Trans. Image Process. **23**(4), 1694–1705 (2014)

22. Naderahmadian, Y., Hosseini-Khayat, S.: Fast watermarking based on QR decomposition in wavelet domain. In: 2010 Sixth International Conference on Intelligent Information Hiding and Multimedia Signal Processing (IIH-MSP), pp. 127–130 (2010)

23. Rassem, T.H., Makbol, N.M., Khoo, B.E.: Performance evaluation of wavelet SVD-based watermarking schemes for color images. In: Anbar, M., Abdullah, N., Manickam, S. (eds.) ACeS 2019. CCIS, vol. 1132, pp. 89–103. Springer, Singapore (2020). https://doi.org/10.1007/978-981-15-2693-0_7

24. Su, Q., Niu, Y., Liu, X., Zhu, Y.: Embedding color watermarks in color images based on Schur decomposition (2012)

25. Su, Q., Niu, Y., Wang, G., Jia, S., Yue, J.: Color image blind watermarking scheme based on QR decomposition (2014)

26. Tsai, H.H., Lo, S.C.: JND-based watermark embedding and GA-based watermark extraction with fuzzy inference system for image verification (2014)

27. Wu, L., Zhang, J., Deng, W., He, D.: Arnold transformation algorithm and anti-Arnold transformation algorithm. In: First International Conference on Information Science and Engineering, pp. 1164–1167 (2009)

28. Zheng, P., Huang, J.: Walsh-Hadamard transform in the homomorphic encrypted domain and its application in image watermarking. Information Hiding, pp. 240–254 (2013)

Role of Distance Measures in Approximate String Matching Algorithms for Face Recognition System

B. Krishnaveni[(✉)] and S. Sridhar

Information Science and Technology Department, Anna University, Chennai, India
krishnaveni1116@gmail.com

Abstract. This paper is based on the recognition of faces using string matching. The approximate string matching is a method for finding an approximate match of a pattern within a string. Exact matching is impracticable for a larger amount of data as it involves more time. Those issues can be solved by finding an approximate match rather than an exact match. This paper aims to experiment with the performance of approximation string matching approaches using various distance measures such as Edit distance, Longest Common Subsequence (LCSS), Hamming distance, Jaro distance, and Jaro-Winkler distance. The algorithms generate a near-optimal solution to face recognition system with reduced computational complexity. This paper deals with the conversion of face images into strings, matching those image strings by using the approximation string matching algorithm that determines the distance and classifies a face image based on the minimum distance. Experiments have been performed with FEI and ORL face databases for the evaluation of approximation string matching algorithms and the results demonstrate the utility of distance measures for the face recognition system.

Keywords: Approximate · String matching · Face recognition · Edit distance · LCSS · Hamming · Jaro distance

1 Introduction

Face recognition is the identification or verification of a person's face from a database of different person's faces [1]. Nowadays, face recognition has become a widely used application in mobile phones, and robotics, and security systems. It has also received attention in applications such as surveillance, human-computer interaction, access control, a criminal investigation, border control, and smart cars. Certain applications such as text searching, DNA subsequence searching, and signal processing and object recognition require faster and near-optimal solutions. Face recognition is also one such application that may require a near-optimal and faster solution.

Several face recognition approaches deal with controlled face recognition [2–8], but only a few approaches address the face recognition under uncontrolled conditions include partial occlusion, pose changes, illumination changes, and expression changes

© IFIP International Federation for Information Processing 2020
Published by Springer Nature Switzerland AG 2020
A. Chandrabose et al. (Eds.): ICCIDS 2020, IFIP AICT 578, pp. 157–169, 2020.
https://doi.org/10.1007/978-3-030-63467-4_12

[9–11]. Uncontrollable face recognition is a difficult and challenging task. An efficient face recognition system should address those conditions.

String matching [12] is a high level structural and syntactic technique for finding similarity between two strings. In other words, a string matching algorithm finds the location of a certain pattern in a larger amount of text or string. The applications of string matching include search engines, object recognition, speech recognition, Information retrieval systems, and molecular biology.

An approximation algorithm [13] is designed to solve optimization problems in polynomial time. A method based on approximation yields a near-optimal solution with a guarantee on the solution's quality. An approximation algorithm for string matching [14, 15] is designed for determining strings that match a pattern approximately instead of exactly.

In this paper, the face recognition problem is modeled as the optimization problem since the approximation algorithm tries to find the distances between the test face image and all the database face images and select the minimum of those distances. Thus, the algorithm performs image matching based on the minimum distance obtained which becomes a minimization problem and any minimization or maximization problem is considered as the optimization problem. The overall goal of the work presented in this paper is to handle the challenges in the uncontrolled face recognition with different distance measures using approximation matching algorithms.

The rest of the paper is ordered as follows. Section 2 presents literature work on face recognition techniques, the methodology is described in Sect. 3, the experimental results are discussed in Sect. 4, and the last section gives the conclusion of the paper.

2 Related Work

Conventional methods for face recognition such as PCA (principal component analysis) [2, 3], LDA (linear discriminant analysis) [4, 5], ICA (independent component analysis) [6], and HMM [7] perform face recognition under controlled conditions. These methods do not perform well when there are uncontrolled conditions such as lighting, pose, expression variations, and occlusion.

The different approaches that deal with the face recognition under uncontrolled conditions include SRC (Sparse representation-based classification) [20], SRC-MRF (SRC with Markov random fields) [21], HQM (half quadratic with multiplicative form) [22], Virtual samples and SRC based algorithm [23] and GSR (group sparse representation-based) [24]. The computational cost of these sparse representation based approaches is more and require a larger number of training samples. Local feature matching face recognition methods such as metric learned extended robust point set matching [25], dynamic image-to-classwarping for face recognition [26] handles uncontrolled face recognition using local features such as image patches.

The research towards face recognition using string matching is very less. One of the related work is an ensemble string matching [27], performs string-to-string matching for the recognition of faces. The human face profile [28] is another work based on representing a face as a sequence of strings.

The latest research face recognition systems based on deep learning [29–33] are succeeded for controlled face recognition, but not producing successful results in uncontrolled conditions. The algorithms presented in this paper make use of different string distance measures for performing face recognition and yield a good recognition rate in case of face recognition application.

3   Methodology

3.1   Face Recognition System Model

Figure 1 demonstrates the architecture of the face recognition system. The various phases in the proposed system are feature extraction, clustering, generation of strings, matching test and database faces. In the feature extraction phase, small patches of m × n pixels are extracted from database gallery images of size p × q pixels. Patches are converted into patch vectors. Here, patch vectors are considered as feature vectors.

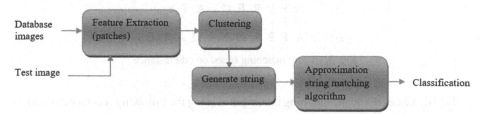

Fig. 1. The architecture of the proposed face recognition system

In the clustering phase, a clustering algorithm is applied on patch vectors to group related patch vectors to the same cluster. Clustering associates each patch vector with a cluster label. There are several clustering algorithms for clustering the data. In this work, K-means clustering is used. In the generation of string phase, a string for a database image is formed by the sequence of cluster labels assigned to patch vectors of a database image. A test mage is also split into a smaller number of patches. Test and database image patch vectors are compared by finding the local distance between them. Cluster label of the database image patch vector\with minimum local distance is obtained and cluster labels with a minimum local distance taken in sequence produce a string for a test image. String generation algorithm is presented in the previous work [34]. Figure 2 shows a sample face image and its corresponding string which is a sequence of cluster labels assigned to patches of the image, the length of the string is 64 which is equal to the number of patches in that image.

In the matching phase, the approximated string matching algorithm is applied to compute the distance between the test and the database image strings and to classify the test image based on the minimum distance computed. The algorithm performs classification in addition to the matching of face images. Thus, the approximation algorithm using distance measures for face image matching does not require any classifier. Hence the complexity of the system is less.

"2 69 12 46 20 46 70 8 17 18 46 20 20 20 43 15 17 61 54
44 28 43 61 55 10 38 70 68 19 22 11 15 30 44 46 20 39 28
22 63 3 62 19 13 60 28 12 63 17 3 62 22 11 35 12 48 17
15 3 27 62 22 63 64"

Fig. 2. Example of a face image and its corresponding string

3.2   Approximate String Matching Algorithm Based on Dynamic Edit Distance

This is also known as Levenshtein distance, which counts the number of dissimilarities between two strings [15]. The Edit distance between two strings s_1 and s_2 is Ed (s_1, s_2), defined as the least number of operations required to transform s_1 into s_2. The operations are transformation (substitution), insertion, and deletion.

Two strings s_1 = "appreciate" and s_2 = "approximate", the edit distance between s_1 and s_2 is 3 since it requires 2 substitutions and 1 insertion to transform s_1 into s_2. This is shown in Fig. 3.

s_1:   A P P R E C I A T E
s_2:   A P P R O X I M A T E

Fig. 3. String matching based on edit distance

Edit distance for image matching is computed using the following recurrence relation Eq. 1.

$$ED(i,j) = \begin{cases} i; \ if \ j = 0 \\ j; \ if \ i = 0 \\ ED[i-1, j-1], \ if \ i, j = 0 \ and \ x_i = y_j \\ \min(ED[i-1, j-1]+1, \ ED[i-1, j]+1, \ ED[i, j-1]+1), \\ \qquad if \ i! = j \ and \ x_i \neq y_j \end{cases} \quad (1)$$

Algorithm 1. Dynamic EDIT distance for image matching

Input: TestT and Database D image strings of lengths n, m respectively
Output:
Edit_dist: the edit distance between test and database image;
Set each element in ED to 0
ED [i, 0] = i; ED [0, j] = j;
fori = 1 ton do
forj = 1 to m do
if (T(i-1) = = D(j-1))
ED (i,j)= ED(i-1,j-1);
else
      ED(i,j) =min{ED(i-1,j-1)+1,ED(i-1,j)+1,ED(i,j-1)+1};
end if
end for
end for
Edit_dist= ED (n,m);
returnEdit_dist;

In Algorithm 1, T and D are test and database face image strings. String element of T is compared with the string element of D if matches EDIT distance is set to 0, otherwise, it is set to a minimum of insert, delete, and substitute operation costs. The time complexity of the EDIT distance algorithm is $O(mn)$, where m is a length of database image string and n is the length of the test image string.

3.3 Approximate String Matching Algorithm Based on Dynamic LCSS Distance

It is a measure of similarity between two strings, which is determined by finding common substrings between them [16]. The pattern is said to match a certain string in a string sequence if the pattern has the maximum number of similarities with that string compared to other strings. In the Longest common subsequence algorithm, the goal is to produce their longest common subsequence: the longest sequence of characters that appear left-to-right order in both the strings.

Two string sequences $s_1 = a_1a_2$—a_p and $s_2 = b_1b_2$—b_q of lengths p and q respectively, The Longest common subsequence problem computes the length of the largest string that is common to two sequences s_1 and s_2. LCSS for string matching is implemented through dynamic programming. For example, X = PQPRQNQP and Y = QRRMNQPR. The LCSS of X and Y is QRNQP with length 5. Figure 4 describes this example.

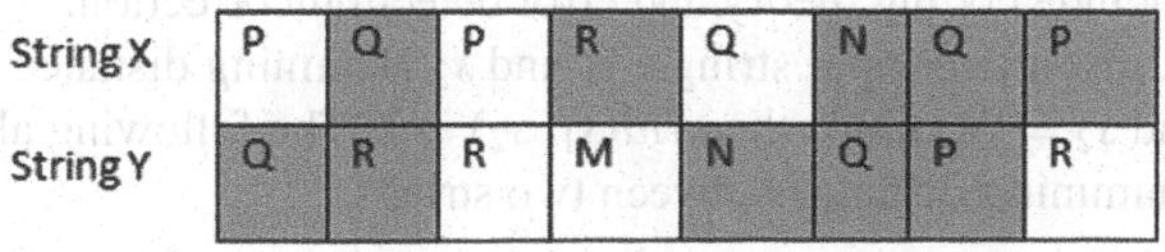

String X	P	Q	P	R	Q	N	Q	P
String Y	Q	R	R	M	N	Q	P	R

Fig. 4. String matching based on LCSS distance

LCSS distance for image matching is computed using the following recurrence relation Eq. 2.

$$LC(i,j) = \begin{cases} 0, \ if \ i = 0 \ or \ j = 0 \\ LC[i-1, j-1] + 1, \ if \ i, j > 0 \ and \ x_i = y_j \\ \max(LC[i,j], \ LC[i,j-1], \ LC[i-1,j]), \ if \ i, j > 0 \ and \ x_i \neq y_j \end{cases}$$

(2)

Algorithm 2. Dynamic LCSS for image matching

---

*Input:*TestT and Database D image strings of lengths n, m respectively
Output:

LCSS_dist: the longest common sub sequence distance between test and database images;
Set each element in LC to 0
LC [:,0] = 0; LC[0,:] = 0;
fori = 1 to n do
for j = 1 to m do
if (T(i-1) = = D(j-1))
LC (i,j) =LC (i-1,j-1)+1;
else
        LC(i,j)= max {LC(i-1,j-1),LC(i-1,j),LC(i,j-1)};
end if
end for
end for
LCSS_dist = LC (n,m);
returnLCSS_dist;

In Algorithm 2, the String element of T is compared with the string element of D, if they are equal LCSS distance is set to diagonal element cost plus 1, otherwise, it is set to a maximum of up, down, and diagonal element costs. If m, n are the lengths of database and test image strings, then the complexity of the LCSS algorithm is $O(mn)$.

3.4 Approximate String Matching Algorithm Based on Hamming Distance

For two similar-length strings X and Y, Hamming distance is the number of character positions at which the corresponding characters in strings are different [17]. Hamming distance determines the least number of replacements or the least number of errors required for transforming a string into another string. The main applications of the hamming code include coding theory and error detection/correction.

For example, given two input strings, s_1 and s_2 Hamming distance is $Hd(s_1, s_2)$. If $s_1 = $ "mouse" and $s_2 = $ "mouth", then $Hd(s_1, s_2) = 2$. The following algorithm is used to calculate the hamming distance between two strings.

Algorithm 3. Hamming distance for image matching

---

*Input:*TestT and Database D image strings of equallength n.
Output:
Hd_dist: the Hamming distance between test and database images;
i=0; replace_char=0; //Intialization
while (T(i) !='\0')
if (T(i) != D(j))
replace_char=replace_char+1;
i=i+1;
end if
endwhile
Hd_dist = replace_char;
returnHd_dist;

In Algorithm 3, if the test image string element (T(i)) does not match with the database image string element (D(j)), then the hamming distance is set to the substitution operation cost plus 1. The final hamming distance is based on the number of substitutions or replacements. The time complexity of the hamming distance is $O(n)$, where n is the length of the database as well as test image strings.

3.5 Approximate String Matching Algorithm Based on Jaro Distance

Jaro similarity measure between two strings is determined based on common characters and character transpositions [18]. The maximum Jaro distance results in more similarity between strings. Jaro distance is a normalized score with value 0 for no match and 1 for the correct match. This distance is computed in two steps; the first is the matching and the second is the transposition.

Matching. In this, a score is computed based on the common character occurrences in the strings. The algorithm finds the common characters in two strings within a specified match range around the character's position in the first string with the consideration of unmatched characters in the second string. If s_1 and s_2 are the two strings, then the match range is computed using the following equation.

$$match\_range = \frac{max(length(s_1), length(s_2))}{2} - 1$$

Transposing. This evaluates the characters in the first string that do not match with the character at the same position in the second string. The transposition score is obtained by dividing this result by 2.

The Jaro distance $Jaro(i, j)$ between 2 strings s_1 and s_2 is

$$Jaro(i, j) = \begin{cases} 0; & if\ m\_c = 0 \\ \frac{1}{2}\left(\frac{m\_c}{n} + \frac{m\_c}{m} + \frac{m\_c - T}{m\_c}\right), \\ otherwise \end{cases}$$

Here m_c is the total number of matches.

T is half of the total transpositions.

n, m are the lengths of two strings.

The strings s_1 and s_2 have two characters that match when they are the same and not beyond the match range which is equal to $max(n, m)/2 - 1$.

Algorithm 4. Jaro distance for image matching

*Input:*TestT and Database D image strings of lengths n, m respectively
Output:
Jaro_dist: the Jaro distance between test and database image;
Matc=0; trans=0; //Intialization
Dist_match=(max(n,m)/2)-1
fori = 0 to n do
    f=max(1,i-Dist_match);
    e=min(i+Dist_match+1,m)
forj = f to e do
if(D_matches(j)) continue;
if (T(i) ! = D(j)) continue;
I_matches(i)=1;
D_matches(j)=1;
matc++;
break;
end if
end for
end for
k1=0;
for k = 0 ton do
if(!I_matches(k)) continue;
while(!D_matches(k1) k1++;
if (T(k) ! = D(k1)) continue;
I_matches(i)=true;
D_matches(j)=true;
trans++;
k1++;
end if
end for
Jaro_dist = ((matc/n) + (matc/m)+((matc − transposition) /match))/2.0
returnJaro_dist;

Jaro-Winkler Distance. This similarity measure between two strings is based on character transpositions to a degree adjusted upwards for common prefixes [19]. The Winkler's modification to Jaro distance acts as a boost on the matching score based on the comparison of the string prefixes. Jaro-Winkler algorithm yields a matching score between 0.0 and 1.0.

4 Experiments and Analysis of Results

The approximation string matching algorithms using distance measures on the face recognition system is evaluated using the face databases such as FEI, and AT&T. Matlab is used for implementing the algorithms. The face databases are stored as mat files. All the experiments were run on a 3.6.5 GHz IntelCore i7 CPU with 16 GB RAM.

The Brazilian face database FEI [35] consists of 2800 face images with varied poses and expressions. A subset of this database [36] has cropped and frontal faces of 360 × 260 pixels size and 200 subjects, each with two frontal face images with two different expressions include neutral and smiling expression. FEI sample images with these expressions are shown in Fig. 5 and Fig. 6. AT&T database [37] has a total of 400 images from 40 subjects (10 images in each subject). This database face images have different pose, lighting, and expression changes. Figure 7 represents sample images from the database.

Fig. 5. Face images with neutral expression from the FEI database

Fig. 6. Face images with smile expression from the FEI database

Fig. 7. Sample face images of the AT&T database with different expressions, poses, and lighting

In all the algorithms, a patch size of 8 × 8 pixels is used, the face images are resized to 64 × 64 pixels and the total number of patches in each image 64 and the number of cluster centers (value of K) is chosen to be the nearest value equal to the number of patches in an image. By empirical observation, the number of clusters is fixed to 70 which has given a good performance.

4.1 Face Recognition with FEI Database

The approximation algorithms are evaluated with the FEI database of 400 images of 200 subjects, each subject has one neutral expression face and a smile expression face. A gallery set of 200 neutral face images and a testing set of 200 smiling face images are used for performance evaluation. The images are resized to 64×64 and divided into patches of size 8×8. The following table describes the comparison between different approximation string matching algorithms and traditional methods Eigen-faces and ICA. The results can conclude that syntactic approaches such as string matching algorithms work well for the face recognition system. All the algorithms except Jaro-Winkler have superior performance over the methods Eigen-faces and ICA. Thus, the approximation string matching algorithms give a better recognition rate than traditional methods for face recognition system (Table 1).

Table 1. Face recognition using the approximation string matching algorithms for the FEI database

Method	No. of gallery images	No. of test images	No. of correct matches	Recognition rate (%)
EDIT	200	200	191	95.5
LCSS	200	200	190	95
Hamming	200	200	192	96
Jaro	200	200	186	93
Jaro-Winkler	200	200	176	88
Eigen-faces	200	200	176	88
ICA	200	200	149	74.5

4.2 Face Recognition with AT&T Database

The performance of the algorithms is evaluated with the AT&T database, which consists of 400 images of 40 subjects. The first five images of all the subjects are used for the gallery database set and the next five images of the subjects are chosen for the testing set. The comparison of the approximation string matching algorithms and the methods Eigen-faces and ICA is described in Table 2. EDIT distance algorithm exhibits a higher recognition rate than other algorithms. Thus, an approximation string matching algorithm using Edit distance performs better for face recognition applications than all other algorithms. Approximation algorithm using hamming distance performs well for recognizing faces with smile expression, but not yields the same performance for the recognition of faces with different poses, lighting, and expression variations. LCSS, hamming, Jaro, and Jaro-Winkler algorithm's performance is average when compared to the EDIT distance algorithm. Face recognition with the AT&T dataset shows that approximation string matching algorithms outperform traditional methods.

Table 2. Comparison between the approximation string matching algorithms and traditional methods for AT&T database

Method	No. of gallery images	No. of test images	No. of correct matches	Recognition rate (%)
EDIT	200	200	187	93.5
LCSS	200	200	176	88
Hamming	200	200	178	89
Jaro	200	200	175	87.5
Jaro-Winkler	200	200	168	84
Eigen-faces	200	200	156	78
ICA	200	200	151	75.5

5 Conclusion and Future Work

In this work, Face recognition is addressed using approximate string matching algorithms using different distance measures such as EDIT, LCSS, hamming, and Jaro. String sequences for the faces are generated and approximated string-based distance is calculated. The string algorithms find the minimum matching path based on the minimum distance and finally, classifies the input face image. The approximation string matching using edit distance yields better results over the other distance metrics such as LCSS, Hamming, Jaro, and Jaro-Winkler for the face recognition system. This work can be used in security, forensic, and other applications. The work presented in this paper concludes that syntactic string approaches perform well for face recognition applications.

References

1. Zhao, W., Chellappa, R., Rosenfeld, A., Phillips, P.: Face recognition: a literature survey. ACM Comput. Surv. **35**, 399–458 (2003)
2. Kirby, M., Sirovich, L.: Application of the Karhunen-Loève procedure for the characterization of the human face. IEEE Trans. Pattern Anal. Mach. Intell. **12**, 103–108 (1990)
3. Turk, M.A., Pentland, A.P.: Face recognition using eigenfaces. In: Proceedings of the IEEE Conference on CVPR, pp. 586–591 (1991)
4. Yu, H., Yang, J.: A direct LDA algorithm for high-dimensional data with application to face recognition. Pattern Recognit. **34**, 2067–2070 (2001)
5. Belhumeur, P.N., Hespanha, J.P., Kriegman, D.J.: Eigenfaces vs. fishy faces: recognition using class specific linear projection. IEEE Trans. Pattern Anal. Mach. Intell. **19**, 711–720 (1997)
6. Bartlett, M.S., Movellan, J.R., Sejnowski, T.J.: Face recognition by independent component analysis. IEEE Trans. Neural Netw. **13**, 1450–1464 (2002)
7. Miar-Naimi, H., Davari, P.: A new fast and efficient HMM-based face recognition system using a 7-state HMM along with SVD coefficients. Iran. J. Electr. Electron. Eng. **4** (2008)
8. Geng, X., Zhou, Z., Smith-Miles, K.: Individual stable space: an approach to face recognition under uncontrolled conditions. IEEE Trans. Neural Netw. **19**(8), 1354–1368 (2008)
9. Gaston, J., Ming, J., Crookes, D.: Matching larger image areas for unconstrained face identification. IEEE Trans. Cybern. **49**(8), 3191–3202 (2019)

10. Qiangchang, W., Guodong, G.: LS-CNN: characterizing local patches at multiple scales for face recognition. IEEE Trans. Inf. Forensics Secur. **15**, 1640–1652 (2020)
11. Alhendawi, K.M.A., Baharudin, S.: String matching algorithms (SMAs): survey & empirical analysis. J. Comput. Sci. Manag. **2**(5), 2637–2644 (2013)
12. Navarro, G.: A guided tour to approximate string matching. ACM Comput. Surv. **33**(1), 31–88 (2001)
13. Ukkonen, E.: Algorithms for approximate string matching. Inf. Control **64**(1–3), 100–118 (1985)
14. Williamson, D.P., Shmoys, D.B.: The Design of Approximation Algorithms, 1 edn. Cambridge University Press, Cambridge (2011)
15. Masek, W.J., Paterson, M.: A faster algorithm computing string edit distances. J. Comput. Syst. Sci. **20**(1), 18–31 (1980)
16. Aho, A.V., Hopcroft, J.E., Ullman, J.D.: The Design and Analysis of Computer Algorithms. Addison-Wesley, Reading (1974)
17. Hamming, R.W.: Error detecting and error-correcting codes. Bell Syst. Tech. J. **29**(2), 147–160 (1950)
18. Jaro, M. A.: Advances in record linkage methodology as applied to the 1985 census of Tampa Florida. J. Am. Stat. Assoc. **84**(406), 414–420 (1989)
19. Winkler, W.E.: Overview of record linkage and current research directions (PDF). Research Report Series, RRS (2006)
20. Wright, J., Yang, A.Y., Ganesh, A., Sastry, S.S., Ma, Y.: Robust face recognition via sparse representation. IEEE Trans. Pattern Anal. Mach. Intell. **31**(2), 210–227 (2009)
21. Zhou, Z., Wagner, A., Mobahi, H., Wright, J., Ma, Y.: Face recognition with contiguous occlusion using Markov random fields. In: IEEE International Conference Computer Vision (ICCV), pp. 1050–1057, October 2009
22. He, R., Zheng, W.S., Tan, T., Sun, Z.: Half-quadratic-based iterative minimization for robust sparse representation. IEEE Trans. Pattern Anal. Mach. Intell. **36**(2), 261–275 (2014)
23. Peng, Y., Li, L., Liu, S., Li, J., Cao, H.: Virtual samples and sparse representation based classification algorithm for face recognition. IET Comput. Vis. **13**(2), 172–177 (2018)
24. Fritz, K., Damiana, L., Serena, M.: A robust group sparse representation variational method with applications to face recognition. IEEE Trans. Image Process. **28**(6), 2785–2798 (2019)
25. Weng, R., Lu, J., Hu, J., Yang, G., Tan, Y.P.: Robust feature set matching for partial face recognition. In: Proceedings of the IEEE International Conference on Computer Vision (ICCV), pp. 601–608, December 2013
26. Wei, X., Li, C.-T., Lei, Z., Yi, D., Li, S.Z.: Dynamic image-to-class warping for occluded face recognition. IEEE Trans. Inf. Forensics Secur. **9**(12), 2035–2050 (2014)
27. Chen, W., Gao, Y.: Face Recognition Using Ensemble String Matching. IEEE Trans. Image Process. **22**(12), 4798–4808 (2013)
28. Gao, Y., Leung, M.K.H.: Human face profile recognition using attributed string. Pattern Recognit. **35**(2), 353–360 (2002)
29. Schmidhuber, J.: Deep learning in neural networks: an overview. Neural Netw. **61**, 85–117 (2013)
30. Mehdipour Ghazi, M., Kemal Ekenel, H.: A comprehensive analysis of deep learning-based representation for face recognition. In: Proceedings of the IEEE Conference on Computer Vision and Pattern Recognition Workshops, pp. 34–41 (2016)
31. Lu, J., Wang, G., Zhou, J.: Simultaneous feature and dictionary learning for image set based face recognition. IEEE Trans. Image Process. **26**, 4042–4054 (2017)
32. Hu, G, Peng, X.Y., Hospedales, Y., Verbeek, T.M., Frankenstein, J.: Learning deep face representations using small data. IEEE Trans. Image Process. **27**, 293–303 (2018)
33. Rawat, W., Wang, Z.: Deep convolutional neural networks for image classification: a comprehensive review. Neural Comput. **29**, 2352–2449 (2017)

34. Krishnaveni, B., Sridhar, S.: Approximation algorithm based on greedy approach for face recognition with partial occlusion. Multimed. Tools Appl. **78**, 27511–27531 (2019)
35. Tenorio, E.Z., Thomaz, C.E.: Analisemultilinear discriminate deformas frontalis de imagens 2D de face. In: Proceedings of the X Simposio Brasileiro de Automacao Inteligente, SBAI, Universidade Federal de Sao Joao del Rei, Sao Joao del Rei, Minas Gerais, Brazil, pp. 266–271, September 2011
36. https://fei.edu.br/~cet/facedatabase.html
37. The Database of Faces, AT&T Laboratories Cambridge (2002). http://www.cl.cam.ac.uk/research/dtg/attarchive/facedatabase.html

Detection of Human Faces in Video Sequences Using Mean of GLBP Signatures

S. Selvi[1(✉)], P. Ithaya Rani[1], and S. Muhil Pradahnji[2]

[1] CSE, Sethu Institute of Technology, Virudhunagar, India
sselvi@sethu.ac.in, muhilrani@gmail.com
[2] CSE Department, AAA College of Engineering and Technology,
Virudhunagar, India

Abstract. Machine analysis of detection of the face is robust research topic in human-machine interaction today. The existing studies reveal that discovering the position and scale of the face region is difficult due to significant illumination variation, noise and appearance variation in unconstrained scenarios. We designed work is spontaneous and vigorous method to identify the location of face area using recently developed You Tube Video face database. Formulate the normalization technique in each frame. The frame is separated into overlapping regions. The Gabor signatures extracted on each region by Gabor filters with different scale and orientations. The Gabor signatures are averaged and then local binary pattern histogram signatures are extracted. The Gabor local binary pattern signatures are passed to Gentle Boost categorizer with the assistance of face and non-face signature of the gallery images for identifying the portion of the face region. Our experimental results on YouTube video face database exhibits promising results and demonstrate a significant performance improvement when compared to the existing techniques. Furthermore, our designed work is uncaring to head poses and sturdy to variations in illumination, appearance and noisy images.

Keywords: Ensemble categorizer · Gabor wavelet · Human computer interaction · Local binary pattern · Normalization

1   Introduction

One of the most interesting fields of image analysis is the automatic identify the area of the human faces. The major applications of finding the face areas are face recognition, facial expression recognition, gender identification, face registration, human-machine interaction, surveillance, etc. Face discovery methods identify the faces in the video clips and provide the location and scale of all faces. But finding the region of the human face is a interesting task as the human face appearances are non-rigid and they appear in different backgrounds (simple,

© IFIP International Federation for Information Processing 2020
Published by Springer Nature Switzerland AG 2020
A. Chandrabose et al. (Eds.): ICCIDS 2020, IFIP AICT 578, pp. 170–184, 2020.
https://doi.org/10.1007/978-3-030-63467-4_13

clutter) and have a high variability of different location, poses, expressions and illuminations (good and bad) [1,2].

To overcome these problems, the planned work is a new approach to identify the face region by Normalized mean of Gabor LBP signatures. The planned methodology is insensitive to head poses and strong to variations in lighting condition and noisy images. The residual portion of the paper is ordered as follows: Sect. 2 briefly evaluate the survey works. Section 3 defines the designed NGLBP signatures details. Section 4 shows the experimental results. Section 5, offers conclusion and plan for future task.

2   Survey Work

Detection of face techniques have been examined immense in the earlier. The methodology for identifying the face area utilizing skin color and the Maximum Morphological Gradient Combination image was exhibited [3,4]. The system failed when it manages with skin color areas including similar color background and region of dress. H. Sagha et.al designed a methodology for discovering sparse signatures using a genetic algorithm for multi view face detection. Notwithstanding, discovering these signatures was time intensive and wasteful by utilizing their strategies [5]. The Gabor Filter (GF) catches the properties of different orientation and spatial localization in the space and frequency domains was utilized in face detection [6–8]. The techniques using the LBP and Local Gradient Pattern (LGP) based signatures for detecting the faces was existed [9–11]. These techniques are sensitive to noise as the signatures at each location compare a central pixel with neighboring pixels. The detection of the facial components utilizing speeded up robust signatures presented in [12] could achieve only moderate performance. The extracts of Haar signature and a learning algorithm (Adaboost) are utilized in [13], where the methods suffer from global illumination variations. Kyungjoong Jeong et al. [14] carried out the work Semi - LBP (SLBP) signatures for face detection. These signatures are robust against noise. Though, higher detection rate could not be achieved.

A lot of existing detection systems utilized one type of signature. Though, for difficult works such as discovering the area of human face, a single signature set is not rich enough to capture all of the information required to detect the face. The robust detection always requires appropriate information on illumination, face appearance variation, and discriminating power of the signature set demanding more than one type of signature set. Finding and fusing relevant signature sets have thus become an energetic research theme in machine learning. Combining the GF and LBP signatures for face recognition is motivated for the work reported in [15]. We plan to combine GF and LBP signatures for discovering the portion of face.

The work considers the local appearance descriptors by Gabor Wavelet (GW) utilized in [6–8] and fusing it with Local Binary Pattern (LBP) signatures as used in [14] rather than working on individual signature set. The GF signatures convert facial shape and appearance information over a broader range of scales. The detection of LBP signature captures little appearance details and tolerance to illumination changes. Local spatial invariance is accomplished by locally pooling (histogramming) the resulting texture codes. The advantage of NGLBP signatures are utilized to capture the local structure corresponding to spatial frequency (scale), spatial localization, and orientation selectivity which are proved to be discriminative the face/non-face and robust to illumination, noise and appearance changes (Fig. 1).

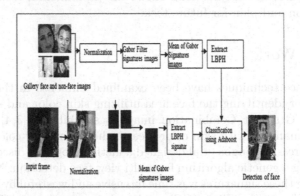

Gallery face and non-face images

Input frame   Normalization   Mean of Gabor signatures images   Detection of face

Fig. 1. Overview of the system diagram for identifying the location of face area where red color box (red color frame size is 30 × 30 pixels) (Color figure online)

3 Planned Work

The video with single subject contains multiple frames depicting the temporal variations in different poses, expressions and varying lighting conditions of the individual. The following steps describe the planned approaches:

1. Initially, normalization techniques are applied on each frame which adjusts the image intensity.
2. Subsequent, each frame is separated into intersecting regions and then local signatures are separated by using GF with different scale and orientation in each region. The Gabor filter signatures are most appropriate for face/non-face classification. The Gabor filter signatures are changed into mean of Gabor signatures. The Gabor signatures have the facial shape and appearance information over a range of coarser scales.
3. 59-LBPH signatures are separated from each region contains Gabor signatures.
4. The Gabor LBP signatures are distributed through the AdaBoost categorizer for the pixel wise classification with well-trained face and non-face signature.

The performance of the planned work NGLBP signatures is compared with the signatures extracted by conventional GF [14], LBP [9] and GLBP by deploying the Ensemble categorizers. For conducting and evaluating the work, YouTube (YT) video face databases [15] are taken. The succeeding sub-divisions reveal the technique in point.

3.1   Normalization

Due to the fact that variant light condition certainly reasons low finding rates and can be removed by illumination normalization, normalization techniques should be well measured in an automatic detection system. So the histogram normalization technique [16–18] was applied on each frame to compensate for different lighting conditions. As the little- contrast image's histogram is narrow and centered toward the middle of the gray scale, if we distribute the histogram to a wider range, the quality of the image will be improved. So we can do it by adjusting the probability density function of the original histogram of the image so that the probability spread equally. It is utilized to produce an image with distributed brightness levels over the image. Initially, each frame is extracted and represented as set of frames f1, f2, ... f_k from video V, where k is the number of frames. The gray levels of the k_{th} frame is first equalized by

$$f_k = E(r_k) \sum_{i=0}^{m} \frac{n_i}{n} \quad m = 0,1,...L-1 \tag{1}$$

where E denote the equalization function, n is the total number of pixels, n_i is the number of pixels with gray level r_i and L is the number of discrete gray levels.

3.2   Signature Extraction of Gabor Filter

Following the intensity normalization, Gabor filters offer the greatest simultaneous localization of spatial and frequency information. The Gabor wavelet (GW) that catches the properties of orientation selectivity, spatial localization and optimally localized in the space and frequency domains was used in face detection [16–18,21]. The extracts of Haar feature and a learning algorithm (Adaboost) are proposed in [22], where the methods suffer from global illumination variations. KyungjoongJeong et al. [14] carriedout the work Semi - LBP features for face detection. Among all the features, the Gabor features [19] are good for solving the computer vision problem such as face detection to provide better accuracy with head poses and appearance variations. Hence, we concentrate on Gabor features. However the Gabor features are limited due to their sensitivity to illumination variations and noisy images. But the performance, quality of a face detection system can be vulnerable to variations in illumination levels; which maybe correlated to the conditions of their surroundings. Therefore, we propose the use of a novel method known as NGF features which is insensitive to variations in lighting condition and noisy images. The 2D Gabor filter is agreed [8] and it could be mathematically stated as:

$$\psi(a,b,\sigma,\theta) = \exp\left(-\frac{(A^2 + \gamma^2 B^2)}{2\sigma^2}\right)\cos\left(\frac{2\Pi}{\lambda}A\right) \tag{2}$$

$$\mathbf{A} = \mathrm{acos}\,\Theta + \mathrm{bsin}\,\Theta;\ \mathbf{B} = -\mathrm{asin}\,\Theta + \mathrm{bcos}\,\Theta$$

where orientation Θ, the effective width σ, the wavelength λ is the spacing factor between filter in the frequency domain, the aspect ratio $\ddot{Y}$. We propose the GF procedure by dividing the kth frame f_k into overlapping regions 'B' represented as. The number of regions are $(m - 30) \times (n - 30)$ and m and n are the number of rows and columns in each frame respectively. Typically, each 'B' size is 30×30 within a frame, its convolution with a Gabor filter Ψ is stated as follows.

$$GF_k, B, \sigma, \Theta(\mathbf{a}, \mathbf{b}) = f_{k,B}(\mathbf{a}, \mathbf{b})\ \Theta\ \Psi\ (\mathbf{a}, \mathbf{b}, \sigma, \Theta) \tag{3}$$

where Θ is the sign for convolution. Eight scales σ, ε (5 to 19 with increments by 2) and four orientations θ $(-45°, 90°, 45°, 0°)$ are utilized in the Gabor filters. The given 'B' region within a input frame f_k is filtered with the Gabor filters as in Eq. (3), ensuing in a series of Gabor filtered images with signatures such as bars and edges usefully emphasized for improved identifying the location of the face. Then extracted signatures are converted into Mean (M) of Gabor signatures in each region. The performance of the planned mean of GF signatures is compared with the Standard deviation (S) and Variance (V) of GF signatures in each region as

$$G_{k,B,M}(a,b) = \frac{1}{\sigma * \theta}\left(\sum_{\sigma=1}^{8}\sum_{\theta=1}^{4} GF_{k,B,\sigma,\theta}(a,b)\right) \tag{4}$$

$$G_{k,B,V}(a,b) = \left(\sum_{\sigma=1}^{8}\sum_{\theta=1}^{4}\left(GF_{k,B,\sigma,\theta}(a,b) - \left(\frac{1}{\sigma * \theta}\left(\sum_{\sigma=1}^{8}\sum_{\theta=1}^{4} GF_{k,B,\sigma,\theta}(a,b)\right)\right)\right)^{2}/\sigma * \theta\right) \tag{5}$$

$$G_{k,B,V}(a,b) = \left(\sum_{\sigma=1}^{8}\sum_{\theta=1}^{4}\left(GF_{k,B,\sigma,\theta}(a,b) - \left(\frac{1}{\sigma * \theta}\left(\sum_{\sigma=1}^{8}\sum_{\theta=1}^{4} GF_{k,B,\sigma,\theta}(a,b)\right)\right)\right)^{2}/\sigma * \theta\right)^{1/2} \tag{6}$$

As a result, each region contains M, S, and V of Gabor signatures.

3.3   Signature Extraction of LBP

The Gabor signatures of each region size are 30×30 resolutions. Then GLBP is defined by a binary coding function [19] to the obtain Gabor signatures in each region. Let $G_{k,B}$ (a, b) be the Gabor signatures in 'B' region within k^{th} frame around pixel (a, b). The center value of 3×3 matrixes is compared with another eight values and an 8bit code is coined, which will be the value at each pixel position(a, b). Let M to represent the matrix as:

$$M = G_{K,B}(\mathbf{a}, \mathbf{b})$$

The value by using the planned method GLBP is obtained as:

$$f_{GLBP}(a,b) = \sum_{a=0}^{2}\sum_{b=0}^{2} T(a,b)2^8$$

$$where \quad a \neq 1 \quad and \quad b \neq 1 \tag{7}$$

and

$$T(a,b) = \begin{cases} 1 & M(a,b) \geq M(1,1) \\ 0 & else \end{cases} \tag{8}$$

After fixing the value using GLBP technique for each pixel related with a region, a 59-bin histogram is applied to capture the signature for the each region. A histogram (H) of the region $f_{G,L,B,P}$ can be defined as:

$$h_L = h_L + I(lower_L < f_{GLBP}(a,b) \geq higher_L)$$

$$1 \leq a \leq m - 30, \qquad 1 \leq b \leq n - 30, \qquad 1 \leq L \leq 59 \tag{9}$$

$$I(A) = \begin{cases} 1 & A \quad is \quad true \\ 0 & A \quad is \quad false \end{cases} \tag{10}$$

where L is the number of bins for the values formed by the GLBP function. The interval of each bin is represented by the range lower L and higher L.

The GLBP histogram holds data about the report of the local micro-patterns such as edges, spots and flat areas, over the whole image, so could be utilized to statistically define image characteristics. We gained 59 - GLBP histogram bins for each region in the frame.

3.4   Classification

Adaboost algorithm as utilized in [19,23,24] for object detection reveals a low false positive performance and hence organized in our work for classification in face/non-face region. Initially the gallery set is formed using the NGLBP signatures from the collection of gallery images having both face and non-face images and stored in database (DB1). The signatures of the each frame are classified in face/non-face region with DB1 utilizing pixel wise classification of boost algorithm. The signature of the each block with in a frame are classified into face and non-face region with training set using boost algorithm [25,26]. Finally the location of face region is obtained from each frame.

$$\mathbf{F = 2 * P * R/P + R} \tag{11}$$

The detection of human face is described in the algorithm as given below.

Algorithm for detection of face region

Input: Given an input video (V), Set scale $\sigma = 8$ and orientations $\theta = 4$, DB - training database

Output: F - face region
V = f1, f2, ..., f_k Collection of frames, K is the number of frames
f(a,   b)   Resize   the   input   framefinto   110   ×   110   pixels.

$$f_k = E(r_k) = \sum_{i=0}^{m} \frac{n_i}{n} \quad m = 0,1,\ldots\ldots, L-1$$

(Apply normalization)

Block $\leftarrow$ 1 represent block, where r and c are the rows and columns of input image respectively, input image is divided into overlapping blocks, (block_height, block_width) = size (block), each block size is of 30 * 30 pixels

Iterate 1:
a = 1:r; b = 1:c f_b(a, b) = f (a:a + block_height -1, b:b + block_width -1)

Iterate 2: s = 1 to σ ; o = 1 to θ

$$\psi(a,b,\sigma,\theta) = \exp\left(-\frac{(A^2 + \gamma^2 B^2)}{2\sigma^2}\right)\cos\left(\frac{2\Pi}{\lambda}A\right)$$

A = acos Θ + bsin Θ
B = $-$asin Θ + bcos Θ
$GK_{k,B,\sigma,\Theta}$(a,b) = $f_{K,B}(a,b)\Theta\Psi$(a, b, σ, Θ)

$$G_{k,B,M}(a,b) = \frac{1}{\sigma * \theta}\left(\sum_{\sigma=1}^{8}\sum_{\theta=1}^{4} GF_{k,B,\sigma,\theta}(a,b)\right)$$

$$G_{k,B,V}(a,b) = \left(\sum_{\sigma=1}^{8}\sum_{\theta=1}^{4}\left[GF_{k,B,\sigma,\theta}(a,b) - \left(\frac{1}{\sigma * \theta}\left(\sum_{\sigma=1}^{8}\sum_{\theta=1}^{4} GF_{k,B,\sigma,\theta}(a,b)\right)\right)\right]^2 / \sigma * \theta\right)$$

$$G_{k,B,\gamma}(a,b) = \left(\sum_{\sigma=1}^{8} \sum_{\theta=1}^{4} \left(GF_{k,B,\sigma,\theta}(a,b) - \left(\frac{1}{\sigma * \theta} \left(\sum_{\sigma=1}^{8} \sum_{\theta=1}^{4} GF_{k,B,\sigma,\theta}(a,b) \right) \right) \right)^2 / \sigma * \theta \right)^{1/2}$$

$M = G_{K,B}(a,b)$
Extract LBP feature

$$f_{GLBP}(a,b) = \sum_{a=0}^{2} \sum_{b=0}^{2} T(a,b) 2^8$$

$$where \quad a \neq 1 \quad and \quad b \neq 1$$

$$T(a,b) = \begin{cases} 1 & M(a,b) \geq M(1,1) \\ 0 & else \end{cases}$$

$$h_L = h_L + I(lower_L < f_{GLBP}(a,b) \geq higher_L)$$

$$1 \leq a \leq m - 30, \qquad 1 \leq b \leq n - 30, \qquad 1 \leq L \leq 59$$

$$I(A) = \begin{cases} 1 & A \quad is \quad true \\ 0 & A \quad is \quad false \end{cases}$$

End iterate 2
F = Classify the face or non-face region using Adaboost classifier algorithm
(f_{GLMP}, DB)

End Iterate 1

4   Experimental Results and Discussion

To evaluate the performance of our planned method, YT video datasets were utilized for the experiment. YT video clips contain 47 celebrities. Some of the videos are low resolution and recorded at high compression rates. This leads to noisy, low-quality image frames. The dataset consists of about 1910 video clips, each containing hundreds of frames. Out of the 1910 video sequence studies, 1870 of them consists of only one person and the remaining have more than one person. For gallery purpose, 805 face images and 1023 non-face images are collected from ORL, Yale databases and background imaged respectively. Figure 2 shows some samples of the gallery images.

(a) face images   (b) non-face images

Fig. 2. Sample of gallery images a) face, non-face images

4.1 Performance of Signature Extraction

We collected the gallery images, which comprise of face images and non-face images. The gallery images are rescaled into three types of resolution such as 25 × 25, 30 × 30 and 35 × 35 pixels for finding the location of face. Initially signatures are extracted such as NGLBP, GLBP, LBP and GF from each gallery image. In our experiments, the orientation and scale of Gabor filters imposed on images are two key parameters that determine the effectiveness of the extracted texture signatures. Figure 3 compares the detection results obtained using 2, 4, 6 and 8 orientations of Gabor filters (the number of the scales is fixed at 8) in each

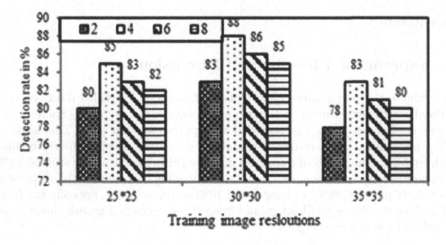

Fig. 3. Four sets of orientation in YT database

resolution gallery images using GA categorizer. The four sets of orientations are (90°, 0°), (−45°, 90°, 45°, 0°), (−45°, −22.5°, 0°, 22.5°, 45°, 90°) and (90°, 67.5°, 45°, 22.5°, 0°, −22.5°, −45°, −67.5°), correspondingly. It can be seen that utilizing the default 4 orientations makes highest accuracy of 89percent among the four sets of orientation values. This result suggests that Gabor filters require at least 4 orientations to be able to capture most of the discrimination information.

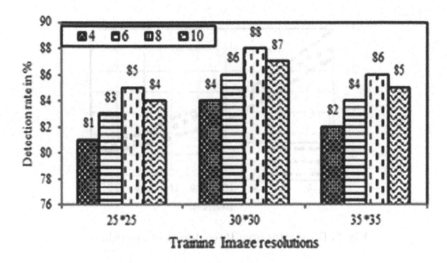

Fig. 4. Four sets of scales in YT database

Figure 4 compares the accuracies obtained using 4, 6, 8, and 10 scales of Gabor filters (the number of orientations of Gabor filters is fixed at four) in each resolution gallery images using GA categorizer. The four sets of scales are composed of (5:2:11), (5:2:15), (5:2:19), and (5:2:23) pixels correspondingly. The results also confirm that the default 4 orientations and 8 scales of Gabor filters are the optimal parameters for detecting the face area. In both Fig. 3 and Fig. 4, 30 × 30 resolution gallery images shows the best result.

Table 1. Compares the accuracies obtained using GW and NGW

Extracting Signatures	Detection rate in %				
	Mean	Std	Variance	Skeweness	Kurtosis
GW	80	79	81	82	81
NGW	85	84	84	84	83

Table 1 compares the accuracies obtained using mean, standard deviation, skewness and variance signatures in gallery images of 30 * 30 resolutions. From the result, it can be observed that mean of Gabor signatures result in better detection of face in 30 × 30 resolutions gallery images.

4.2   Performance of Categorizer

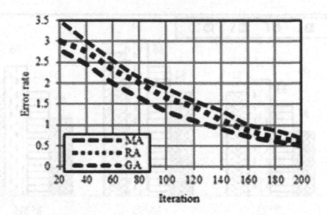

Fig. 5. Error rates from three Adaboost algorithms

The NGLBP signatures are trained through variant Adaboost categorizers such as RA, GA and MA in 30 × 30 resolution gallery images. They are compared for error checking with 100 boosting iterations as shown in Fig. 5. From the analysis GA returns the lowest error rate and is selected as the detection algorithm for our system.

4.3   Performance of NGLBP

Figure 6 shows the receiver operating characteristic curves (ROC). The curve is made by YT databases with GF, LBP, GLBP and NGLBP signatures are tested in GI, BI, N and MS. Figure 6 depicts the relationship between a number of false positives and the detection rate. The NGLBP signatures highlight higher performance of 3 videos respectively. The LBP signatures higher performance of 3 lower performance by 2 performance by 1.5 individual signature under all types of videos. The averages of all types of videos for identifying the location of face region rate considerably improved to about 4 reported for utilizing NGLBP signatures over using individual signature set in different video conditions.

Detection of the face includes calculating the Sensitivity, Precision and F measure the results are shown in Table 2, which indicates the number of video sequences with GI, BI, N and MS. It can be seen that the GF and LBP signatures perform poorly owing to their sensitivity to various illumination variations

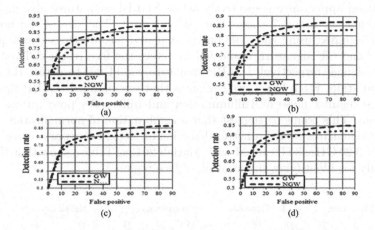

Fig. 6. Roc curve false positive rate vs. Detection rate a) Good Illumination b) Bad Illumination c) Noisy images d) Multiple Subject

Table 2. Result for identifying the location of face

author	Signature Extraction	Categorizer	Sensitivity in (%)				Precision in (%)				F measure in (%)			
			No. of videos				No. of videos				No. of videos			
			625	420	542	35	625	420	542	35	625	420	542	35
			GI	BI	N	MS	GI	BI	N	MS	GI	BI	N	MS
	GF	Gentleboost	84	80	78	79	85	80	82	81	84	80	79	80
	LBP	Gentleboost	85	83	80	82	87	82	83	84	85	82	82	83
proposed	GLBP	Gentleboost	87	85	83	85	88	84	85	86	87	84	85	86
	NGLBP	Gentleboost	90	88	86	87	90	85	87	88	90	86	87	87

Table 3. Time complexity for comparing our designed approach with existing approach

Signature Extraction	Signature extraction for gallery Images (compilation time in ms)	Testing Videos for face detection (compilation time in ms)			
		GI	BI	N	MS
LBP [9]	25	974000	789300	1213450	632211
GF[14]	28	982120	801342	1238691	637625
GLBP	28.5	983450	812420	1256302	638823
NGLBP	29.5	985629	816458	1259000	639000

and common appearance respectively, while NGLBP signatures give much better performance. Figure 7 shows the sample result. The size of the bounding box is determined using the scale on the detected face on the video sequences. The LBP and GF signature would fail to detect the face in the different poses and noise image respectively. GLBP signature would fail to detect the face in the BI variation. From Fig. 7, we can show that our plannedNGLBP signatures are robust against noise and illumination and differences pose variations and expressions. This seems to suggest that a combination of normalization, Gabor and LBP signatures result in better detection of face. In our work by using Intel Core i5 @ 3.20 GHz, 8 GB RAM with Matlab 2013a. Table 3 show that the time complexity for identifying the location of face region.

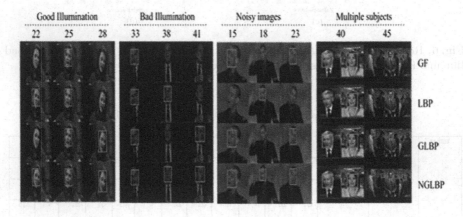

Fig. 7. Sample detection of face result in YT databases

5   Conclusion and Future Work

This paper investigated the benefits of NGLBP signatures for strong discovery of face area in video sequences under uncontrolled scenario. The experimental results showed that the planned method is successful when compared to the existing methods. The advantage of NGLBP signatures is amazingly uncaring in appearance varieties through illumination, expression, and noise in the images. NGLBPsignatures are not just robust to the varieties of image condition, additionally in encoding discriminate information, i.e. face/non-face area in spatial and frequency domains. The NGLBP test results exhibit that signature finds the best execution of revelation of face applications on the YT databases. In future work on the face regions are utilized for facial expression recognition.

References

1. Yang, M., Kriegman, D., Ahuja, N.: Detecting faces in images: a survey. IEEE Trans. Pattern Anal. Mach. Intell. **24**(1), 34–58 (2002)

2. Hielm, E., Low, B.: Face detection: a survey. Comput. Vis. Image Underst. **83**(3), 236–274 (2001)
3. Borah, S., Konwar, S., Tuithung, T., Rathi, R.: A human face detection method based on connected component analysis. In: International Conference on Communications and Signal Processing, pp. 1205–1208 (2014)
4. Byung-Hun, O., Kwang-Seok, H.: A study on facial components detection method for face-based emotion recognition. In: International Conference on Audio, Language and Image Processing (ICALIP), pp. 256–259 (2014)
5. Sagha, H., Kasaei, S., Enayati, E., Dehghani, M.: Finding sparse signatures in face detection using genetic algorithms. In: IEEE International Conference on Computational Cybernetics, pp. 179–182 (2008)
6. Xiaohua, L., Lam, K.-M., Lansun, S., Jiliu, Z.: Face detection using simplified Gabor signatures and hierarchical regions in a cascade of categorizers. Pattern Recognit. Lett. **30**, 717–728 (2009)
7. Huang, L., Shimizu, A., Kobatake, H.: Robust face detection using Gabor filter signatures. Pattern Recognit. Lett. 1641–1649 (2005)
8. Yun, T., Guan, L.: Automatic face detection in video sequences using local normalization and optimal adaptive correlation techniques. Pattern Recognit. **42**(9), 1859–1868 (2009)
9. Ojala, T., Pietikainen, M., Harwood, D.: A comparative study of texture measures with classification based on signature distributions. Pattern Recognit. **29**(1), 51–59 (1996)
10. Froba, B., Ernst, A.: Face detection with the modified census transform. In: Proceedings of the IEEE International Conference on Automatic Face and Gesture Recognition, pp. 91–96 (2004)
11. Jun, B., Kim, D.: Robust face detection using local gradient patterns and evidence accumulation. Pattern Recognit. **45**(9), 3304–3316 (2012)
12. Kim, D., Dahyot, R.: Face components detection using SURF descriptors and SVMs. In: Machine Vision and Image Processing International Conference, pp. 51–56 (2008)
13. Viola, P., Jones, M.: Robust real-time object detection. IJCV **57**(2), 137–154 (2004)
14. Jeong, K., Choi, J., Jang, G.-J.: Semi-local structure patterns for robust face detection. IEEE Signal Process. Lett. **22**(9), 1400–1403 (2015)
15. Kim, M., Kumar, S., Pavlovic, V., Rowley, H.: Face tracking and recognition with visual constraints in real-world videos. In: IEEE Conference Computer Vision and Pattern Recognition (2008)
16. Ithaya Rani, P., Muneeswaran, K.: Facial emotion recognition based on eye and mouth regions. Int. J. Pattern Recognit. Artif. Intell. **30**(6) (2016)
17. Ithaya Rani, P., Muneeswaran, K.: Recognize the facial emotion in video sequences using eye and mouth temporal Gabor features. Multimed. Tools Appl. **76**, 10017–10040 (2017). https://doi.org/10.1007/s11042-016-3592-y. Impact factor 1.53
18. Ithaya Rani, P., Muneeswaran, K.: Gender identification using reinforced local binary pattern. IET Comput. Vis. (2017). https://doi.org/10.1049/iet-cvi.2015.0394. ISSN 1751–9632. Impact Factor 963
19. Ithaya Rani, P., Muneeswaran, K.: Emotion recognition based on facial components. Sadhana **43**, 48 (2018). https://doi.org/10.1007/s12046-018-0801-6. Impact factor 1.53
20. Viola, P., Jones, M.: Rapid object detection using a boosted cascade of simple signatures. In: Proceedings of IEEE Computer Society Conference on Computer Vision and Pattern Recognition, vol. 1, pp. 511–518 (2001)

21. Ithaya Rani, P., Muneeswaran, K.: Fusing the facial temporal information in videos for face recognition. IET Comput. Vis. (2016). ISSN 1751–9632 https://doi.org/10.1049/iet-cvi.2015.0394
22. Ithaya Rani, P., Hari Prasath, T.: Ranking, clustering and fusing the normalized LBP temporal facial features for face recognition in video sequences. Multimed. Tools Appl. **77**(5), 5785–5802 (2017). https://doi.org/10.1007/s11042-017-4491-6
23. Ithaya Rani, P., Muneeswaran, K.: Robust real time face detection automatically from video sequence based on Haar features. In: IEEE International Conference on Communication and Network Technologies, pp. 276-280 (2014). https://doi.org/10.1109/CNT.2014.7062769
24. Ithaya Rani, P., Hari Prasath, T.: Detection of human faces in video sequences using normalized Gabor LBP histograms. Int. J. Comput. Math. Sci. **6**(9) (2017). ISSN 2347–8527
25. Ithaya Rani, P., Hari Prasath, T.: Detection of human faces in video sequences using normalized Gabor LBP histograms. In: International Conference on New Frontiers of Engineering, Science, Management and Humanities (ICNFESMH-2017) at The Institution of Electronics and Telecommunication Engineers, Osmania University Campus, Hyderabad, India 2017 (2017). ISBN: 978-81-934288-3-2
26. Ithaya Rani, P., S. Selvi: skewness Gabor features used for detection of human faces automatically in different poses and illuminations. In: The National Conference on Hi-Tech Trends in Emerging Computational Technologies on 22nd: at Sethu Institute of Technology, Kariapatti, Virudhunagar (2019)

Malware Family Classification Model Using User Defined Features and Representation Learning

T. Gayathri [iD] and M. S. Vijaya[(✉)]

PSGR Krishnammal College for Women, Coimbatore, India
Gayathrithangamuthu73@gmail.com, msvijaya@psgrkc.ac.in

Abstract. Malware is very dangerous for system and network user. Malware identification is essential tasks in effective detecting and preventing the computer system from being infected, protecting it from potential information loss and system compromise. Commonly, there are 25 malware families exists. Traditional malware detection and anti-virus systems fail to classify the new variants of unknown malware into their corresponding families. With development of malicious code engineering, it is possible to understand the malware variants and their features for new malware samples which carry variability and polymorphism. The detection methods can hardly detect such variants but it is significant in the cyber security field to analyze and detect large-scale malware samples more efficiently. Hence it is proposed to develop an accurate malware family classification model contemporary deep learning technique. In this paper, malware family recognition is formulated as multi classification task and appropriate solution is obtained using representation learning based on binary array of malware executable files. Six families of malware have been considered here for building the models. The feature dataset with 690 instances is applied to deep neural network to build the classifier. The experimental results, based on a dataset of 6 classes of malware families and 690 malware files trained model provides an accuracy of over 86.8% in discriminating from malware families. The techniques provide better results for classifying malware into families.

Keywords: Malware classification · Machine learning · Representation learning · Deep neural network

1 Introduction

Malware is a malicious code, and it is harmful when executed on a computing device or system. Software that is specifically designed to disrupt and damage, or gain unauthorized access to a computer system. These are proposed to gain access to computer systems and network resources, disturb computer operations, and gather personal information without taking the consent of system's owner, thus creating a menace to the availability of the internet, integrity of its hosts, and the privacy of its users.

The threats posed by malware are perceived as too numerous and agile to be managed by humans in a meaningful way. Generic detection of malware comes at the cost of

© IFIP International Federation for Information Processing 2020
Published by Springer Nature Switzerland AG 2020
A. Chandrabose et al. (Eds.): ICCIDS 2020, IFIP AICT 578, pp. 185–195, 2020.
https://doi.org/10.1007/978-3-030-63467-4_14

precision; leading to information that is often limited to the fact that malware has been detected. This is useful for much of the network defense community in incident response and remediation activities often need correct and concise identification of malware-related threats. Such identification provides incident responders with the information necessary to understand what threats they are actually facing and to allocate resources accordingly.

Web usage mining is discovery of meaningful pattern from data generated by client server transaction on one or more web localities. Several web transactions automatically make the data which gets gathered in server access logs, refers logs, agent logs, client side's cookies, user profile, metadata, page attribute, page content and site structure. Search engines began to understand their unintended contribution in malware distribution. Web services enable the detection of malware with a huge partner's data. But even a massive malware database does not guarantee detection of recent ones. Most of anti-malware software products, such as Kaspersky, Symantec, and MacAfee normally use the signature based method to recognize threats. This malicious software can perform heterogeneity of functions such as encrypting and destroy data, hijacking core computing functions and accessing user system activity without their permission.

The damage affected by a virus that corrupts a computer or a corporate network can be different from an irrelevant increase in outgoing traffic to the complete network breakdown or the loss of hypercritical data. The scale of the damage depends on the purpose of the virus, and sometimes the results of its activity are undetectable for the users of a compromised machine. A virus on a commercial network can be considered a force mature and the damage affected by it as being equal to the loss associated with the network downtime essential for disinfection.

Malwares come in wide range of variations like Virus, Worm, Trojan horse, Ransomware, Backdoor, Zeus, Key loggers, Adware and six malware families such as allaple, cryptolocker, agent, Trojan generic, wannacry and zbot are taken into account for implementation.

2   Literature Survey

Several research works have been carried out currently using machine learning. Various features such as static and dynamic have been used to build models. In few cases signature based methods, image processing methods have been used. Based on the study of various literatures available on malware family classification, a brief report is presented in this section about the developments in the respective area in the last several years.

Schultz et al. [1] were the first to introduce the concept of data mining for detecting malwares. They applied three different static features for malware classification: Portable Executable (PE), strings and byte sequences. They used a data set consisted of 4266 files including 3265 malicious and 1001 benign programs. The Naive Bayes algorithm, taking strings as input data, gave the highest classification accuracy of 97.11%. The authors claimed that the rate of detection of malwares using data mining method was twice as compared to signature based method.

Nari et al. [2] presented a framework for automated malware classification into their respective families based on network behavior. Network traces were taken as input to the framework in the form of pcap files, from which the network flows were extracted. From these behavior graphs, the features like graph size, root out-degree, average out-degree, maximum out-degree, number of specific nodes were extracted. These features were used to classify malwares using classification algorithms available in WEKA library and it was concluded that J48 decision tree performs better than other classifiers.

Nataraj et al. [3] were the first to explore the use of byte plot visualization for automatic malware classification. They converted all the malware samples to grayscale byte plot representations and extracted texture-based features from the malware image. They used an abstract representation technique, GIST (global image descriptors), for computing texture features from images. The dataset consisted of 9,458 malware samples belonging to 25 different classes, collected from the Anubis system. They used the global image-based features to train a K-Nearest Neighbour model, with Euclidean distance as distance measure, to classify malware samples into their respective classes and an accuracy of 97.18% was obtained.

Rieck et al. [4] suggested a new method for automated identification of new classes of malware with similar type of behavior using clustering and classifying previously unseen malware to these discovered classes by classification using machine learning. They used more than 10,000 malware samples, belonging to 14 different families, in their experiment. These malware samples were collected using honeypots and spam-traps. They reported an accuracy of 88% on family classification using simple SVM based classifier.

Tian et al. [5] used function length frequency to classify Trojans. Function length was measured by the number of bytes in the code. Their results specify that the function length along with its frequency was important in classifying malware family and can be grouped with another feature for quick and scalable malware families' classification. They applied machine learning algorithms available in the WEKA library for classifying malware. They used a data set of 1368 malware to demonstrate their work and achieved an accuracy of over 97%.

In the existing work, the classifications were performed by training malware executable files or images or PE (portable executable) files. The features based on texture were extracted from executable files that assisted in recognizing the malwares or in classification of malware. In a few cases, malware image datasets were used and the corresponding binary values were trained to build classifiers. This motivated to carry out research work for building the malware family classification model by deriving an array of binary features and training the feature dataset through deep learning classifiers.

3 Malware Family Classification Model

The problem of malware family identification is formulated as multiclass classification task and solved using machine learning technique. The methodology of proposed malware family classification includes four different stages. In the first stage, malware family corpus development is performed wherein the virus executable files corresponding to six families of malware are collected. The second stage is malware family dataset creation. In this stage, the executable files are converted into images which are then converted into binary arrays to form the feature vectors of malware family dataset. In the third stage, the training dataset is used to develop the malware family classification models by implementing supervised learning and deep learning algorithms. Finally, the classification models are evaluated in terms of precision, recall, F-measure, and accuracy. The methodology of the malware family classification model is shown in Fig. 1.

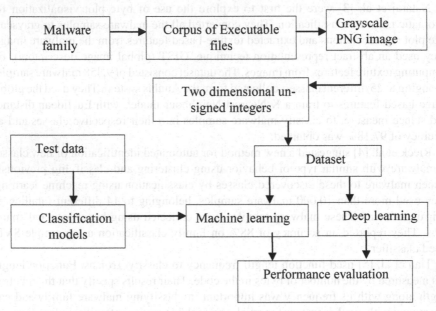

Fig. 1. Methodology of malware family classification model

3.1 Corpus Development and Dataset Creation

Initially a corpus consisting of 690 malware executable files related to six types of malware family is developed. Here allaple, cryptolocker, agent, Trojan generic, wannacry and zbot are six different types of malware family considered. These viruses corresponding to six malware families are collected from the das malwark, virustotal and a malwr.com malware samples. These portals have to collect malware using honeypots and also users around the world submit files over them for analysis and sharing malware samples. A set of 100 files of allaple malware, 150 exe files of cryptolocker, 100 files of agent family, 150 files of Trojan generic, 90 files of wannacry and 100 files of zbot malware family has been collected and a malware family corpus of size 690 is developed. Malware executable file in hexadecimal view is shown in Fig. 2.

Malware coders change small parts of the original source code to produce a new variant. Images can capture small changes yet retain the global structure. Images give more information about the structure of the malware. The malware files in the corpus are converted into a grayscale PNG (Portable Network Graphics) image of size 64×64 using python code. Conversion of executable file to image results are shown in Fig. 3.

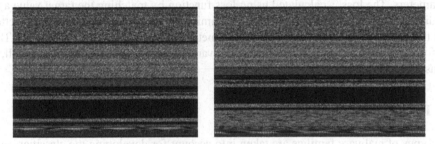

```
            yitaly.exe.zip        29c7e8735
            □ ■ ■ ■ ■ ■ ■ ■ ☑ Apply this colorset to all tabs
0000000   37 7A BC AF 27 1C 00 04 09 C5 AF 42 D0 1C 01 00   7z..'......B....
0000010   00 00 00 00 9A 00 00 00 00 00 00 00 5C E7 D5 37   ............\..7
0000020   6B CB 1E ED A7 0C CF 47 9D 5D 2D 31 42 8D 2F 46   k......G.]-1B./F
0000030   FA FA C9 BC AE 60 AD CB 76 77 49 5D FF A5 BF 8E   .....`..vwI]....
0000040   B0 3E 51 F5 78 4E E1 82 30 2F 31 10 5C 17 3A 1C   .>Q.xN..0/1.\.:.
0000050   00 77 06 DC 33 FF 56 FC 0C D3 2C 9D 73 FE A2 E3   .w..3.V...,.s...
0000060   D0 AE 61 58 FF A5 C6 A0 06 D0 05 32 1A 8A 6A 91   ..aX.......2..j.
0000070   EC 62 E6 E8 FD 9C 94 B1 7A F0 73 28 A2 CD 1F 66   .b......z.s(...f
0000080   43 7B E0 26 F5 7E E5 1A 62 C1 AC A8 8D 98 12 55   C{.&.~..b.......U
0000090   5D E9 3C 65 CA 2E 37 31 58 81 3A 3C 6A 73 CD EF   ].<e..71X.:<js..
00000A0   8E 61 A2 F9 CB 69 6F 2D C0 91 FB E0 05 6F BA 54   .a...i o-.....o.T
00000B0   6D 2B 23 20 DE 7D D4 14 B0 90 8D F9 FB A0 88 09   m+#..}..........
00000C0   62 84 DE 2C 17 4D 3C 10 A9 16 FB 7D 5B 93 B6 D5   b..,.M<....}[...
00000D0   FB 58 B7 DB 2D F0 3F 4B 2C AC 6E 46 75 24 01 89   .X.-.?K,.nFu$..
00000E0   27 7C 01 A9 F2 FF 53 A1 0C D6 B8 12 48 CE 54 1B   '|....S.....H.T.
00000F0   1D B2 B8 C8 33 58 67 B9 DA 01 77 C2 68 13 34 9C   ....3Xg...w.h.4.
0000100   F9 A0 E2 1A 24 19 85 B5 EE 61 1A 15 F5 DD 84 DF   ....$....a......
0000110   FC 2F AD 61 48 F9 5C 90 D8 40 FF B4 42 99 CC B4   ./.aH.\.@..B...
0000120   D0 F4 D3 35 A0 9E CC D5 C6 4B 2F 0D 6E C8 E1 65   ...5.....K/.n..e
0000130   57 0A 0D 27 4B 56 D5 D6 48 BE 2A 22 1A B3 29 F7   W..'KV..H.*"..).
0000140   0F 92 77 C3 4D C7 5D 6D EF 32 94 37 66 23 4A 71   .w.M.]m.2.7f#Jq
0000150   E4 89 8A FF 01 76 D7 4C 4F A4 DD CF 9C 17 A8 79   .....v.LO......y
```

Fig. 2. Malware executable files in hexadecimal view

Fig. 3. Image of zbot exe file

A sequence of pixel values corresponding to each byte is stored in the binary file as a sequence of ones and zeros. Width of the image is set according to the file size. Height of the image varies depending on the file size. Each image is resized to 28 x 28 and represented in binary form. The binary files are difficult to read as its dimension is high. Hence, it is mapped into a one dimensional array of integers between 0 and 255. It is a vector of 8-bit unsigned integers. This vector is then reshaped into a two dimensional array of unsigned integers. The resizing is done based on the height and width of the file size. The height of the file is the total length of the one dimensional array and the width of the array is nothing but the file size. Finally, the two dimensional feature matrix of size 28 × 28 amounting to 784 feature values is derived.

The rows of 2D matrix is organized as a feature vector of size 784 and the dimensionality reduction technique namely PCA (Principle Component Analysis) is used to reduce the size of the feature vectors. This dimensionality reduction method efficiently represents interesting parts of an array as a compact feature vector of size 480. Data transformation and dimensionality reduction are performed to achieve maximum prediction accuracy.

In this manner all the 690 virus exe files are converted into feature vectors which are then assigned class labels 1 to 6 as 1 for alleple, 2 for cryptolocker, 3 for agent, 4 for Trojan generic, 5 for wannacry and 6 for zbot. Finally a malware family dataset with 690 instances of dimension 480 is created and stored in a csv file.

3.2 Model Building

Malware family classification model is built with the above dataset as input and a deep learning architecture namely deep neural network is employed to build the malware family classification model. Deep neural network performs the representation learning from the features and self-extracts hidden patterns during training. Various hyper parameters such as learning rate, epochs, loss function, activation function and optimizers are defined to improve the efficiency of learning and to build the accurate malware family classification model. The learning rate is used to adjust the weights and back propagate the weights to make correct predictions. Epochs defines the number of iterations, the learning algorithm work through the entire training dataset for leaning the patterns efficiently. The learning algorithms train the patterns through the layers such that the error rate of the model is sufficiently minimized. Loss function helps in optimizing the parameters of the neural networks. This is minimizing the loss for a neural network by optimizing its parameters. The loss is calculated using loss function by matching the target value and predicted value by a neural network. The optimization is used to produce slightly better and faster results by updating the model parameters such as weights and bias values. The performance of the model is evaluated using various metrics such as precision, recall, F measure and accuracy.

4 Experiments and Results

Six types of malware families are taken into account for developing the classifiers and hence malware recognition problem turn out to be multi classification. In these experiments, the sequential model is developed to build DNN classifier. The hyper parameters used to build model are learning rate is 0.001 and dropouts is 0.2, 0.3. The input layers have been given 480 attributes and similarly the hidden layer has been given as a 240 in sequential model. The output layers have been specified with class labels 1 to 6 for recognizing six families of malwares. The activation functions namely softmax and relu with adam optimizers are used in this work. A softmax function is used in output layer and relu functions used for hidden layer. Adam optimizer is used with value of epochs is 200 and batch size is 10 to increase the prediction accuracy. Deep neural network is prepared to assign the layers and input of one layer is passed as output to the other layer. The functioning of DNN based malware family classification method is determined based on classification metrics. Prediction accuracy of 86.8% is achieved by DNN classifier and evaluated with respect to six class labels i.e. are allaple, cryptolocker, agent, Trojan generic, wannacry and zbot. The precision is high for allaple, agent and zbot with the value of 1.00 and recall value of 1.00 is maximum for class allaple and zbot. The F-measure with the value of 1.00 is excessive for class allaple and zbot. The average values of DNN methods with precision of 0.82, recall of 0.80 and F-measure of 0.85 is obtained for six malware family types. Class-wise performance of DNN model is shown in Table 1 and results obtained for various dropouts and epochs are illustrated in Table 2.

Table 1. Performance results of deep neural network classifier by class

Class labels	Precision	Recall	F-measure
Allaple	1.00	1.00	1.00
Cryptolocker	0.97	0.95	0.96
Agent	1.00	0.96	0.98
Trojan generic	0.78	0.64	0.68
Wannacry	0.48	0.58	0.53
Zbot	1.00	1.00	1.00
Accuracy	0.78	0.86	0.81

The malware classification models based on deep neural network is built with performance of the model evaluated using classification metrics such as precision, recall, F-score and accuracy. The performance results of the deep neural network classification based on various metrics are shown Table 2 (Fig. 4).

Table 2. Performance results of deep learning

Dropout	Epochs	Precision	Recall	F-measure	Accuracy
0.2	50	0.58	0.68	0.60	0.63
	100	0.78	0.77	0.77	0.77
	150	0.80	0.81	0.80	0.79
	200	**0.75**	**0.81**	**0.77**	**0.81**
0.3	50	0.75	0.83	0.78	0.80
	100	0.81	0.80	0.79	0.83
	150	0.84	0.83	0.83	0.85
	200	**0.82**	**0.80**	**0.85**	**0.86**

The results of DNN classifier is compared with the implementation results of supervised learning algorithms such as decision tree, neural networks, support vector machine which have been implemented using the same dataset. It is discovered that deep neural network had achieved highest prediction accuracy of 86.8% was acquired by trained supervised models. DNN classifier obtained a precision of 0.82, recall value of 0.80 with F-measure of 0.85 and accuracy 86.8%. The comparative predictive performances of malware family classifiers are shown in Table 3 (Fig. 5).

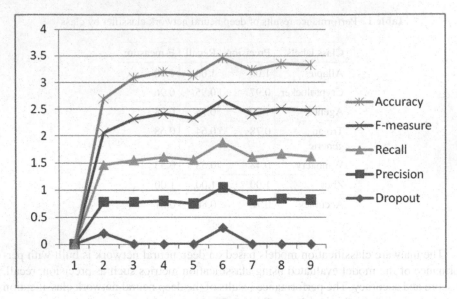

Fig. 4. Performance results of deep learning at various dropouts

Table 3. Comparative results of DNN and supervised classification algorithms

Performance evaluation	Classifiers			
	Neural network (%)	Decision tree (%)	Support vector machine (%)	Deep neural network (%)
Precision	75	80	76	82
Recall	82	86	81	80
F-measure	78	86	78	85
Accuracy	85	82	85	86

The deep neural network model have capability to modify the weights in deep neural network so the error rate is minimized which gives accurate prediction. The significant result of deep neural network based classification model is attained by maximizing hidden layers and number of epochs in sequential model. The performance of the deep neural network classifier is validated using its measures with high accuracy and least error rate for malware family classification. Deep neural network classifier achieves better performance through image features and outperforms with supervised learning classification models.

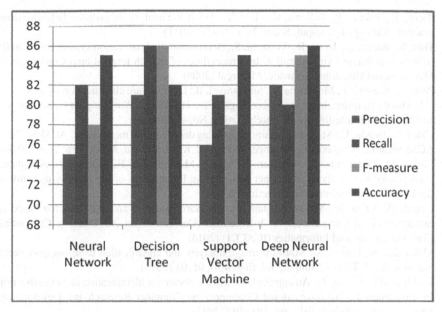

Fig. 5. Comparison of classifiers based on various evaluation metrics

5 Conclusion

This paper demonstrated the implementation of deep neural network based malware family classification model built with the aim to predict the malware family through DNN classifier using malware family dataset. DNN method automatically learns high level features from the user defined features in the malware family dataset. The executable malware files are collected from various malware resources and malware family corpus has been developed. The performance of the supervised and deep learning classifiers is evaluated in terms of accuracy, precision, recall, F-measure and the results are compared. The DNN based malware family classification model has shown high classification accuracy compared with supervised classifiers. Finally, it is concluded that deep neural network based classification method performs accurate prediction in classifying the malware families.

References

1. Schultz, M., Eskin, E., Zadok, F., Stolfo, S., Data mining methods for detection of new malicious executables. In: Proceedings of 2001 IEEE Symposium on Security and Privacy, Oakland, 14–16 May 2001, pp. 38–49 (2001)
2. Nari, S., Ghorbani, A.: Automated malware classification based on network behavior. In: Proceedings of International Conference on Computing, Networking and Communications (ICNC), San Diego, pp. 642–647 (2013)
3. Nataraj, L., Karthikeyan, S., Jacob, G., Manjunath, B., Malware images: visualization and automatic classification. In: Proceedings of the 8th International Symposium on Visualization for Cyber Security, Article No. 4 (2011)

4. Rieck, K., Trinius, P., Willems, C., Holz, T.: Automatic analysis of malware behavior using machine learning. J. Comput. Secur. **19**, 639–668 (2011)
5. Tian, R., Batten, L., Islam, R., Versteeg, S.: An automated classification system based on the strings of trojan and virus families. In: Proceedings of the 4th International Conference on Malicious and Unwanted Software, Montréal (2009)
6. Park, Y., Reeves, D., Mulukutla, V., Sundaravel, B.: fast malware classification by automated behavioral graph matching. In: Proceedings of the 6th Annual Workshop on Cyber Security and Information Intelligence Research, Article No. 45 (2010)
7. Cakir, B., Dogdu, E.: Malware classification using deep learning methods. In: ACM SE 2018: ACM SE 2018: Southeast Conference, Richmond, KY, USA, 5 p. ACM, New York (2018)
8. Bailey, M., Oberheide, J., Andersen, J., Mao, Z.M., Jahanian, F., Nazario, J.: Automated classification and analysis of internet malware. In: Proceedings of the 10th Symposium on Recent Advances in Intrusion Detection (2007)
9. Uppal, D., Sinha, R., Mehra, V., Jain, V.: Malware detection and classification based on extraction of API sequences. In: International Conference on Advances in Computing, Communications and Informatics (ICACCI) (2014)
10. Makandar, A., Patrot, A.: Malware image analysis and classification using support vector machine. Int. J. Trends Comput. Sci. Eng. **4**(5), 01–03 (2015)
11. Zolkipli, M.F., Jantan, A.: An approach for malware behavior identification and classification. In: Proceeding of 3rd International Conference on Computer Research and Development, Shanghai, 11–13 March 2011, pp. 191–194 (2011)
12. Biley, M., Oberheid, J., Andersen, J., Morley Mao, Z., Jahanian, F., Nazario, J.: Automated classification and analysis of internet malware. In: Proceedings of the 10th International Conference on Recent Advances in Intrusion Detection, vol. 4637, pp. 178–197 (2007)
13. Islam, R., Tian, R., Battenb, L., Versteeg, S.: Classification of malware based on integrated static and dynamic features. J. Network Comput. Appl. **36**, 646–656 (2013)
14. Bhodia, N., Prajapati, P., Di Troia, F., Stamp, M.: Transfer learning for image-based malware classification. In: 3rd International Workshop on Formal Methods for Security Engineering (ForSE 2019), in Conjunction with the 5th International Conference on Information Systems Security and Privacy (ICISSP 2019), Prague, Czech Republic (2019)
15. Gandotra, E., Bansal, D., Sofat, S.: Malware analysis and classification: a survey. J. Inf. Secur. **5**, 56–64 (2014)
16. Maksood, F.Z.: Analysis of data mining techniques and its applications. Int. J. Comput. Appl. (0975 – 8887) **140**(3), 6–14 (2016)
17. Babu, S.I., Chandra Sekhara Rao, M.V.P., Nagi Reddy, G.: Research methodology on web mining for malware detection. Int. J. Comput. Trends Technol. (IJCTT) **12**(4) (2014)
18. Gavrilut, D., Cimpoeşu, M., Anton, D., Ciortuz, L.: Malware detection using machine learning. In: Proceedings of the International Multi conference on Computer Science and Information Technology, pp. 735–741 (2009)
19. Saxe, J., Berlin, K.: Deep neural network based malware detection using two dimensional binary program features. In: 10th International Conference on Malicious and Unwanted Software: "Know Your Enemy" (MALWARE) (2015)
20. Dychka, I., Chernyshev, D.: Malware detection using artificial neural networks. In: ICCSEEA, Advances in Computer Science for Engineering and Education II, vol. 938, pp. 3–12 (2019)
21. Islam, R., Altas, I.: A comparative study of malware family classification. In: Chim, T.W., Yuen, T.H. (eds.) ICICS 2012. LNCS, vol. 7618, pp. 488–496. Springer, Heidelberg (2012). https://doi.org/10.1007/978-3-642-34129-8_48
22. Saxe, J., Berlin, K.: Deep neural network based malware detection using two dimensional binary program features. In: Proceedings of the 10th International Conference on Malicious and Unwanted Software (MALWARE), IEEE, pp. 11–20 (2015)

23. Nazario, J., Oberheid, J., Andersen, J., Morley Mao, Z., Jahanian, F., Biley, M.: Automated classification and analysis of internet malware. In: Proceedings of the 10th International Conference on Recent Advances in Intrusion Detection, vol. 4637, pp. 178–197 (2007)
24. Mohaisen, A., Alrawi, O.: Unveiling Zeus: automated classification of malware samples. In: Proceedings of the 22nd International Conference on World Wide Web, pp. 829–832 (2013)
25. Kong, D., Yan, G.: Discriminate malware distance learning on structural information for automated malware classification. In: Proceedings of the ACM SIGMETRICS/International Conference on Measurement and Modeling of Computer Systems, pp. 347–348 (2013)
26. Rieck, K., Holz, T., Willems, C., Düssel, P., Laskov, P.: Learning and classification of malware behavior. In: Zamboni, D. (ed.) DIMVA 2008. LNCS, vol. 5137, pp. 108–125. Springer, Heidelberg (2008). https://doi.org/10.1007/978-3-540-70542-0_6
27. Pascanu, R., Stokes, J.W., Sanossian, H., Marinescu, M., Thomas, A.: Malware classification with recurrent networks. In: 2015 IEEE International Conference on IEEE Acoustics, Speech and Signal Processing (ICASSP), pp. 1916–1920 (2015)

Fabric Defect Detection Using YOLOv2 and YOLO v3 Tiny

R. Sujee(✉), D. Shanthosh, and L. Sudharsun

Department of Computer Science and Engineering, Amrita School of Engineering,
Amrita Vishwa Vidyapeetham, Amrita University, Coimbatore, India
r_sujee@cb.amrita.edu, shanthoshdevaraj@gmail.com, sudhas1728@gmail.com

Abstract. The paper aims to classify the defects in a fabric material using deep learning and neural network methodologies. For this paper, 6 classes of defects are considered, namely, Rust, Grease, Hole, Slough, Oil Stain, and, Broken Filament. This paper has implemented both the YOLOv2 model and the YOLOv3 Tiny model separately using the same fabric data set which was collected for this research, which consists of six types of defects, and uses the convolutional weights which were pre-trained on Imagenet dataset. Observed and documented the success rate of both the model in detecting the defects in the fabric material.

Keywords: YOLOv2 · YOLOv3 Tiny · Fabric defect detection

1   Introduction

In the growing trend of Artificial Intelligence, machine learning, and deep learning, many man-made tasks are automated and developed. But still, there are many industries to be revolutionized in these technologies. In that an important industry is textile. In textile industries, quality inspections are one of the major problems for fabric manufacturers as it requires lots of man-hours and wages. Currently, in textile industries, inspection is carried out manually, due to this there are possibilities of human errors with high inspection time. To overcome this problem, automating the process of finding defects is necessary.

In this paper, two object detection algorithms YOLOv2 and YOLOv3 Tiny are used to solve the problem of finding defects in fabric material with better accuracy and minimum loss with less computational time to suite the industrial rapidness. Six different defects:-Oil Stain: Discoloration of fabric due to oil; Holes: A small opening in fabric; Broken Filament: Individual filaments constituting the main yarn are broken; Rust: Presence of red or yellow color coating made of iron oxide formed by oxidation, especially in the presence of moisture in the fabric; Grease: A black thick oily substance in fabric; and Slough: A thick bundle of yarn is wrapped into the fabric in the direction of the weft due to the slip of the yarn coils from the pirn during weaving; are classified and localized.

Supported by Amrita School of Engineering, Coimbatore.

2   Literature Survey

Paper [1], proposed a new descriptor based on an mutual information, which is then used as input to neural networks. One and two hidden layer of neural networks were used. The result revealed that the recognition rates were 100% for training and 100% for generalization.

In paper [2] it was shown that image quality can be enhanced by improving contrast using a histogram matching and thus texture classification can be carried out in a better and efficient way.

Paper [3], suggested a method based on local homogeneity and mathematical morphology. The spatial homogeneity of each pixel was calculated to create a new homogeneity image. This image was placed in a series of morphological and Structuring Element operations to determine the feature of the existing fabric. The functionality of the system was evaluated extensively using various fabric samples, which varied in defect type, size and shape, text background, and image resolution.

In paper [4], it was proposed to provide a diagnostic procedure for detecting and classifying defects in warp and weft using a computer program developed by MATLAB that analyzed images of fabric samples obtained using a flat scanner. Three were determined to detect defects. Finally, a neural network was introduced to identify the class of defects present in the fabric.

Paper [5], described a system of pre-processing train images collected by a web crawler. The result showed before developing the system, two important factors have to be taken care of. First of all, the size of the training image and the acquisition image should be the same, second, the size of the object's location should be the same as the training image and the detection image. Six classes were used in the study and that satisfied the YOLO training value.

In paper [6], the 276 images associated with the textile fabrications were collected, pre-processed and labeled. After that, YOLO9000, YOLO-VOC, and Tiny YOLO were used to build models for fabric detection. Through a comparative study, YOLO-VOC was chosen for further development by optimizing the super-parameters of the deep neural network. In this paper, only 3 types of defects are identified, namely, Belt yarn, Knot tying, and Hole, with less number of defect images. In the proposed paper, we try to identify 6 defects as mentioned above, using YOLOv2 and YOLOv3.

In paper [7], a working model was developed for portable devices such as a laptop or mobile phone which lacks the Graphics Processing Unit (GPU). The model was first trained in the PASCAL VOC dataset then it was trained in the COCO database. It was seen that the YOLO-LITE speed was 3.8x faster than the SSD MobilenetvI.

Paper [8] suggested several improvements to the YOLO adoption method, both novel and based on previous work. The improved model, YOLOv2, performed better while compared to Faster RCNN with Resent and SSD. Also, YOLO9000 was trained in the COCO acquisition and ImageNet classification data and it was shown that YOLO9000 was able to predict more than 9000 different object classes.

3   Proposed System

3.1   Choice of YOLO

From the literature survey, it was clear that YOLO outperforms other object
detection frameworks. The result of the comparative Performance of YOLO2 and
other state-of-the-art detection systems from paper [8] is shown in the Table 1.
As mAP and FPS is better for YOLOv2 we have chosen the YOLO model to
detect defects in the fabric. For optimal FPS and mAP, we chose YOLOv2 with
input size 416 × 416.

Table 1. Performance comparison

Detection frameworks	Train	mAP	FPS
Faster R-CNN VGG-16	2007 +2012	73.2	7
Faster R-CNN ResNet	2007 + 2012	76.4	5
YOLOv2 288 × 288	2007 + 2012	69.0	91
YOLOv2 416 × 416	2007 + 2012	76.8	67
YOLOv2 544 × 544	2007 + 2012	78.6	40

Table 2. Sample annotation data

Class	x1	y1	x2	y2
1	0.24609	0.70361	0.41601	0.59277
1	0.77197	0.75976	0.12793	0.46679

3.2   Data Set Creation

First step in proposed system is to collect fabric material with and without
defects used in real factories and capture the images of those fabric materials
from different orientations and angles. Then the images should be pre-processed
and made to meet the dimensions of 416 * 416 pixels.

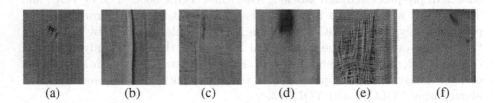

(a)   (b)   (c)   (d)   (e)   (f)

Fig. 1. Dataset (a) Hole (b) Slough (c) Rust (d) Grease (e) Broken Filament (f) Oil
Stain

After preprocessing images, in order to feed them to YOLOv2 it is necessary
to annotate the images with rectangle boxes specifying their defect area in the
following format. Table 2 shows the format and sample values of center of defect,
width and height of the defect with respect to the image. Figure 1 displays the
sample defect images from the dataset used in training YOLOv2 and YOLOv3
Tiny.

Each of the fields in Table 2 represent following data (Table 3):

Table 3. Number of images used for each class

Class	Hole	Rust	Broken filament	Oil stain	Slough	Grease	Total
Defect images	80	160	100	85	75	100	600

- class: Represents the type of defect present in the corresponding box
- x1, x2: Center of defect with respect to the image
- y1, y2: Width and Height of the defect with respect to the image coordinates.

3.3 Object Detection Algorithm

In this proposed system, it is not only necessary to classify the images as defects and not defects but it is also important to identify their place of occurrence for improving the manufacturing process in the future. In order to achieve this, paper uses YOLO object detection algorithm.

You Look At Once (YOLO), is a real-time object detection system, and is a neural network that can see what is in the image and where content is, just in one passing. It offers bounding boxes around discoveries and can detect multiple objects at a time, providing a huge advantage when comparing it to other neural network models.

To train neural networks, a framework called Darknet is used, it is open source and serves as the basis for YOLO written in C/CUDA. The original repository, by J Redmon. Darknet is used as the framework for training YOLO, i.e it sets the architecture of the network. Figure 2 shows the flow of the proposed method for training starting from capturing the defect, annotating the defect image, training the model using the annotated image obtained from the above step. During test phase when image is given to the trained model it gives the output as show in the last step of flow diagram.

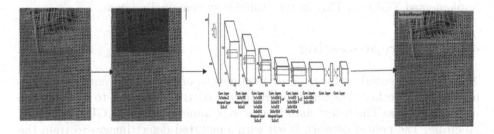

Fig. 2. Flow diagram

YOLOv2 Model was mainly aimed at improving recall and localization while maintaining the accuracy of the classification. For better performance, techniques such as batch normalization, high resolution classification, convolution with anchor boxes were incorporated. The architecture of Darknet-19 is the backbone of YOLOv2.

YOLOv3 Tiny Model. In YOLOv2 there are No residual blocks, no skip connections, and no up-sampling. YOLO v3 incorporates all of the above features. YOLOv3 use a new network for performing feature extraction. The latest network is a hybrid approach of YOLOv3, and residual network, so it has some shortcut connections. The Darknet-53 architecture is the backbone of YOLOv3. Tiny-yolov3 is a simplified version of YOLOv3, which contains a smaller number of convolution layers than YOLOv3, in turn the neural network doesn't need to take up much of the memory, reducing the need for hardware. And it also greatly speeds up detection, as a result it is well suited to use in a resource-constrained and rapid computing environment.

3.4 System Specification

For the current implementation of this paper, the camera and system with the following specification are used to capture and test the textile material. Table 4 specifies system specification.

Table 4. System specification

Component	Specification
Camera	DSLR
RAM	8 GB
Processor	Intel Core i7-7700HQ
Graphic card	NVIDIA GeForce GTX 1050Ti 4 GB

With this system environment, we were able to implement YOLOv2 for fabric defect detection. Due to the limitation of the RAM and Graphics card, where YOLOv3 requires 16 GB and NVIDIA GeForce GTX 1080Ti 8 GB GPU, we implemented YOLOv3 Tiny in the available system environment.

4   Data Preprocessing

The dataset created had a total of 1000 images of both defect and non-defect images put together. The images were resized to 416 × 416 × 3 to feed into the neural networks. The defect data images were annotated with a GUI tool called labelImg. The neural network is fed with annotated defect images to train the model. After running K means clustering separately for YOLOv2 and YOLOv3 Tiny on the manually labelled boxes, it turns out that most of the manually labelled boxes which is referred as actual bounding box have certain height-width ratios. So instead of directly predicting a rectangle to localize the defects, YOLOv2 and YOLOv3 predict off-sets from a pre-established set of boxes with particular height-width ratios to overcome the difficulty of detecting the objects having midpoints in the same grid we call these as anchors which are set before training the neural network using the actual bounding box. In YOLO v2, 5 anchor boxes are used and in YOLOv3 Tiny, tiny 6 anchor boxes are used. While training, the model uses this anchor box to localize the defect accurately especially when there are multiple defects in a single image and while predicting the defects in the test dataset bounding boxes are predicted relative to these predefined anchors.

5 Implementation

The dataset was split as 80-20. The YOLOv2 and YOLOv3 tiny model was trained separately using the 80% of the images from the fabric defect dataset and tested on the randomly picked 20% images of fabric defect dataset.

Table 5. Layer information - YOLOv2

Number	Layer	Input	Output
0	conv	416 × 416 × 3	416 × 416 × 32
1	max	416 × 416 × 32	208 × 208 × 32
2	conv	208 × 208 × 32	208 × 208 × 64
3	max	208 × 208 × 64	104 × 104 × 64
4–6	conv	104 × 104 × 64	104 × 104 × 128
7	max	104 × 104 × 128	52 × 52 × 128
8–10	conv	52 × 52 × 128	52 × 52 × 256
11	max	52 × 52 × 256	26 × 26 × 256
12–16	conv	26 × 26 × 256	26 × 26 × 512
17	max	26 × 26 × 512	13 × 13 × 512
18–24	conv	13 × 13 × 512	13 × 13 × 1024
25	route		
26	conv	26 × 26 × 512	26 × 26 × 64
27	reorg	26 × 26 × 64	13 × 13 × 256
28	route		
29–30	conv	13 × 13 × 1280	13 × 13 × 30

Table 6. Layer information - YOLOv3 Tiny

Number	Layer	Input	Output
0	conv	416 × 416 × 3	416 × 416 × 16
1	max	416 × 416 × 16	208 × 208 × 16
2	conv	208 × 208 × 16	208 × 208 × 32
3	max	208 × 208 × 32	104 × 104 × 32
4	conv	104 × 104 × 32	104 × 104 × 64
5	max	104 × 104 × 64	52 × 52 × 64
6	conv	52 × 52 × 64	52 × 52 × 128
7	max	52 × 52 × 128	26 × 26 × 128
8	conv	26 × 26 × 128	26 × 26 × 256
9	max	26 × 26 × 256	13 × 13 × 256
10	conv	13 × 13 × 256	13 × 13 × 512
11	conv	13 × 13 × 512	13 × 13 × 512
12–15	conv	13 × 13 × 512	13 × 13 × 33

YOLOv2 model consists of 23 Convolution Layers, 5 Max Pooling Layers, 2 Route Layers, and 1 Reorg layer. Before training the YOLOv2 model for fabric dataset the followings parameters were tuned; the number of classes was set as 6, the batch parameter and subdivisions were fixed as 64 and 16 respectively, the maximum batches was set as 12000 using the formula - number of classes * 2000, steps were set as 0.8 * maximum batches, 0.9 * maximum batches which equals to 9600 and 10800. The number of filters used was 33 calculated using the formula - (Number of Classes + 5) * 3. The five anchor box values used are [[1.3221, 1.73145], [3.19275, 4.00944], [5.05587, 8.09892], [9.47112, 4.84053], [11.2364, 10.0071]]. The Table 5 explains YOLOv2 architecture and its layers in detail.

YOLOv3 model consists of 10 Convolution Layers, and 6 Max Pooling Layers. In YOLOv3, the six anchor values used are [[10, 14], [23, 27], [37, 58], [81, 82], [135, 169], [344, 319]] and rest of the parameters used are same as YOLOv2. The Table 6 explains YOLOv3 Tiny architecture and its layers in detail. The proposed model is implemented using both YOLOv2 and YOLOv3 Tiny separately with the architecture mentioned in the Table 5 and Table 6 respectively.

6   Results and Discussion

In defect detection, the actual bounding boxes are the manually labeled box which specify where the defect is in the image and the predicted bounding is the box the model gives as an output. IOU is used for the evaluation of the model which is computed by Area of Intersection divided over Area of Union of those two boxes. For detection to be accurate this paper use Non Max Suppression (NMS). In NMS all the boxes having probabilities less than or equal to a pre-defined threshold are discarded. For the remaining boxes, box with the highest probability is chosen and taken as the output prediction and boxes having IoU greater than the threshold with the output box from the above step are discarded. Mean Average Precision is used as one of the factor to check the performance of the model. mAP is calculated by taking the mean of average precision over all classes and/or over all Intersection over Union thresholds and Average Precision score calculated by taking the average value of the precision across all recall values. The formula for IOU, Precision and recall are mentioned in the Fig. 3.

The field with greater risk should have the IoU value high. But, in the fabric industry, the defects produced doesn't cause any risk to human lives and our primary focus is to know the type of defect and the prime area where the defect occurred rather than the exact point, therefore IOU is set to 0.5, reducing further will make the model to point totally a different area relative to the defect which will not solve our purpose.

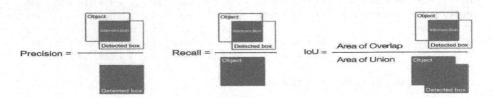

Fig. 3. Formula for calculating Precision, Recall and IOU

The Performance Metrics of YOLOv2 and YOLOv3 Tiny is shown in Table 7. The IoU threshold used for both the models was 50%, used Area-Under-Curve for each unique (Fig. 4 and Table 8).

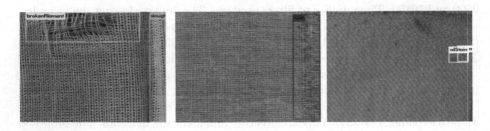

Fig. 4. Experimental test results of YOLOv2

Table 7. Performance metrics comparison

Model	Precision	Recall	F1-score	Accuracy	mAP
YOLOv2	0.77	0.66	0.71	83.12	65.25
YOLOv3	0.58	0.60	0.50	72	60.13

Table 8. Class-wise accuracy of YOLOv2

Class	Hole	Rust	Broken filament	Oil stain	Slough	Grease	Average accuracy
Accuracy	87.5	78.125	85	76.5	86.6	85	83.12

7   Conclusion and Future Work

YOLOv3 Tiny is lighter and faster than YOLOv2, by sacrificing some accuracy for it. Initially, with a small dataset, YOLOv2 and YOLOv3 tiny gave similar same accuracy. But, with an increase in dataset resulted in the greater accuracy for YOLOv2. But in case of YOLOv3 tiny, as the number of layers was less, the model was not able to learn enough about the defects after a certain extent from the available dataset. This paper faces challenges due to evolving patterns and styles in the fabric material as even a hole in certain areas can be due to a new style. And the mixture of colors in the material can also decrease the accuracy of the model as the trained images are in a similar background.

As this model will not address all the defects identified in a cloth. The model can be trained with other type of defects. Due to lacking variety in data set the identification of the defects may be affected by various factors like evolving colours and patterns. A well developed data set can be created and trained. As we were short of dataset, we produced the dataset and processed it in a short period of time. Further improvements to the paper in the upcoming years can be made by including same background, different background, single color fabric, multicolored fabric image datasets. In the future, methods to be discussed to implement the proposed system in a cost-efficient manner in the industries.

References

1. Abdel-Azim, G., Nasri, S.: Textile defects identification based on neural networks and mutual information. In: 2013 International Conference on Computer Applications Technology (ICCAT) (2013)
2. Sujee, R., Padmavathi, S.: Image enhancement through pyramid histogram matching. In: 2017 International Conference on Computer Communication and Informatics (ICCCI)
3. Rebhi, A., Abid, S., Fnaiech, F.: Fabric defect detection using local homogeneity and morphological image processing. In: 2016 International Image Processing, Applications and Systems (IPAS), Hammamet (2016)
4. Jmali, M., Zitouni, B., Sakli, F.: Fabrics defects detecting using image processing and neural networks. In: 2014 Information and Communication Technologies Innovation and Application (2014)

5. Jeong, H., Park, K., Ha, Y.: Image preprocessing for efficient training of YOLO deep learning networks. In: 2018 IEEE International Conference on Big Data and Smart Computing (BigComp), Shanghai (2018)
6. Zhang, H., Zhang, L., Li, P., Gu, D.: Yarn-dyed fabric defect detection with YOLOV2 based on deep convolution neural networks. In: 2018 IEEE 7th Data Driven Control and Learning Systems Conference (DDCLS), Enshi (2018)
7. Huang, R., Pedoeem, J., Chen, C.: YOLO-LITE: a real-time object detection algorithm optimized for non-GPU computers (2018)
8. Redmon, J., Farhadi, A.: YOLO9000: better, faster, stronger (2017)

Automatic Questionnaire and Interactive Session Generation from Videos

V. C. Skanda$^{(\boxtimes)}$ ⓘ, Rachana Jayaram ⓘ, Viraj C. Bukitagar ⓘ,
and N. S. Kumar ⓘ

Department of CSE, PES University, Bengaluru, India
skandavc18@gmail.com, rachana.jayaram@gmail.com, viraj.bukitagar@gmail.com,
nskumar@pes.edu

Abstract. In this paper, we present a tool that interleaves lengthy lecture videos with questionnaires at optimal moments. This is done to keep students' attention by making the video interactive. The student will be presented with MCQ type questions based on the topic covered so far in the video, at regular intervals. The questions are generated based on the transcript of the video lecture using machine learning and natural language processing techniques. In order to have continuity and proper flow of teaching, a LDA-based (Latent Dirichlet Allocation) model has been proposed to insert those generated questions at appropriate points called logical points.

Keywords: Interactive videos · Question generation · Speech to text · Logical points · Latent Dirichlet Allocation · Natural language processing · Machine learning

1 Introduction

According to a study by Microsoft, the average human attention span is decreasing [1]. Decreasing attention span can have a huge impact on the learning outcomes of children. Educational/lecture videos are generally monotonous with little to no interaction. Thus, in order to keep students attentive, the videos must be made interactive. So we present a method in order to generate and insert questions automatically at appropriate points.

2 Literature Survey

Interactive teaching is known to be very effective. In Richard Hake's landmark 1998 study on the effectiveness of lecture-based instruction, he showed that interactive classes outperform traditional classes when it comes to learning effectiveness and concentration retention [2].

In order to generate and insert questions, transcription has to be done. In 2014 Coates et al. introduced a state of the art speech recognition system with a 84% accuracy using end-to-end deep learning [3].

© IFIP International Federation for Information Processing 2020
Published by Springer Nature Switzerland AG 2020
A. Chandrabose et al. (Eds.): ICCIDS 2020, IFIP AICT 578, pp. 205–212, 2020.
https://doi.org/10.1007/978-3-030-63467-4_16

In 2011, Crossno et al. compared topic modelers and found that LDA performed better than LSA especially for smaller document sizes [4]. Our approach uses LDA for topic modelling as a part logical point detection.

In his 2010 study on automated question generation, Heilman [5] delves into the intricacies of generation of factual questions from text. Our approach to cloze question generation relies heavily on machine learning techniques as well natural language processing.

In 2010, Altabe and Maritxalar presented a corpus-based approach to domain-based distractor generation [6] which is quite similar to the approach to MCQ option generation presented in this paper.

3   Current Work

The best way to make the videos interactive is to insert questions based on the video topic at appropriate "logical points". Logical points are those time points in the video which mark the beginning or the end of a topic or a sub topic in the video. Here a topic is defined as a collection of related paragraphs.

If logical points are too sparsely distributed, there will be very few questions in the video. To maintain a balance in the length of time between questions, the logical points are found such that they are evenly distributed throughout the video. But this interval between questions can also be set manually if needed.

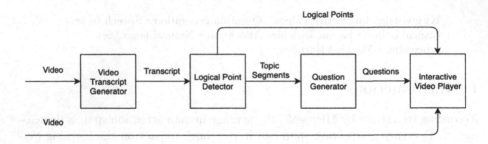

Fig. 1. System architecture

The questions are based on the topics covered so far in the video. They are generated from the video transcript using natural language processing and machine learning techniques. We have proposed a solution wherein the questions and the logical points at which a question must be inserted are extracted from the transcript itself.

The overall architecture of the solution has been shown in Fig. 1. The tool consists of 4 main parts which work in sequence:

1. Transcript Generator
2. Logical Point Detector
3. Question Generator
4. Interactive Video Player

3.1 Transcript Generation

The audio is first extracted from video. The retrieved audio file is split into multiple parts based on silence. Here silence implies that the audio level has fallen below a certain threshold. The threshold is a tunable parameter whose default value is set to 16 dbFS. Splitting the audio file is necessary as transcribing a huge audio file leads to bad transcription. Instead of splitting the audio into equal intervals, the splits are based on silence. A "silent" point is a good indication of a logical point. After splitting, the audio files are transcribed using existing transcription tools (Fig. 2).

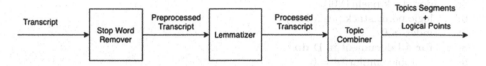

Fig. 2. Logical point detection architecture

3.2 Logical Point Detection

Text preprocessing is done on the generated transcripts before detecting logical points. The first step is to remove stop words. Stop words are articles (a, an, the), verbs (like is, was, were, etc.), pronouns (like he, she, it, they, etc.). Then lemmatization is done in order to remove different forms of the same word.

Next, in order to organize the transcripts (which were earlier generated from the audio) and detect logical points, we first find the topics for each transcript document.

LDA (Latent Dirichlet Allocation) topic modeler was used in order to extract the topics from the transcript documents. LDA is a three-level hierarchical Bayesian model, in which each transcript document is viewed as a mixture of topics. The LDA algorithm maps the topics with the documents such that words in the documents are mostly captured by those topics [7]. It returns a list of topics and their relative importance in the given document.

The extracted topics are used for deciding whether two consecutive paragraphs can be combined together or not. In order to check whether a given paragraph can be combined with its preceding paragraph, common topics among both are searched for. If there are no topics in common, then the paragraphs are not combined. This signals a logical point between the paragraphs.

If there are common topics, then we check the extent of similarity by accumulating the difference in the relative weights assigned to topics in both the paragraphs. If the accumulated difference is greater than zero, then the two paragraphs are taken to be belonging to same topic and combined together.

Otherwise it is considered as a different topic and is taken to be different paragraph. Thus, it is pushed to the stack along with corresponding time stamp as
a logical point. This is summarized by the Topic combiner algorithm described
in Algorithm 1 and Fig. 3.

Algorithm 1 Topic Combiner

1: **function** TOPIC-COMBINER(D, T)
        ▷ // D: list of all pre-processed transcript documents generated
        ▷ // T: list of end timestamps of transcripts
2:     D = LDA(D)   // performs LDA on each transcript documents
3:     doc_stack = [ ]
4:     logic_point_stack = [ ]
5:     doc_stack.push(D[0])
6:     logic_point_stack.push(T[0])
7:     index = 0
8:     **for all** document in D **do**
9:         topic_similarity = 0
10:        **for all** topic in document.topics **do**
11:            **if** topic in doc_stack.top.topics **then**
12:                topic_similarity += (doc_stack.top.topics[topic].weight - docu-
    ment.topics[topic].weight)
13:            **end if**
14:        **end for**
15:        **if** topic_similarity > 0 **then**
16:            doc_stack.top.concat(document)   // concat document to top of stack
17:        **else**
18:            doc_stack.push(document)   // push current document to stack
19:            logic_point_stack.push(T[index])   // push current timestamp to stack
20:        **end if**
21:        index += 1
22:    **end for**
        return logic_point_stack, doc_stack
23: **end function**

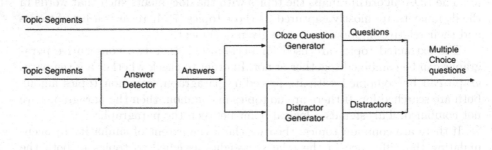

Fig. 3. Question generation architecture

3.3   Question Generation

The question generation consists of three steps:

1. Answer Detection - Given a block of text we first find all the tokens of the document that could potentially be an answer. This is similar to keyword detection. The classification of a token as an answer or not is done using a Naïve Bayes classifier trained on the Stanford Question Answering Dataset [8]. The attributes considered are part-of-speech of the token, whether the token is a named entity, tf-idf of the token, dependency of the token in its abstract syntax tree and shape of the token. Here dependency attribute of a token gives its syntactic dependency on the head token. The shape attribute gives information about capitalization, punctuation, digits in the token. It is in essence a transform on the token's string done in order to learn more about its orthographic features. The transform involves the following mappings –
   a. Lower case alphabetic characters (a-z) are mapped to 'x'.
   b. Upper case alphabetic characters (A-Z) are mapped to 'X'.
   c. Numeric characters are mapped to 'd'.
   d. Post mapping sequences of 5 or more of the same replacement characters are truncated to length 4. For instance, 'Xxxxxxx' becomes 'Xxxxx'.

   As an example, consider the sentence "Clifford is a big red dog.". The attributes of the token "dog" are listed as follows:
   a. Part of speech – Noun
   b. Named entity – False
   c. Dependency – Attribute
   d. Shape – xxx

   Consider the attributes of the token "Clifford" from the same sentence:
   a. Part of speech – Proper Noun
   b. Named entity – True
   c. Dependency – Nominal subject
   d. Shape – Xxxxx

2. Question generation - Given a sentence that contains a word categorized as an "answer", we generate a fill-in-the-blank type question by replacing the occurrences of the answer with a blank. For example, if "dog" was categorized as an answer, a sentence in the input text containing "dog" would be transformed as follows: "Clifford is a big red dog." becomes "Clifford is big red ____ .".

3. Distractor generation - To build an MCQ type question, we have to generate options or distractors. Given an answer to a question, 3 words most similar to it in a relevant vocabulary to use as distractors. This is implemented using word vectors. For example, words similar to "dog" are "cat", "wolf" and "fox". These three distractors would be presented as options along with the correct answers.

Thus, for the sentence "Clifford is a big red dog.", the question generated is: Clifford is big red ____.

a. dog   b. cat   c. wolf   d. fox

Question validation – A cosine similarity check is done between the answer of the question and the topic of the given text (provided in the LDA stage). Questions with the highest similarities are presented to the student and the rest are discarded.

4   Results

With this paper, we could achieve:

1. a. Logical segmentation of lecture videos into topics using LDA.
2. b. Automated generation of questions from the transcripts.
3. c. Generation of distractors to form MCQs.
4. d. Insertion of questionnaires in the lecture at logical points.

In order to know the efficacy of the overall methods proposed, we tested them on two videos. The first test was on a C++ video lecture from NPTEL [9]. 20 students were made to watch the video in a interactive video player with the UI functionality to pause the lecture and display the questionnaires at logical points. The students' responses were collected. The video lecture was 18 min long. 4 logical points were detected. 6 questions were generated by question generator. Some of the questions are as follows:

1. We use 'printf' from the ____ library and print the hello world on to the terminal or which is formally set to with the stdout file.
   a. stdin   b. stderr   c. stdio   d. stdout
2. C strings are actually a collection of ____ in string.h
   a. Functions   b. Objects   c. Class   d. Constructor
3. '212' in ____ will be considered a const int
   a. C98   b. C99   c. C97   d. C96

The statistics of student performance for the first lecture are in Table 1.

Table 1. Student performance statistics for the first video.

Question number	Correctly answered	Incorrectly answered
1	12	8
2	8	12
3	9	11

The second video was on an introduction to literary history from NPTEL [10]. It was a 22 min video. Totally 4 logical points were detected and 10 questions were generated. Some of the questions which were generated for the video are as follows:

1. The hundred years war and the wars of the _____ accordingly had defined the fortunes of the nation.
   - a. roses   b. tulips   c. orchids   d. lilies
2. The Elizabethan period spans over _____ years from the ascension of queen Elizabeth from 1558 to the death of king James 1 1625.
   - a. 67   b. 68   c. 69   d. 66
3. In many different ways England becomes a leader from the time of the reign of queen _____ I.
   - a. Elizabeth   b. Mary   c. Anne   d. Margaret
4. Thomas More's _____ is considered as a significant writing of the times.
   - a. Utopia   b. Utopian   c. Dystopia   d. collectives

The statistics of student performance for the second lecture are in Table 2.

5 Future Work

Currently, the appropriateness of the location of logical points are validated manually. In order to automate the process, a machine learning model can be trained on a dataset created by manual tagging.

Table 2. Student performance statistics for the second video.

Question number	Correctly answered	Incorrectly answered
1	16	4
2	5	15
3	11	9
4	12	8

Logical points are currently detected by using silence and topic clustering. The appropriateness of logical points can be improved by taking into consideration the reason for silence as well. This can be achieved by using Video Analytics.

In order to improve distractor quality, vocabularies relevant to the video topics can be compiled by training GloVe models on data accumulated from online articles specifically related to the video subject [11].

The system currently in place for question validation involves performing a cosine similarity test. A more robust system can be developed by training a machine learning model to validate the questions generated. This would require building a new dataset of manually formulated questions from transcripts and training a machine learning model on the same.

References

1. Attention spans, consumer Insights, Microsoft. http://dl.motamem.org/microsoft-attention-spans-research-report.pdf. Accessed 21 Nov 2019
2. Hake, R.R.: Interactive-engagement versus traditional methods: a six thousand-student survey of mechanics test data for introductory physics courses. Am. J. Phys. **66**, 64 (1998)
3. Hannun, A., Case, C., Casper, J., et al.: DeepSpeech: scaling up end-to-end speech recognition. In: ArXiv e-prints (2014)
4. Blei, D.M., Ng, A.Y., Michael, J.I., Lafferty, J.: Latent Dirichlet allocation. J. Mach. Learn. Res. **3**, 993–1022 (2003)
5. Heilman, M.: Automatic factual question generation from text, 195. Carnegie Mellon University (2011)
6. Aldabe, I., Maritxalar, M.: Automatic distractor generation for domain specific texts. In: Loftsson, H., Rögnvaldsson, E., Helgadóttir, S. (eds.) NLP 2010. LNCS (LNAI), vol. 6233, pp. 27–38. Springer, Heidelberg (2010). https://doi.org/10.1007/978-3-642-14770-8_5
7. Crossno, P.J., Wilson, A.T., Shead, T.M., Dunlavy, D.M.: TopicView: visually comparing topic models of text collections. In: 23rd International Conference on Tools with Artificial Intelligence, Boca Raton, FL, pp. 936–943. IEEE (2011)
8. SQuAD: 100,000+ questions for machine comprehension of text. https://rajpurkar.github.io/SQuAD-explorer. Accessed 21 Nov 2019
9. Module 1: Recap of C (Lecture 01), NPTEL. https://nptel.ac.in/courses/106105151. Accessed 21 Nov 2019
10. Introduction to literary history (Week 1), NPTEL. https://nptel.ac.in/courses/109/106/109106124. Accessed 21 Nov 2019
11. Pennington, J., Socher, R., Manning, R.D.: GloVe: global vectors for word representation. In: Empirical Methods in Natural Language Processing (EMNLP), pp. 1532–1543 (2014)

Effective Emotion Recognition from Partially Occluded Facial Images Using Deep Learning

Smitha Engoor[1] (ID), Sendhilkumar Selvaraju[1] (✉) (ID), Hepsibah Sharon Christopher[2] (ID),
Mahalakshmi Guruvayur Suryanarayanan[2] (ID), and Bhuvaneshwari Ranganathan[2] (ID)

[1] Department of Information Science and Technology, Anna University,
Chennai, Tamil Nadu, India
ssk_pdy@yahoo.co.in
[2] Department of Computer Science and Engineering, Anna University,
Chennai, Tamil Nadu, India

Abstract. Effective expression analysis hugely depends upon the accurate representation of facial features. Proper identification and tracking of different facial muscles irrespective of pose, face shape, illumination, and image resolution is very much essential for serving the purpose. However, extraction and analysis of facial and appearance based features fails with improper face alignment and occlusions. Few existing works on these problems mainly determine the facial regions which contribute towards discrimination of expressions based on the training data. However, in these approaches, the positions and sizes of the facial patches vary according to the training data which inherently makes it difficult to conceive a generic system to serve the purpose. This paper proposes a novel facial landmark detection technique as well as a salient patch based facial expression recognition framework based on ACNN with significant performance at different image resolutions.

Keywords: Emotion recognition · Partial occlusion · Facial features · ACNN

1 Introduction

Facial expression classifiers are successful on analyzing constrained frontal faces. There have been less reports on their performance on partially occluded faces [3]. In this paper, emotion recognition is performed with partially occluded faces by identifying the blocked region in the image and get information from unblocked region or informative region. Facial expressions inherently extend to other parts of the face producing visible patches. If the facial parts are hidden or blocked by hands or spectacles or any other means, symmetric parts of the face or the regions that contribute to the expression shall be involved for facial emotion recognition. Inspired by the intuition, Attention based Convolutional Neural Networks (ACNNs) automatically handles the occluded regions by paying attention to informative facial regions. Every Gate Unit of ACNN associates an importance based weight factor via thorough adaptive learning.

© IFIP International Federation for Information Processing 2020
Published by Springer Nature Switzerland AG 2020
A. Chandrabose et al. (Eds.): ICCIDS 2020, IFIP AICT 578, pp. 213–221, 2020.
https://doi.org/10.1007/978-3-030-63467-4_17

In this work, two versions of ACNN: patch-based ACNN (pACNN) and global–local-based ACNN (gACNN) are deployed. pACNN has a Patch-Gated Unit (PG-Unit) which is used to learn, weigh the patch's local representation by its unobstructedness that is computed from the patch itself. gACNN integrates local and global representations concurrently. A Global-Gated Unit (GG-Unit) is adopted in gACNN to learn and weigh the global representation.

2 Related Work

2.1 Facial Occlusion Method

VGG Net has the image represented as feature maps. ACNN decomposes the feature maps into multiple sub feature maps to obtain local patches. The feature maps are also sent to gg-unit to identify occlusion. The pg-unit and gg-unit are concatenated and softmax loss is used to predict the final output.

Patch based ACNN (pACNN) is decomposed into two schemes: (i) region decomposition (ii) occlusion perception. In region decomposition 69 facial landmarks and selecting 24 points [5] which covers all information. In occlusion perception it deals with pg-unit. In each patch-specific PG-Unit, the cropped local feature maps are fed to two convolution layers without decreasing the spatial resolution, so as to preserve more information when learning region specific patterns. Then, the last set of feature maps are processed in two steps: vector-shaped local features and attention net that estimates the importance based scalar weights. The sigmoid activation of Attention net forces the output as [0,1], where 1 indicates the most salient unobstructed patch and 0 indicates the completely blocked patch

Global-local based ACNN (gACNN) is divide into two schemes: (i) integration with full face region (ii) global-gated unit. In integration with full face region the gACNN takes the whole face region on the one hand, the global-local attention method is used to know the local details and global context cues. The global representation is then weighed by the computed weight. The ACNNs rely on the detected landmarks. It cannot be neglected that facial landmarks will suffer misalignment in the presence of severe occlusions. The existing ACNNs are not sensitive to the landmark misalignment.

2.2 Detecting the Shape of Faces

Regression and Deep regression are proposed in the literature for face detection purposes [7]. Deep regression network aims at characterizing the nonlinear mapping from appearance to shape. For a deep network with m − 1 hidden layers, de-corrupt auto-encoders are used for recovering the occluded faces. Auto-encoder will tackle partial occlusion and de-corrupt auto-encoder will occlude the parts by partitioning the face image x into j components. Therefore, after partitioning the image there will be 68 facial points which has 7 components. The components cover all the information regions. Considering that the face appearance varies under different poses and expressions, it is nontrivial to design one de-corrupt auto-encoder network to reconstruct the details of the whole face.

The third approach, cascade deep regression with de-corrupts auto-encoder, concatenates both deep regression and de-corrupts auto-encoder to get the local patches.

By learning de-corrupt auto-encoder networks and deep regression networks under a cascade structure, they can benefit from each other. On the one hand, with more accurate face shape, the appearance variations within each component becomes more consistent, leading to more compact de-corrupt auto-encoder networks for better de-corrupted face images. On the other hand, the deep regression networks that are robust to occlusions can be attained by leveraging better de-corrupted faces

2.3 Localization of ROI

The eyes and nose localization is detected by Haar cascade algorithm. The Haar classifier returns the vertices of the rectangular area of detected eyes. The eye centers are computed as the mean of these coordinates. Similarly, nose position was also detected using Haar cascades. In case the eyes or nose was not detected using Haar classifiers [2], the system relies on the landmark coordinates detected by anthropometric statistics of face. In summary, Deep Regressive and De-corrupt Auto-encoders does not recover other type of deep architecture for the genuine appearance for occluded parts. In traditional CNN based approaches, registered facial images were handled and partial occlusion is not addressed.

3 Proposed Work: ACNN for Partial Facial Occlusion

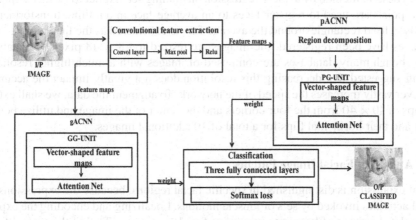

Fig. 1. Working model for ACNN based partial facial occlusion driven emotion detection

In this work, to address the occlusion issue, ACNN endeavors to focus on different regions of the facial image and weighs each region according to its obstructed-ness as well as its contribution to facial emotions. The CNN takes facial image as the input. The image is fed into a convolutional net as feature maps. Then, ACNN decomposes the feature maps of the whole face to multiple sub-feature maps to obtain diverse local patches. Each local patch is encoded as a weighed vector by a Patch-Gated Unit (PG-Unit). A PG-Unit computes the weight of each patch by an Attention Net, considering its obstructedness. The feature maps of the whole face are encoded as a weighed vector by a Global Gated Unit (GG-Unit). The weighed global facial features with local representations are concatenated and serve as a representation of occluded face. Two fully connected layers are followed to classify the facial emotions. ACNNs are optimized by minimizing the softmax loss (refer Fig. 1).

3.1 Convolution and Max-Pooling Layer

The work presented here is inspired by the techniques provided by the GoogLeNet and AlexNet architectures. Our network consists of two traditional CNN modules (a traditional CNN layer consists of a convolution layer and a max pooling layer). Both of these modules use rectified linear units (ReLU) which have an activation function. Using the ReLU activation function [1] allows us to avoid the vanishing gradient problem caused by some other activation functions. Following these modules, we apply the techniques of the network in network architecture and add two "Inception" style modules, which are made up of a $1 \times 1, 3 \times 3$ and 5×5 convolution layers (Using ReLU) in parallel. These layers are then concatenated as output and we use two fully connected layers as the classifying layers.

In yet another approach, bidirectional warping of Active Appearance Model (AAM) and a Supervised Descent Method (SDM) called IntraFace to extract facial landmarks is proposed, however further work could consider improving the landmark recognition in order to extract more accurate faces. IntraFace uses SIFT features for feature mapping and trains a descent method by a linear regression on training set in order to extract 49 points. These points are used to register faces to an average face in an affine transformation. Finally, a fixed rectangle around the average face is considered as the face region. Once the faces have been registered, the images are resized to 48×48 pixels for analysis. Even though many databases are composed of images with a much higher resolution testing suggested that decreasing this resolution does not greatly impact the accuracy, however vastly increases the speed of the network. To augment the data, we shall extract 5 crops of 40×40 from the four corners and the center of the image and utilize both of them and their horizontal flips for a total of 10 additional images.

3.2 Analysing Facial Image Patches

Facial expression is distinguished in specific facial regions, because the expressions are facial activities invoked by sets of muscle motions. Localizing and encoding the expression related parts is of benefit to recognize facial expression. To find the typical facial parts that related to expression, we first extract the patches according to the positions of each subject's facial landmarks. We first detect 68 facial landmark points and then, based on the detected 68 points, select or re-compute 24 points that cover the informative region of the face, including the two eyes, nose, mouth, cheek, and dimple. The selected patches are defined as the regions taking each of the 24 points as the center. It is noteworthy that face alignment method in is robust to occlusions, which is important for precise region decomposition. The patch decomposition operation is conducted on the feature map from convolution layers rather than from the original image. This is because sharing some convolutional operations can decrease the model size and enlarge the receptive fields of subsequent neurons. Based on the $512 \times 28 \times 28$ feature maps as well as the 24 local region centers and get a total of 24 local regions, each with a size of $512 \times 6 \times 6$.

In PG-CNN [4], the idea was to embed the PG-Unit (refer Fig. 1) to automatically percept the blocked facial patch and pay attentions mainly to the unblocked and informative patches. In each patch-specific PG-Unit, the cropped local feature maps are fed to two convolution layers without decreasing the spatial resolution, so as to preserve more information when learning region specific patterns. Then, the last $512 \times 6 \times 6$ feature maps are processed in two branches. The first branch encodes the input feature maps as the vector-shaped local feature. The second branch consist an attention net that estimates a scalar weight to denote the importance of the local patch. The local feature is then weighted by the computed weight. In PG-Unit, each patch is weighted differently according to its occlusion conditions or importance. Through the end-to-end training of overall PG-CNN, PG-Units can automatically learn low weights for occluded parts and high weights for unblocked and discriminative parts.

4 Experimental Results and Discussion

4.1 Dataset

In this work, Facial Expression Dataset with Real Occlusion (FER-RO) is used. The occlusions involved are mostly real-life originating with natural occlusions limited to arising from sunglasses, medical mask, hands or hair (refer Table 1).

Table 1. Total number of images for each category in FER-RO

Dataset	Neutral	Anger	Disgust	Fear	Happy	Sad	Surprise
Total	50	53	51	58	59	66	63
Training	22	17	18	27	25	33	22
Testing	28	36	33	31	34	33	41

4.2 Results

4.2.1 Convolutional Feature Extraction

The input to convolutional layer is of fixed size 224×224 RGB image. The image is passed through a stack of convolutional layers, where the filters were used with a very small receptive field: 3×3 (which is the smallest size to capture the notion of left/right, up/down, center). In one of the configurations, it also utilizes 1×1 convolution filters, which can be seen as a linear transformation of the input channels (followed by non-linearity). The convolution stride is fixed to 1 pixel; the spatial padding of conv. layer input is such that the spatial resolution is preserved after convolution, i.e. the padding is 1-pixel for 3×3 conv. layers. Spatial pooling is carried out by five max-pooling layers, which follow some of the convolutional layers, not all the conv. layers are followed by max-pooling, which is performed over a 2×2 pixel window, with stride 2.

4.2.2 Patch-Based ACNN

After identifying the 68 facial landmark points we select or recompute 24 points that cover the informative regions of the face, including the eyes, nose, mouth, cheeks. Then we extract the patches according to the positions of each subject's facial landmarks. The selection of facial patches follows the procedure below:

- Pick 16 points from the original 68 facial landmarks to cover each subject's eyebrows, eyes, nose, and mouth. The selected points are indexed as 19, 22, 23, 26, 39, 37, 44, 46, 28, 30, 49, 51, 53, 55, 59, 57.
- Add one informative point for each eye and eyebrow. We pick four point pairs around the eyes and eyebrows, then compute midpoint of each point pair as delegation. It is because we conduct patch extraction on convolutional feature maps rather than on the input image, adjacent facial points on facial images will coalesce into a same point on feature maps.

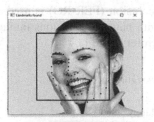

Fig. 2. Facial LandMark detection for images from FFE [6]

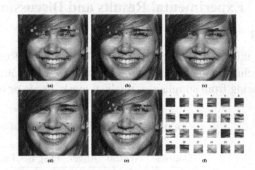

Fig. 3. Patch identification by region decomposition

Based on the $512 \times 28 \times 28$ feature maps as well as the 24 local region centers (refer Fig. 2), we get a total of 24 local regions, each with a size of $512 \times 6 \times 6$. Following this, we embed the Patch-Gated Unit in the pACNN (refer Fig. 3). Under the attention mechanism in the proposed Gate-Unit, each cropped patch is weighed differently according to its occlusion conditions or importance.

4.2.3 Global Local-Based ACNN

gACNN takes global face region into consideration. Global-Local Attention method helps to infer local details and global context cues from image concurrently. On the other hand, gACNN can be viewed as a type of ensemble learning, which seeks to promote diversity among the learned features. We further embed the GG-Unit in gACNN to automatically weigh the global facial representation.

4.2.4 Classification

The output from the convolutional layers represents high-level features in the data. While that output could be flattened and connected to the output layer, adding a fully-connected layer is a deep learning framework. After feature extraction we need to classify the data into various classes, this can be done using a fully connected neural network. The weight of patch-based ACNN and global local-based ACNN is integrated to find the emotion on the given face. The emotions can be classified into seven categories such as: happy, sad, disgust, fear, neutral, anger and surprise. pACNN is capable of discovering local discriminative patches and is much less sensitive to occlusions. Among all the different network structures, pACNN and gACNN are capable of perceiving occlusions and shifting attention from the occluded patches. The proposed approach arrives at a precision of 0.929 for partially occluded faces (refer Fig. 4 and Fig. 5).

Fig. 4. Emotion classification for facial image with partial occlusion

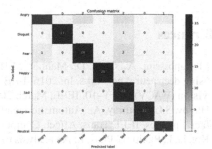

Fig. 5. Confusion matrix

4.2.5 Discussion

Sparse representation classifiers were previously applied for partial facial occlusions [8, 9]. However, Li et al. [5] attempts to apply ACNN for non-occluded and occluded face images. These occluded images attempted by Li et al. [5] are full occlusions. Wang et al. [10] have explored region based attention models for robust expression recognition, however, in this work [10] patches were not involved. Yet another work on the lines of using attention models are discussed in Wang et al. [11]. Here, ACNNs are applied over unconstrained facial expressions but which were shot in the wild. Neither of the above approaches are applied over partial facial occlusions. Retail consumer segment has great potential to leverage into facial expression recognition. Since partial facial occlusion is very natural and common in retail, the proposed work records the usage of ACNN for partial facial occlusions captured with better clarity, which is reportedly first work in this direction. In addition, the proposed work uses less images for training for two reasons: 1. The ACNN model used with patch based approaches has the capability of learning from least inputs 2. The images fed to training were peak images for respective expressions which is enough to govern the handling of partially occluded faces.

5 Conclusion

This paper proposes CNN with attention mechanism (ACNN) for facial expression recognition in the presence of occlusions. The proposed work shall be applied successfully for customer interest detection [12], human behavior analysis [13]. In order to make it more real-time, handling video facial data is essential. For real-time facial emotion recognition with occlusion, the losses have to be very minimal [14]. Further, to maintain a balance [15] between rich and poor facial feature classes, feature augmentation and feature normalization have to be addressed.

Acknowledgements. This Publication is an outcome of the R&D work undertaken in the project under the Visvesvaraya Ph.D. Scheme (Unique Awardee Number: VISPHD-MEITY-2959) of Ministry of Electronics & Information Technology, Government of India, being implemented by Digital India Corporation (formerly Media Lab Asia).

References

1. Mollahosseini, A., Chan, D., Mahoor, M.H.: Going deeper in facial expression recognition using deep neural networks. In: 2016 IEEE Winter Conference on Applications of Computer Vision (WACV), pp. 1–10. IEEE, March 2016
2. Happy, S.L., Routray, A.: Automatic facial expression recognition using features of salient facial patches. IEEE Trans. Affect. Comput. **6**(1), 1–12 (2014)
3. Kotsia, I., Buciu, I., Pitas, I.: An analysis of facial expression recognition under partial facial image occlusion. Image Vis. Comput. **26**(7), 1052–1067 (2008)
4. Li, Y., Zeng, J., Shan, S., Chen, X.: Patch-gated CNN for occlusion-aware facial expression recognition. In: 2018 24th International Conference on Pattern Recognition (ICPR), pp. 2209–2214. IEEE, August 2018
5. Li, Y., Zeng, J., Shan, S., Chen, X.: Occlusion aware facial expression recognition using CNN with attention mechanism. IEEE Trans. Image Process. **28**(5), 2439–2450 (2018)
6. Lyons, M.J., Akamatsu, S., Kamachi, M., Gyoba, J., Budynek, J.: The Japanese female facial expression (JAFFE) database. In: FG, pp. 14–16 (2005)
7. Zhang, J., Kan, M., Shan, S., Chen, X.: Occlusion-free face alignment: deep regression networks coupled with de-corrupt autoencoders. In: Proceedings of the IEEE Conference on Computer Vision and Pattern Recognition, pp. 3428–3437 (2016)
8. S. F. Cotter, Sparse representation for accurate classification of corrupted and occluded facial expressions, Proc. ICASSP, pp. 838–841, Apr. (2010)
9. S. F. Cotter, Weighted voting of sparse representation classifiers for facial expression recognition, Proc. Signal Process. Eur. Conf., pp. 1164–1168, (2010)
10. Wang, K., Peng, X., Yang, J., Meng, D., Qiao, Y.: Region attention networks for pose and occlusion robust facial expression recognition. IEEE Trans. Image Process. **29**, 4057–4069 (2020)
11. Wang, C., et al.: Lossless attention in convolutional networks for facial expression recognition in the wild. arXiv preprint. arXiv:2001.11869 (2020)
12. Yolcu, G., Oztel, I., Kazan, S., Oz, C., Bunyak, F.: Deep learning-based face analysis system for monitoring customer interest. J. Ambient Intell. Hum. Comput. **11**(1), 237–248 (2019). https://doi.org/10.1007/s12652-019-01310-5

13. Sajjad, M., Zahir, S., Ullah, A., Akhtar, Z., Muhammad, K.: Human behavior understanding in big multimedia data using CNN based facial expression recognition. Mob. Netw. Appl. **25**(4), 1611–1621 (2019). https://doi.org/10.1007/s11036-019-01366-9
14. Wei, X., Wang, H., Scotney, B., Wan, H.: Minimum margin loss for deep face recognition. Pattern Recogn. **97**, 107012 (2020)
15. Wang, P., Fei, S., Zhao, Z., Guo, Y., Zhao, Y., Zhuang, B.: Deep class-skewed learning for face recognition. Neurocomputing **363**, 35–45 (2019)

Driveable Area Detection Using Semantic Segmentation Deep Neural Network

P. Subhasree[1]([✉]), P. Karthikeyan[1], and R. Senthilnathan[2]

[1] Division of Mechatronics, Department of Production Technology, MIT Campus,
Anna University, Chennai, India
subhasreesenthilnathan@gmail.com
[2] Department of Mechatronics Engineering, SRM Institute of Science and Technology,
Kattankulathur, India

Abstract. Autonomous vehicles use road images to detect roads, identify lanes, objects around the vehicle and other important pieces of information. This information retrieved from the road data helps in making appropriate driving decisions for autonomous vehicles. Road segmentation is such a technique that segments the road from the image. Many deep learning networks developed for semantic segmentation can be fine-tuned for road segmentation. The paper presents details of the segmentation of the driveable area from the road image using a semantic segmentation network. The semantic segmentation network used segments road into the driveable and alternate area separately. Driveable area and alternately driveable area on a road are semantically different, but it is a difficult computer vision task to differentiate between them since they are similar in texture, color, and other important features. However, due to the development of advanced Deep Convolutional Neural Networks and road datasets, the differentiation was possible. A result achieved in detecting the driveable area using a semantic segmentation network, DeepLab, on the Berkley Deep Drive dataset is reported.

Keywords: Road detection · Road segmentation · Driveable area detection · Semantic segmentation · Autonomous driving

1 Introduction

Advanced Driver Assist Systems (ADAS) and autonomous driving technology have greatly contributed to the explosive growth of the field of deep learning which was fueled by the massive collection and availability of road datasets for public usage. Among the many computer vision tasks involved in ADAS and Autonomous driving, road detection has been considered one of the most important topics of research especially in the scope of level 4 and level 5 of driving automation [1]. The road perception algorithms are the first step for any subsequent path planner which aid in autonomous navigation. Remarkable progress has been achieved in road segmentation in the last decade both in feature-based and learning-based approaches. Automatic driving decisions collision-free on the

© IFIP International Federation for Information Processing 2020
Published by Springer Nature Switzerland AG 2020
A. Chandrabose et al. (Eds.): ICCIDS 2020, IFIP AICT 578, pp. 222–230, 2020.
https://doi.org/10.1007/978-3-030-63467-4_18

segmentation roads from vehicles. Often high-level information about roads such as ego-lane detection [2], object-lane relationships [3], etc. is required for successful cognitive actions ensuring collision-free navigation. With the advent of deep neural networks and publicly available datasets that aid in explicit labelling of road relates features such as drivable area, lane markings, etc. there is a tremendous interest in extending the road detection problem to drivable area detection. In fact, much recent research had focused on the detection of flat areas which are considered to be drivable. Such high-level information allow self-driving cars to act very similarly to human drivers. The paper presents the details of the implementation of driveable area detection using a deep neural network developed for semantic segmentation. Driveable area detection involves precise segmentation of ego-lane i.e. the lane of travel of the vehicle on which the camera is mounted (ego-vehicle). The driveable area is segmented based on the availability of the lane markings. The dataset used for training in the research work presented is Berkeley Deep Drive (BDD) Dataset [4].

2 Related Work

Many researches and works have been going on to solve road segmentation problem since the development of deep neural networks. Even before the recent evolution of deep neural networks, road segmentation has been carried out using other computer vision algorithms, but the features learned by DCNNs were incomparably good. In a work, Road images were segmented into parking lanes, sidewalk and into a road using ariel and ground views of the image, with inputs from Camera, GPS, and IMU systems [5]. Caltagirone L et al., in their work, detected road using DCNN with data from camera and LIDAR [6]. They have applied FCN on a subset of the KITTI dataset for the experiment. In another work, a different approach was carried out to detect roads with minimal amounts of labeled data. The work used semi-supervised learning applied to inputs from camera and LIDAR data combined together. Yang X et al. extracted lane marking and detected road using a combination of Recurrent Neural Network and U-net to reduce the propagation error and improve the accuracy of detection [7]. A model based on the conditional random field, which takes in inputs from both camera and LIDAR as random variables, was experimented by Xiao L [8]. Various research has been carried out for road detection and lane marking. The introduction of the Berkley DeepDrive dataset gave rise to the experimentations on driveable area related research.

Semantic segmentation has evolved since the evolution of CNNs had taken heights. First of such state-of-the-art architectures, was developed by Long J et al. [9]. Using the classification network's learned features, the model adapted the classification network into a fully convolutional network and fine-tuned it for semantic segmentation. Unlike [9], which has done end-to-end learning, the model proposed by Noh H et al., which was built on the classification network VGG-16, integrated deep deconvolution network and proposals prediction [10]. R-CNN model developed by Girshick R et al. applied convolutions on each of the region proposals extracted from the image and classified the regions into labels [11]. This work gave rise to a new branch of research on combining region proposals and convolutions. Convolution and deconvolution on the image during segmentation, due to downsampling and pooling, misses the high-resolution information required for segmentation. In the model developed by Lin G et al., to retain the high-level spatial information, residual connections between the convolution and deconvolution layers were given like identity mapping [12]. An interesting work by Luc P combined the sematic segmentation network with an adversarial network. Using adversarial training, the inconsistencies between ground truth and the predicted map were detected and corrected [13]. Then came the set of models that concentrated on dilated convolutions, that helped in getting the local information along with the global information, by increasing the receptive field of the network [14–17]. One such model is DeepLab which combines deep convolutions, dilated convolutions at different range, and Conditional random fields. This model gained 79.7% mIOU on PASCAL VOC-2012 dataset.

3 Details of Dataset

To find a suitable dataset for the intended purpose of driveable area detection, the datasets available for road data were referred. PASCAL-VOC 2012 dataset [18], Microsoft COCO dataset [19], ADE20K dataset are some of the general semantic segmentation datasets. These datasets contain different objects of vast categories, thus helpful in training the network to identify any object present in the image. For some specific objects of interest to be identified by the network (e.g. road, car, signal, etc. for the autonomous driving scenario), the weights of the network can be fine-tuned by training it with objects specific dataset (e.g. Datasets with road objects). Some road-specific datasets available are Cityscapes dataset [20], KITTI road dataset [21], Apolloscape dataset [22], Oxford RobotCar Dataset [23], Berkley Deep Drive dataset [4]. Among these, the BDD dataset was chosen based on the merit of large scale annotations for driveable area segmentation. This information will be helpful in determining vehicle localization on the road with image data and in turn be helpful in decision making. Though few other datasets have annotations done for road segmentation, these were non-standard and very small in size. BDD dataset is promising in the scope of its attributes such as diverse nature, a sheer large number of annotations of images specifically made for driveable area detection. The dataset has images taken from a variety of geographical locations, environmental and weather conditions. The dataset has labels, "Directly Drivable Area", Alternatively Drivable Area". "Directly Drivable Area" is the area that the driver is currently driving on. The driver has priority of this area over the other cars. "Alternately Drivable Area" is the area that the driver is not currently driving on, but can drive on it by changing lane. The dataset contains 91626 instances of directly drivable area and 88392 instances of Alternatively drivable area. The annotation is given as a mask image which contains pixel level labels for drivable area, alternative area and the background. The dataset is also annotated for lane markings, object detection and instance segmentation. For all the work reported in this paper, the BDD dataset images with annotations for driveable area detection alone are utilized. The driveable area annotations consist of a pair of images for each example. One image is the raw RGB image of resolution 1280 × 720 pixels and the other image is the annotation image of same size which is also an RGB image. The annotation image has the ego-lane pixels in red, other lanes marked in blue and all other regions of the image in black. A sample image pair is shown in Fig. 1.

Fig. 1. A sample example of annotation for driveable area detection in BDD dataset (Color figure online)

4 Deep Learning Network

The work presented in the paper is about the implementation details of applying a popular sematic segmentation network for the purpose for driveable area detection. Though conceptually driveable map inferencing is similar to the general semantic segmentation problem, the challenges involved in the former is very high. This is mainly because the classification is between two regions which are exactly identical in terms of all visual properties. The only differentiating element is the lane marking which is assumed to be of contrasting the color of the driveable area (roads). Deep convolutional neural networks have been the choice for various classification tasks in the computer vision system mainly attributed to the built-in invariance to local image transformations. But this property doesn't meet the requirements of dense inferences such as semantic segmentation where abstraction of information in the image is not a requirement. Among the Deep CNN based architectures available for semantic segmentation in the current work reported in the paper, DeepLab [24] is identified to be suitable for the intended purpose of driveable area detection. This is mainly because of its key property such as the good resolution of the features, good ability to tackle the problem of objects' existence at multiple scales and high localization accuracy which, in a deep CNN framework is difficult to obtain due to the invariance property. DeepLab enjoys these merits by strategically incorporating significant changes in the architecture. Some of the important changes introduced in the DeepLab which not only has benefited the semantic segmentation problem but also to a higher level of complexities such as the driveable area detection problem as follows:

- Atrous convolution instead of convolution with downsampled filters: Generally CNNs would have a series of convolutional layers interleaved with pooling layer resulting in downsampling of filters. Atrous convolution does convolution after the last few pooling layers with upsampled filters instead of downsampled filters as in the case of regular convolution. This strategy basically improves the feature resolution which is very valuable in pixel-level inferences as in the case of driveable area detection. It may be noticed that the atrous convolution serves as a valuable replacement for the deconvolutional layers which is commonly found in most semantic segmentation networks [25].
- Atrous spatial pyramid pooling: To address the problem of scale invariance, generally images of objects acquired at many scales are used. DeepLab utilizes a pyramid

pooling strategy which is a computationally efficient way of ensuring scale invariance. The strategy involves the usage of multiple filters that have complementary fields of view. This ensures capturing of objects as well the context in multiple scales resulting in the scale invariance property.

- Fully-connected CRFs: The property of accurate localization is incorporated through the use of fully connected pairwise conditional random fields.

A high-level block diagram of the network architecture is shown in Fig. 2.

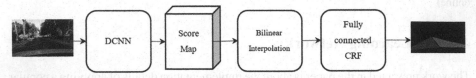

Fig. 2. Abstract block diagram of DeepLab network

Beyond the specified advantages that DeepLab specifically addresses some important advantages that are relevant to the driveable detection problem from a practical standpoint are speed, accuracy, and simplicity of the network. Since the intended application is for a driving scenario real-time nature, accuracy and speed are primary concerns for any segment of algorithms. The detailed block diagram of the DeepLab network is presented in Fig. 3.

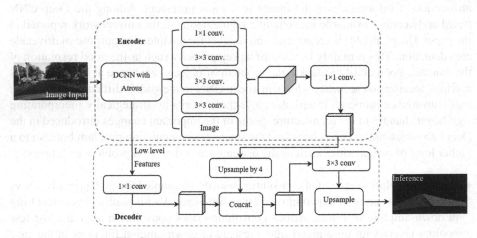

Fig. 3. Detailed block diagram of DeepLab network

The implementation of DeepLab network for the driveable area detection is carried out with Tensorflow API. Tensorflow provides pre-trained models of DeepLab which are meant for semantic image segmentation of a wide variety of classes. The network is pre-trained on PASCAL-VOC 2012 dataset. The weights of the pre-trained model which are available in a frozen inference graph is used before the training is initiated on the

BDD driveable dataset. All the frozen inference graphs by default use an output stride of 8. The backbone of the DeepLab utilized in this work is Xception_65 [26]. The block diagram of a single block of a DeepLab network is presented in Fig. 4. This network is pre-trained on datasets such as ImageNet, MS-COCO part for VOC. As mentioned earlier all the training had been only for generic semantic image segmentation. To ensure robustness the network is initialized with pre-trained weights from PASCAL-VOC. The training is carried out in a phased manner to understand the convergence of the model The key specifications of the training are listed in Table 1.

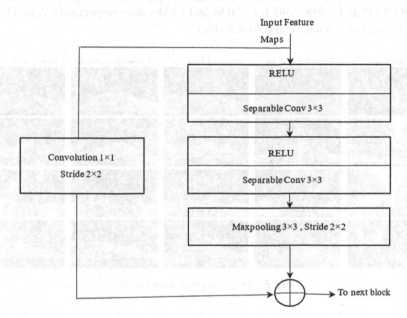

Fig. 4. Xception-65 block at entry flow

Table 1. Key training specifications

Parameter	Specification
Number of iterations	2000
Atrous rates	6, 12, 18
Output stride	16
Decoder output stride	4
Train crop size	513
Train batch size	4
Fine tuning of batch normalization	Yes

5 Results and Conclusion

In the lines of most semantic segmentation benchmarking, mIoU (mean intersection of union) is used as the evaluation metric to understand the performance of the network. Other evaluation metrics are not presented within the scope of this paper. Sample images from the validation set and the corresponding inference images, containing three classes viz., driveable area, alternately drivable area, background, are presented in Fig. 5. The intersection of union is measured between the inference images and the ground truth in the dataset [19]. IoU for the first 48 images are presented in Fig. 6. The mean IoU are 0.9379145 and 0.93885483 for 3 lakhs and 4 lakhs steps respectively. A step here is (Total number of training images/Batch size).

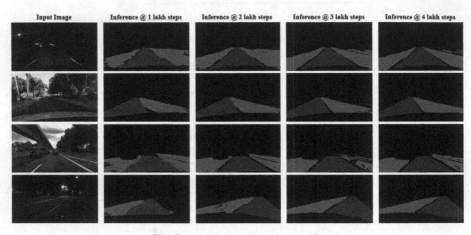

Fig. 5. Sample segmentation results

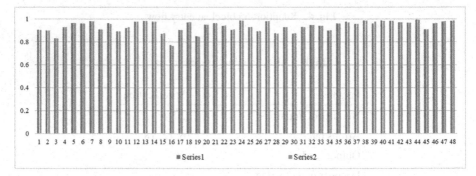

Fig. 6. IoU for first 48 images from validaion set

The IoU for driveable area detected using DeepLab semantic segmentation network is given in the paper. As part of the future work, driveable area detection would be carried out on state-of-the-art semantic segmentation networks and compared. And extraction of high level semantics useful for autonomous driving would be carried out using the detected driveable area.

References

1. Levinson, J., et al.: Towards fully autonomous driving: systems and algorithms. In: IEEE Intelligent Vehicles Symposium (IV). IEEE (2011)
2. Kim, J., Park, C.: End-to-end ego lane estimation based on sequential transfer learning for self-driving cars. In: Proceedings of the IEEE Conference on Computer Vision and Pattern Recognition Workshops (2017)
3. Ohn-Bar, E., Trivedi, M.M.: Are all objects equal? Deep spatio-temporal importance prediction in driving videos. Pattern Recogn. **64**, 425–436 (2017)
4. Yu, F., et al.: BDD100K- a diverse driving video database with scalable annotation tooling. arXiv (2018)
5. Máttyus, G., Wang, S., Fidler, S., Urtasun, R.: HD maps: fine-grained road segmentation by parsing ground and aerial images. In: Proceedings of the IEEE Conference on Computer Vision and Pattern Recognition, pp. 3611–3619 (2016)
6. Caltagirone, L., Bellone, M., Svensson, L., Wahde, M.: LIDAR–camera fusion for road detection using fully convolutional neural networks. Robot. Auton. Syst. **111**, 25–31 (2019)
7. Yang, X., Li, X., Ye, Y., Lau, R.Y., Zhang, X., Huang, X.: Road detection and centerline extraction via deep recurrent convolutional neural network U-Net. IEEE Trans. Geosci. Remote Sens. **57**(9), 7209–7220 (2019)
8. Xiao, L., Wang, R., Dai, B., Fang, Y., Liu, D., Wu, T.: Hybrid conditional random field based camera-LIDAR fusion for road detection. Inf. Sci. **432**, 543–558 (2018)
9. Long, J., Shelhamer, E., Darrell, T.: Fully convolutional networks for semantic segmentation. In: Proceedings of the IEEE Conference on Computer Vision and Pattern Recognition, pp. 3431–3440 (2015)
10. Noh, H., Hong, S., Han, B.: Learning deconvolution network for semantic segmentation. In: Proceedings of the IEEE International Conference on Computer Vision, 1520–1528 (2015)
11. Girshick, R., Donahue, J., Darrell, T., Malik, J.: Rich feature hierarchies for accurate object detection and semantic segmentation. In: Proceedings of the IEEE Conference on Computer Vision And Pattern Recognition, 580–587 (2014)
12. Lin, G., Milan, A., Shen, C., Reid, I.: RefineNet - multi-path refinement networks for high-resolution semantic segmentation. In: Proceedings of the IEEE Conference on Computer Vision and Pattern Recognition, 1925–1934 (2017)
13. Luc, P., Couprie, C., Chintala, S., Verbeek, J.: Semantic segmentation using adversarial networks. arXiv preprint arXiv:1611.08408 (2016)
14. Wang, P., et al.: Understanding convolution for semantic segmentation. In: IEEE Winter Conference on Applications of Computer Vision, pp. 1451–1460 (2018)
15. Yu, F., Koltun, V.: Multi-scale context aggregation by dilated convolutions. arXiv preprint arXiv:1511.07122 (2015)
16. Wei, Y., Xiao, H., Shi, H., Jie, Z., Feng, J., Huang, T.S.: Revisiting dilated convolution: a simple approach for weakly-and semi-supervised semantic segmentation. In: Proceedings of the IEEE Conference on Computer Vision and Pattern Recognition, pp. 7268–7277. (2018)
17. Zhao, H., Qi, X., Shen, X., Shi, J., Jia, J.: ICNET for real-time semantic segmentation on high-resolution images. In: Proceedings of the European Conference on Computer Vision, pp. 405–420 (2018)
18. Everingham, M., Eslami, S.A., Van Gool, L., Williams, C.K., Winn, J., Zisserman, A.: The Pascal visual object classes challenge: a retrospective. Int. J. Comput. Vis. **111**(1), 98–136 (2015). https://doi.org/10.1007/s11263-014-0733-5
19. Lin, T.-Y., et al.: Microsoft COCO: common objects in context. In: Fleet, D., Pajdla, T., Schiele, B., Tuytelaars, T. (eds.) ECCV 2014. LNCS, vol. 8693, pp. 740–755. Springer, Cham (2014). https://doi.org/10.1007/978-3-319-10602-1_48

20. Cordts, M., et al.: The cityscapes dataset for semantic urban scene understanding. In: Proceedings of the IEEE Conference on Computer Vision and Pattern Recognition, pp. 3213–3223 (2016)

21. Geiger, A., Lenz, P., Stiller, C., Urtasun, R.: Vision meets robotics: the KITTI dataset. Int. J. Robot. Res. 32(11), 1231–1237 (2013)

22. Huang, X., Wang, P., Cheng, X., Zhou, D., Geng, Q, Yang, R.: The apolloscape open dataset for autonomous driving and its application. arXiv preprint arXiv:1803.06184 (2018)

23. Maddern, W., et al.: 1 year, 1000 km - the Oxford robot car dataset. Int. J. Robot. Res. 36(1), 3–15 (2017)

24. Chen, L.-C., et al.: DeepLab: semantic image segmentation with deep convolutional nets, atrous convolution, and fully connected CRFs. IEEE Trans. Pattern Anal. Mach. Intell. 40(4), 834–848 (2018)

25. Chen, L.C., Papandreou, G., Schroff, F., Adam, H.: Rethinking atrous convolution for semantic image segmentation. arXiv preprint arXiv:1706.05587 (2017)

26. Chollet, F.: Xception: deep learning with depthwise separable convolutions. arXiv preprint arXiv:1610–02357 (2017)

Data Science

Data Science

WS-SM: Web Services - Secured Messaging Framework with Pluggable APIs

Kanchana Rajaram(✉) and Chitra Babu

Department of Computer Science and Engineering, Sri Sivasubramaniya Nadar College of Engineering, Anna University, Chennai 603110, Tamil Nadu, India
{rkanch,chitra}@ssn.edu.in

Abstract. Dynamic composition of web services is important in B2B applications where user requirements and business policies change and new services get added to the service registry frequently. In a dynamic composition environment, ensuring the security of messages communicated among the web services becomes challenging since, several attacks are possible on SOAP messages in the public network due to their standardized interfaces. Most of the existing works on web services security provide solutions to ensure basic security features such as confidentiality, integrity, authentication, authorization, and non-repudiation. Few existing works that provide solutions such as schema validation and schema hardening for attacks on web services do not provide attack-specific solutions. The web services security standard and all the existing works have addressed only the security of messages between a client and a single web service but not the security for messages between two services which is quite challenging. Hence, a security framework for secured messaging among web services has been proposed to provide attack-specific solutions. Since new types of web service attacks are evolving over time, the proposed security solutions are implemented as APIs that are pluggable in any server where the web service is deployed. The proposed framework has been tested for compliance with WSI-BP to demonstrate its interoperability and subjected to vulnerability testing which proved its immunity to attacks. The stress testing results revealed that the throughput decreased only by 35% achieving a good trade-off between performance and security.

Keywords: Web services · Composition · SOAP messages · Security · Threats

1 Introduction

Service Oriented Architecture (SOA) [1] represents an open, agile and composable architecture comprised of autonomous, interoperable and potentially reusable services, implemented as web services [2]. The use of web services implies

© IFIP International Federation for Information Processing 2020
Published by Springer Nature Switzerland AG 2020
A. Chandrabose et al. (Eds.): ICCIDS 2020, IFIP AICT 578, pp. 233–247, 2020.
https://doi.org/10.1007/978-3-030-63467-4_19

crossing of trust boundaries and the involvement of software with uncertain reliability that leads to mitigation of risks. The security mechanisms that aim to mitigate risks are applied at four different levels in an SOA, namely user, message, service, and transport. At the service level, the objective is to ensure the availability and correct functioning of a service. Due to standardized interfaces, web services are prone to security attacks, since attackers know more about the format of these interfaces. At the transport level, the goal is to guarantee a seamless and reliable communication between parties.

The web services communicate using SOAP protocol. In General, SOAP message is transmitted over HTTP [3], which can flow freely through a firewall and it cannot protect SOAP messages that transmit in application layer. Secure Socket Layer (SSL)/Transport Layer Security (TLS) [4] is inadequate for protecting SOAP messages since it is designed to operate between two endpoints. Most of the times, SOAP messages might be processed by SOAP intermediaries. If the SOAP intermediaries are compromised, the security provided by SSL/TLS becomes insufficient to ensure the end-to-end integrity and confidentiality of SOAP messages. Hence, SOAP message communication in the application layer is prone to attacks.

Services can be composed in two ways viz., statically or dynamically. In a dynamic composition scenario, the concrete workflow is created during run-time while the abstract workflow is created during design-time. In such scenario, the web service is discovered and composed at run-time and SOAP messages are exchanged between web services. When these messages travel through a public network, they become vulnerable to attacks and hence, it is essential to protect these messages.

Lemos et al. [5] surveyed a variety of techniques and tools for web service compositions as well as provided a systematic analysis of the most representative service composition approaches by evaluating and classifying them against the proposed taxonomy. Mouli et al. [6] presented a systematic review on the studies of web service security and observed that the solutions were mainly proposed using dynamic analysis, closely followed by static analysis. Masood et al. [7] review techniques and tools to improve services security by detecting vulnerabilities and discuss the potential static code analysis techniques to discover these vulnerabilities. Singhal et al. [4] describe various web service threats and the basic security standards that provide solution to few of the web service attacks. The various attacks possible on SOAP messages are tabulated in Table 1. Jensen et al. [8] surveyed the vulnerabilities in the context of web services. Their methods provide solutions based on accessing WSDL of the domain service for each attack which is a time-consuming process.

In a dynamic composition environment, a composer service in the middleware discovers and invokes services based on the user requirements specified at run-time. The composer communicates with client programs as well as domain web services. It is quite challenging to provide security for the messages exchanged between the composer and domain services since, the composer handles multiple communication channels. None of the existing security standards and solutions

Table 1. Attacks on SOAP message communication among web services

Attack	Description
SOAPAction Spoofing[a]	Changes the operation in SOAPAction header that leads to execution of unintended operation
WS-Addressing Spoofing[b]	Changes the address of addressing field which leads to flooding of web service[c]
Replay of Messages[d]	Resending the message to the same web service
Message Alteration Attack	Modifying the content of SOAP message
Loss of Confidentiality	Discloses the information to the unauthorised person
XML Injection Attack	Changes the structure of SOAP message
Principal Spoofing	Changes the credentials of SOAP message
Forged Claims	Construct SOAP Message using false credentials
Falsified Messages	Sends fictitious message to receiver

[a]SOAPAction Spoofing, (2017), http://www.ws-attacks.org/SOAPAction_Spoofing
[b]WS-Addressing Spoofing, (2015), https://www.ws-attacks.org/WS-Addressing_sp oofing
[c]Web Services Addressing, (2004), http://www.w3.org/Submission/ws-addressing/
[d]Replay of Messages Attack, (2010) http://msdn.microsoft.com/en-us/library/ff649 371.aspx

address security of messages exchanged between web services during dynamic composition and execution of services. Hence, a novel **Web Services - Secured Messaging (WS-SM)** framework is proposed in this paper, with pluggable APIs for attack-specific security solutions in a dynamic composition scenario.

The rest of the paper is organized as follows, Sect. 2 describes existing works on web service security. Section 3 explains the architecture and methodology of the proposed system. Testing of the proposed solution and the results are discussed in Sect. 4 and Sect. 5 concludes the paper.

2 Existing Work

The defacto standard for providing security on SOAP message communication, WS-Security [9] provides basic security features such as confidentiality, integrity, authentication, authorization, and non-repudiation. Moreover, it detects the presence of a few web service attacks on the communicated SOAP message. However, it does not provide solutions to overcome any of these attacks and does not address some of the attacks like SOAPAction Spoofing, WS-Addressing Spoofing, etc. Alotaibi [10] implemented a secure Web Service using WS-Security specifications such as Signature, Timestamp and Username Token. Thelin et al. [11] proposed a security framework to achieve end-to-end propagation of security credentials throughout the SOAP processing stack.

Hua Yue et al. [12] discussed about the security issues of Web based services on heterogeneous platforms. This approach uses WS-Security to provide security solution with asymmetric cryptography algorithms. Since it is provided as a wrapper in Axis2 platform, it is difficult to provide the solution for upcoming attacks without changing the functionality of rampart.

Layer 7 SecureSpan and CloudSpan[1] provide APIs against attacks that occur in SOA. Layer 7 XML Gateway implements WS-* standards to ensure integrity between the transactions. Layer 7 SecureSpan XML Firewall provide developers the ability to define and enforce security policy through a simple graphical policy language. The web services Security Programming Application Programming Interfaces (WSS API)[2] is used for securing the SOAP messages. However, the functionality of the application needs to be changed for providing security. All the above solutions consider SOAP message communication between a client program and a web service.

Kishore Kumar et al. [13] proposed an API based solution that protects the SOAP messages that are communicated between web services from Message Alteration Attack (MAA). However, this solution is not generic and pluggable.

3   WS-SM Framework

In a dynamic composition environment, user requirements from clients are submitted to the composer which dynamically selects and composes services according to user requirements. While invoking domain web services, the composer sends back SOAP request message and in turn the domain web services send SOAP response message that contains the output or the fault information. During the exchange of these messages via a public network, they are prone to attacks. The security gateway adds (removes) basic security functionality to (from) SOAP request (response) messages irrespective of the attack type. The secured SOAP message is intercepted by different APIs to protect the message from the particular web service attack. When a specific attack occurs, the respective API in the composer or in the domain server side detects it and overcomes. Certain web service attacks are prevented by the corresponding APIs installed on both sides. The proposed framework is depicted in Fig. 1. The design of security gateway is detailed in the next subsection and the design of proposed APIs are described in the subsequent subsections.

3.1   Security Gateway

The security gateway implemented as a web service is deployed in composer as well as in the domain server side. The security gateway in composer side remembers a copy of the SOAP request and redirects it to the security gateway

---

[1] Layer7 Technology, (2013), http://www.layer7tech.com/solutions/web-api-attack-protection.

[2] IBM WSSAPI, (2014), https://www.ibm.com/support/knowledgecenter/SSEQTP_8.5.5/com.ibm.websphere.base.doc/ae/cwbs_wss_api.html.

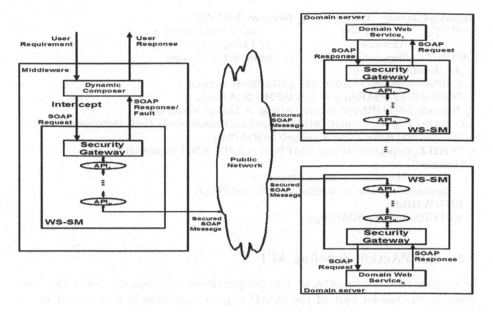

Fig. 1. Architecture of WS-SM framework

in the domain server. The security gateway in domain server side receives the SOAP request and invokes the domain service. The response from domain service is redirected by the gateway to its counterpart in the composer side after remembering a copy. The security gateway in the composer side forwards the response to the composer. In case of an attack on the communicated message in the network, the receiving gateway generates a fault message and in response, the sender gateway sends the remembered copy of the original message. If the communication channel is subject to an attack repeatedly, resending the copy of the original message several times would incur increased overhead. Hence, a *threshold* for maximum number of times the message copy can be re-transmitted is designed and it is made configurable. After the message is resent a specific number of times equal to the pre-defined *threshold*, the security gateway notifies the composer. Based on this, the composer selects an alternative service with an alternative communication channel so that the attack can potentially be avoided. The design of security gateway service is described in Algorithm 1.

SecurityGateway (InSOAPMsg, ResponseSOAPMsg)
INPUT: InSOAPMsg   // Input SOAP message
              Threshold   // Configured by WS-SM
OUTPUT: ResponseSOAPMsg   // Output SOAP message
1. NumberofResends = 0
2. EndPointReference ← InSOAPMsg.SOAPBody.endpoint
3. OutSOAPMsg.SOAPBody ← InSOAPMsg.SOAPBody
// Redirect *InSOAPMsg* to Service Gateway in Domain Server side
4. OutSOAPMsg.To ← concat(GetIPAddress(EndPointReference), GetPath(SecurityGateway))
5. ResponseSOAPMsg ← Sendmessage(OutSOAPMsg)
6. **WHILE**(ResponseSOAPMsg.SOAPFault = TRUE **AND** NumberofResends
<Threshold)
    NumberofResends = NumberofResends + 1
    ResponseSOAPMsg ← Sendmessage(OutSOAPMsg)
  ENDWHILE
7. RETURN(ResponseSOAPMsg)

3.2 SOAPAction Spoofing API

In a public network, an attacker can tamper the SOAP request, change the operation in the header part of the SOAP request and allow it to transmit in the same communication channel. In effect, an unintended wrong operation would be invoked. In order to avoid this SOAPAction Spoofing attack, the SOAPAction Spoofing API (SAS API) is proposed as a part of WS-SM framework. It is installed in the domain server end. It intercepts the SOAP request and compares the content in the header field and the operation in the SOAP Body. If they are not the same, it responds with a SOAP fault message. On receiving the fault message, the security gateway service at the composer end resends a copy of the request. If the attacker hacks the communication channel more than the *threshold* number of times that is pre-configured, the composer will be notified regarding the attack. The message communication sequence is depicted in Fig. 2.

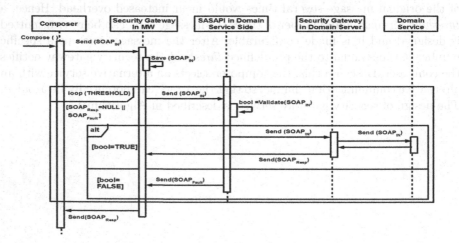

Fig. 2. Securing message communication from SAS attack

3.3 WS-Addressing Spoofing API

An attacker can tamper the SOAP message in transit and add an invalid address in *ReplyTo, FaultTo* or *To* fields which causes redirection of SOAP message to an unintended receiver. This WS-Addressing Spoofing attack is addressed by proposing WS-Addressing Spoofing API (WSAS API) in WS-SM framework. The WSAS API is installed in domain server end to intercept the SOAP request and validate the *RelyTo/FaultTo/To* fields against the *whitelist,* a list of valid URLs registered with the domain service that can communicate with it. If the address is not present in the *whitelist,* the WSAS API sends a fault response to the composer as shown in Fig. 3.

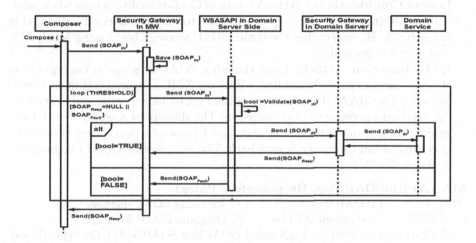

Fig. 3. Securing message communication from WSAS attack

3.4 Replay of Messages API

Every SOAP message has a unique *MessageId* to differentiate it from another message. An attacker can tamper this *MessageId* and send the same message several times as if it is a request from a valid client. The proposed Replay of Messages API (RoM API) intercepts and encrypts the *MessageId* field before sending the request through the network so that, the attacker cannot change it. The RoM API at the receiver end decrypts the *MessageId* value of the received message and validates with *MessageIdList* file that contains all the *MessageIds* of previously executed SOAP requests. If the *MessageId* is not present in the list, then the RoM API allows the request to invoke the domain service. Otherwise, it drops the request.

3.5   Message Alteration Attack API

The Message Alteration Attack (MAA) API is proposed in WS-SM framework to ensure the confidentiality of SOAP messages. This API intercepts the outgoing SOAP message from security gateway and encrypts[3] its entire body so that the attack is prevented. If the attacker changes the encrypted content of the message during transit, decryption of the SOAP message by MAA API in the receiver end will result in an error. Due to this, the request will not be executed in the domain server. To detect this attack, the MAA API at the domain server end intercepts the request and decrypts the content of the incoming request. If this decrypted content is some junk value, then it is sent back to the security gateway to resend the original message. The MAA API also addresses the following attacks.

- **Loss of Confidentiality Attack:** Loss of Confidentiality occurs when information within a SOAP Message is disclosed to an unauthorized individual in the network. By encrypting the entire SOAP message body using MAA API, this attack is prevented.
- **XML Injection Attack:** Even though a SOAP message is encrypted, an attacker can tamper it by adding additional XML tags or deleting encrypted contents. The MAA API can detect this attack by intercepting and decrypting the received encrypted SOAP message. The decrypted message is validated against the schema of the SOAP message. In case of a decryption or a validation error, a fault message is sent back. The design of MAA API is described in Algorithm 2.

MAAAPI(InSOAPMsg, ResponseSOAPMsg)
INPUT: InSOAPMsg   // Incoming SOAP message
OUTPUT: ResponseSOAPMsg   // Outgoing SOAP Message
1. **IF** Equals(getElementsByTagName(InSOAPMsg.SOAPBody), EncryptedData)
THEN
       ResponseSOAPMsg ← Decrypt(InSOAPMsg.SOAPBody)
       IF DECRYPTERROR **OR NOT** Validate(ResponseSOAPMsg, SOAP-Schema) **THEN**
                THROW SOAPFault   // If MAA is detected, a fault is thrown
       ELSE   ResponseSOAPMsg ← Encrypt(InSOAPMsg.SOAPBody)
2. **RETURN** (ResponseSOAPMsg)

3.6   Principal Spoofing Attack API

Principal Spoofing Attack (PSA)[4] occurs when an attacker hacks the SOAP message and sends it as if it is sent by another authorized client, by stealing its credentials. The attack results in the execution of an unintended service. In order to avoid this attack, the proposed PSA API intercepts the message and adds the *username* token containing *UserName*, *Password*, *Created* and *Nonce*

---

[3] Encryption Algorithm, (2014), https://www.princeton.edu/~ota/disk2/1987/8706/870612.PDF.

[4] Principal Spoofing, (2014), https://capec.mitre.org/data/definitions/195.html.

tags. Then, a digital signature[5] is added to the message and is sent through the network. The PSA API in the other end verifies the incoming SOAP message and sends a fault message in case it is tampered during transit. The PSA API also addresses the following attacks.

- Forged Claims Attack
- Falsified Messages Attack

Forged Claims Attack. This type of attack is a variation of PSA where an attacker constructs a new SOAP message using credentials of a different authorized client. The attacker sends the SOAP message as if it is sent by an authorized client. The PSA API prevents this attack by creating the digest of the password of an identity. The password digest is calculated using plain password, nonce and created time. It is not possible for the attacker to obtain the password from the password-digest and hence the credentials.

Falsified Messages Attack. This type of attack is also another variation of PSA where a fictitious message is created by an attacker with the same credentials so that the receiver believes that the SOAP message comes from the original authorized client. The PSA API prevents this attack by making digest of sender password so that the attacker will not be able to extract password of a valid sender. Thus, the PSA API prevents principal spoofing attack. The design of PSA API is described by Algorithm 3.

PSAAPI(InSOAPMsg, ResponseSOAPMsg)
INPUT: InSOAPMsg   //Incoming SOAP message
OUTPUT: ResponseSOAPMsg   //Outgoing SOAP Message
//Check if header of InSOAPMsg contains UserNameToken
1. **IF** $InSOAPMsg.SOAPHeader.getChildrenWithLocalName() \notin UsernameToken$
THEN
    InSOAPMsg.SOAPHeader.addChild(UserName)
    InSOAPMsg.SOAPHeader.addChild(Password)
    InSOAPMsg.SOAPHeader.addChild(Nonce)
    InSOAPMsg.SOAPHeader.addChild(CreatedTime)
    InSOAPMsg.SOAPHeader.addChild(Signature) //Add Digital Signature to
InSOAPMsg
    ResponseSOAPMsg ← InSOAPMsg
  ELSE   Validate PasswordDigest and Digital Signature
    IF NOT(Validate(InSOAPMsg.SOAPHeader.UserNameToken) **AND**
          Validate(InSOAPMsg.SOAPHeader.Signature)) **THEN**
      THROW SOAPFault //Principal Spoofing is detected
2. **RETURN** (ResponseSOAPMsg)

---

[5] XML Signature, (2013), http://www.xml.com/pub/a/2001/08/08/xmldsig.html.

4   Testing and Discussion

The proposed WS-SM Framework has been tested with the following three objectives:

- To test the compliance of WS-SM framework for interoperability to prove suitability of WS-SM for SOA based applications.
- To test the immunity of WS-SM approach against web service attacks on communication among web services
- To analyze the trade-off between providing security using WS-SM approach and degradation of application performance.

4.1   Compliance Testing

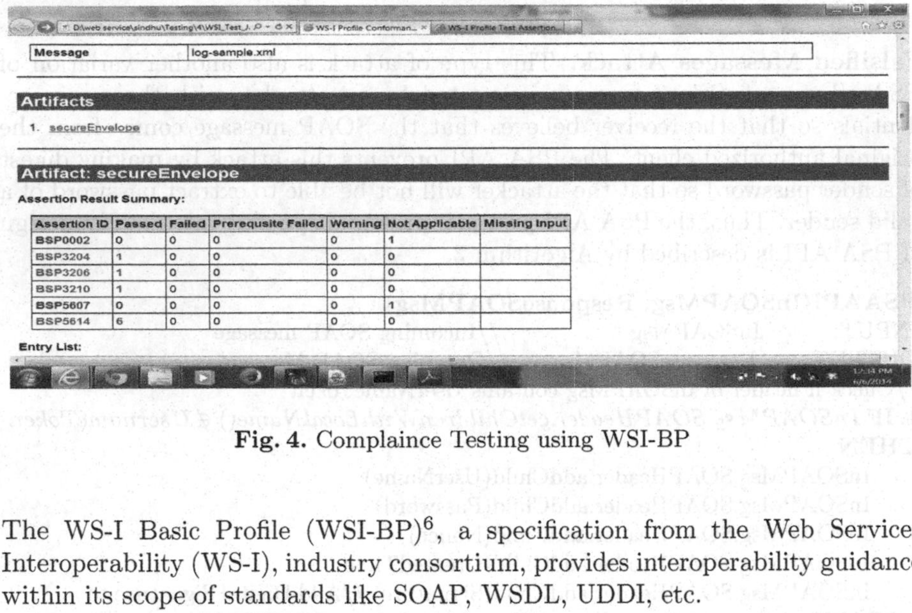

Fig. 4. Complaince Testing using WSI-BP

The WS-I Basic Profile (WSI-BP)[6], a specification from the Web Services Interoperability (WS-I), industry consortium, provides interoperability guidance within its scope of standards like SOAP, WSDL, UDDI, etc.

Conformance to WSI-BP is defined by adherence to the set of requirements for a specific target, within the scope of WSI-BP. Requirements state the criteria for conformance to WSI-BP and consist of refinements, interpretations and clarifications that improve interoperability. Targets allow for the description of conformance in different contexts, to allow conformance testing and certification of artifacts such as SOAP messages and WSDL descriptions, web service instances, and web service consumers. WSI-BP makes requirement statements about artifacts such as SOAP messages, WSDL Descriptions, and registry elements. A web service is allowed to advertise conformance to WSI-BP by annotating these

---

[6] WS-I Basic Security Profile, (2007), http://ws-i.org/Profiles/BasicProfile-2.0-2010-11-09.html.

artifacts with conformance claims, which use a URI to assert conformance with a particular profile. An instance of an artifact is considered conformant when all of the requirements associated with it are met.

The proposed WS-SM Framework was adapted to an example case study of banking application where dynamic composition of services is involved. The compliance of WS-SM framework to WSI-BP has been checked using an existing tool[7] by submitting the SOAP messages communicated in the network as input. The snapshot of the conformance report from the tool is depicted in Fig. 4. It indicates that the assertions related to the artifact of SOAP messages have been passed. Hence, it is established that WS-SM Framework is compliant with WSI-BP and thus interoperable.

4.2 Vulnerability Testing

Most of the existing methodologies and tools for vulnerability testing either do not work properly, are poorly designed, or do not fully test for real world web service vulnerabilities[8]. Typically, web application penetration tests are not scoped properly to include the related web services. Testing methodologies should include not only technical details on how to test web services, but also non-technical information such as proper scoping as well as pre-engagement requirements, which often are overlooked by penetration testers. In addition, current methodologies lack information on a complete threat model for web services. Depending on the data being exchanged, threats need to be carefully identified. SOAPUI [14], WS-Attacker[9] [15] and WSBang[10] are some of the testing tools used to test the functionality and few of the security features like SQL Injection, Cross Site scripting, etc. However, these tools are used for testing the attacks against a single web service [16] and they do not work in a dynamic composition environment. Hence, a specific penetration testing tool has been designed based on ATLIST (Attentive Listener) [17] methodology. ATLIST is a vulnerability analysis method developed during and for the analysis of SOA service orchestrations. It facilitates the detection of known vulnerability types and enables derivation of vulnerability patterns for tool support. ATLIST is applicable to business processes composed of services as well as single services. ATLIST offers better transferability by guiding the analysis with a set of analysis elements such as Point of View, Attack Effect, Active Component, Involved Standard, and Triggering Property.

---

[7] WS-I Basic Security Profile Tool, (2009), http://www.ws-i.org/deliverables/workinggroup.aspx?wg=testingtools.

[8] Eston, T., J. Abraham, and K. Johnson.: Dont Drop the SOAP: Real World Web Service Testing. Retrieved July 6, 2013.

[9] WS-Attacker, (2013), http://sourceforge/p/ws-attacker/wiki/Home.

[10] WSBang Testing Tool, (2014), https://www.isecpartners.com/tools/application-security/wsbang.aspx.

The proposed ATLIST based penetration tool considers banking application as a business process (POV), domain web server as an active component, and SOAP as an involved standard. For each of the nine web service attacks that have been addressed by the proposed WS-SM framework, the triggering property (Precondition) and test case(s) have been identified and tabulated in Table 2. Each attack was simulated in the dynamic composition environment in accordance with the test case. For the input SOAP message generated according to the test case, the attack effect (Postcondition) was observed. The banking application was then enabled with WS-SM approach, the vulnerability testing was repeated with the same test case, and the outcome (Protection measure) of the proposed approach was observed and tabulated. It is found that the proposed WS-SM approach is not vulnerable to any of the specified nine attacks. It detects and overcomes five of the attacks and prevents four of the attacks.

4.3 Stress Testing

A prototype banking application involving dynamic composition of services was implemented on a 2.4 GHz Intel Core2 Duo processor with 4 GB RAM. The domain web services, composer and client were developed on different machines with different platforms such as Linux Ubuntu 10.04, Ubuntu 11.04, and Fedora 12. The services were deployed in different web servers like Glassfish, Apache Tomcat, etc. The composer and various domain services with WS-SM framework were all deployed in different LAN terminals.

The performance of the banking application was assessed in terms of throughput as the number of requests per second by increasing the number of concurrent user requests from 25 to 200. The throughput was measured 10 times at different workloads of network and the maximum values have been plotted as a graph shown in Fig. 5. Then, the banking application was enabled with WS-SM framework by installing the Security APIs and security gateway in both composer as well as in each of the domain server end. The throughput readings were taken and the impact of security APIs over the performance of the application was analyzed. It is found that even with 200 concurrent user requests, the throughput decreased by merely 35% when compared to not providing the security using the proposed approach.

In general, the attacks are generated automatically and repeatedly. The proposed security APIs prevent or detect the attacks every time they affect the communication. In order to decrease the turnaround time of the user request, the proposed WS-SM approach allows to configure the maximum number of times a particular instance of an attack is prevented or overcome. When this threshold reaches, the composer is notified so as to select an alternate service and thus an alternate communication channel to potentially avoid the attack. The impact of this threshold over performance of the WS-SM enabled banking application prototype has been analyzed. The threshold was configured as 1 and 3. For each of these configuration, the number of concurrent user requests was varied from 25 to 200 and the throughput in terms of number of requests per

Table 2. ATLIST based vulnerability testing of proposed security APIs

Attack	Triggering properties	TestCase	Attack effect	Outcome from proposed security API
SOAP Action Spoofing	Action field in soapenv:Header	Different operations in action field of soapenv:Header and operation field of soapenv:Body	Execution of incorrect SOAP operation	Prevention of invoking incorrect service and responding with soapenv:Fault message
WS-Addressing Spoofing	1. ReplyTo addressing field in soapenv:Header 2. FaultTo addressing field in soapenv:Header 3. To addressing field in soapenv:Header	1. Blacklisted address in ReplyTo 2. Blacklisted address in FaultTo 3. Blacklisted address in To	1. Redirection of SOAP Response to blacklisted address	Preventing the redirection of SOAP message and responding with soapenv:Fault message
Replay of Messages	MessageId in soapenv:Header	SOAP Messages with same/different MessageId and same soapenv:Body	Unintended execution of same service more than once	All resent SOAP Messages are dropped except the original message
Message Alteration Attack	Plaintext in soapenv:Body	soapenv:Body with unintended contents in plaintext	SOAP Message with unintended contents	Testcase not allowed
Loss of Confidentiality		SOAP Message with plaintext	Exposure of SOAP Message to unintended users	
XML Injection	Encrypted/ Unencrypted Message in soapenv:Body	Encrypted/ Unencrypted Message in soapenv:Body not conforming to XML Schema	soapenv:Fault or SOAP Message with unintended contents	Invalidating SOAP Message and responding with soapenv:Fault message
Principal Spoofing	1. Username in soapenv: Header 2. Password in soapenv: Header 3. Nonce in soapenv: Header 4. Created in soapenv: Header	UsernameToken with unregistered credentials in soapenv:Header	Service provisioning to user with false credentials	Preventing service provision to user with false credentials and responding with soapenv:Fault
Forged Claims				Testcase Not allowed
Falsified Messages				Testcase Not allowed

second was measured. Three samples were taken for each reading at different workloads of network and the maximum values have been plotted as a graph shown in Fig. 6. It is observed that the throughput decreased by 46 % when the threshold is increased from 1 to 3.

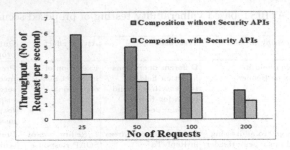

Fig. 5. Security vs performance

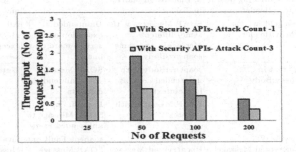

Fig. 6. Impact of repeated attacks on performance

5 Conclusion

A framework for secured messaging among web services that either prevents or detects and overcomes some of the web service attacks has been implemented in the context of dynamic web service composition. Compliance testing of WS-SM for WSI-BP revealed that it is interoperable. From the vulnerability testing of WS-SM, it was found that the proposed WS-SM approach is not vulnerable to any of the specified nine attacks. It detects and overcomes five of the attacks and prevents four of the attacks. A banking application involving dynamic composition of services was adapted with WS-SM framework and subjected to stress testing. Even with 200 concurrent requests, the throughput decreased by merely 35% when compared to not providing the security using the proposed approach demonstrating a good tradeoff between security and performance.

The proposed WS-SM framework with pluggable APIs can be adapted for any application involving dynamic composition by installing it in the web servers where the composer as well as domain services are deployed. The proposed solution is generic since new APIs for other kinds of attacks can be added on both sides as and when new types of attacks emerge. Moreover, the solution is entirely transparent to the client.

The future work involves creating APIs for addressing and overcoming the other web service attacks such as denial of service.

Acknowledgments. Authors would like to thank S.M. Sindhu, postgraduate student for her coding efforts.

References

1. Erl, T.: Service-Oriented Architecture concept, Technology, and Design. Pearson Education, London (2006)
2. Schmelzer, R., Vandersypen, T.: XML and Web Services Unleashed. Sams Publication, Chennai (2002)
3. Cerami, E.: Web Services Essentials: Distributed Applications with XML-RPC, SOAP, UDDI & WSDL. O'Reilly Media, Inc., Sebastopol (2002)
4. Singhal, A., Winograd, T., Scarfone, K.: Guide to secure web services. Technical report of National Institute of Standards and Technology, Special Publication 800-95 (2007)
5. Lemos, A.L., Daniel, F., Benatallah, B.: Web service composition: a survey of techniques and tools. ACM Comput. Surv. (CSUR) 48(3), 1–41 (2016). Article No. 33
6. Mouli, V.R., Jevitha, K.P.: Web services attacks and security - a systematic literature review. Procedia Comput. Sci. 93, 870–877 (2016)
7. Masood, A., Java, J.: Static analysis for web service security - tools & techniques for a secure development life cycle. In: IEEE International Symposium on Technologies for Homeland Security (HST), pp. 1–6 (2015)
8. Jensen, M., Gruschka, N., Herkenhoner, R.: A survey of attacks on web services - classification and countermeasures. Comput. Sci. Res. Dev. (CSRD) 24(4), 189–197 (2009). https://doi.org/10.1007/s00450-009-0092-6
9. Nordbotten, N.A.: XML and web services security standards. IEEE Commun. Surv. Tutorials 11(3), 4–21 (2009)
10. Alotaibi, S.J.: Toward a secure web service by using WS-security specifications. J. Comput. Theoret. Nanosci. 14(8), 3837–3842 (2017)
11. Thelin, J., Murray, P.J.: A public web services security framework based on current and future usage scenarios. In: International Conference on Internet Computing, pp. 825–833 (2002)
12. Yue, H., Tao, X.: Web services security problem in service-oriented architecture. In: International Conference on Applied Physics and Industrial Engineering, vol. 24, no. 9, pp. 1635–1641 (2001)
13. Kumar, R.K., Kanchana, R., Babu, C.: Security for SOAP based communication among web service. In: IJCA Proceedings on International Conference on Science. Engineering and Management (ICSEM 2013), pp. 46–51. Foundation of Computer Science, USA (2013)
14. Altaani, N.A., Jaradat, A.S.: Security analysis and testing in service oriented architecture. Int. J. Sci. Eng. Res. 3(2), 1–9 (1981)
15. Mainka, C., Somorovsky, J., Schwenk, J.: Penetration testing tool for web service security. In: IEEE 8th World Congress on Services, pp. 163–170 (2012)
16. Salas, M.I.P., Martins, E.: Security testing methodology for vulnerabilities detection of XSS in web services and WS-security. Electron. Notes Theoret. Comput. Sci. 302, 133–154 (2014)
17. Lowis, L., Accorsi, R.: Vulnerability analysis in SOA-based business processes. IEEE Trans. Serv. Comput. 4(3), 230–242 (2011)

Online Product Recommendation System Using Multi Scenario Demographic Hybrid (MDH) Approach

R. V. Karthik[(⊠)] [iD] and Sannasi Ganapathy [iD]

School of Computer Science and Engineering,
Vellore Institute of Technology, Chennai, Chennai 127, India
karthikr.v2016@vitstudent.ac.in, sganapathy@vit.ac.in

Abstract. The recommendation system plays very important role in the ecommerce domain to recommend the relevant products/services based on end user preference or interest. This helps end users to easily make the buying decision from a vast range of product and brands are available in the market. A lot of research is done in recommendation system, aim to provide the relevant product to the end user by referring end user past purchase history, transaction details etc. In our Multi scenario demographic hybrid (MDH) approach, important demographic influence factors like the user age group and located area are considered. The products are also ranked with associated age group category. The experimental results of the proposed recommendation system have proven that it is better than the existing systems in terms of prediction accuracy of relevant products.

Keywords: Ecommerce · User reviews · Recommendation

1 Introduction

Recommendation systems are applied in almost all the applications like Movie recommendation, book recommendation, product recommendation in online shopping sites, travel package recommendation etc. The recommendation system helps to propose a subset of relevant item/service that user are interested from the large amount of datasets. In online shopping sites, Recommendation system understands the user behavior and purchase interest and gives the customized recommendation based on each user.

A recommendation system is commonly classified in to a content based recommendation system and collaborative filtering approach. Content based recommendation system considers user past purchased data to understand the user interest and propose new product list, which are relevant or similar to the user past purchased product. Another recommendation system collaborative Filtering method focus on the similarities of other users. Based on similarity between the users, product purchased by one user is recommended to another user if both users are having same interest. Both these approaches has advantages and drawbacks. Content based recommendation system requires past

© IFIP International Federation for Information Processing 2020
Published by Springer Nature Switzerland AG 2020
A. Chandrabose et al. (Eds.): ICCIDS 2020, IFIP AICT 578, pp. 248–260, 2020.
https://doi.org/10.1007/978-3-030-63467-4_20

history of the user. On other hand, collaborative filtering face challenge to recommend the product for some new customers.

To overcome this problem, the hybrid recommendation system is developed which combines both content based and collaborative filtering method. The results of the hybrid recommendation system also give better result when compared to individual recommendation systems. Now a day deep learning techniques are applied for recommendation system to improve the prediction accuracy and also the performance of the recommendation system.

In this paper, a new recommendation system is proposed to recommend suitable products by considering demographic information. The major contribution of this paper are 1) identify the end user age group category from customer review 2) computes the rating score for each user category 3) Apply fuzzy logic to get the relevant product. The remainder of this research article is formulated as below: The related works in this area are elaborated in Sect. 2. The proposed recommendation working model is described in Sect. 3. Section 4 explains the performance metrics and experimental results of the new MDH approach. Conclusions on the proposed work with future focus are described in Sect. 5.

2 Literature Survey

There are various recommendation systems are developed to predict the user interested item/ service from a large amount of items. Soe-Tsyr Yuan et al. [1] proposed customized product recommendation by considering clustering algorithm. Jae Kyeong Kim et al. [2] used associate rule mining and decision tree to improve the recommendation system by selecting the relevant taxonomy and the preference of the user.

Duen-Ren Liu et al. [3] proposed Hybrid based approach by combining preference based collaborative filtering and Weighted RFM methods. Clustering is done based on the similar user preference and interest. Finally, recommendation is performed by grouping users with similar weighted RFM value.

Sang Hyun Choi et al. [4] proposed utility based recommendation system. In this approach multi attribute decision making procedure is used to predict the best product using utility value among other similar products. Utility value is computed by considering all the multi influence attributes. This approach shows better results and predicting accuracy when compared to distance based similarity product identification.

Kun Chang Lee et al. [5] proposed new casual mapping approach which consider qualitative characteristics that influence the quantitative factors. By considering this relationship helps predict the interest level of product and which in turn consider for later product recommendation. Fuzzy based product recommendation is proposed by Yukun Cao et al. [6] where the user needed features are noted. Based on the importance, raking is done. Fuzzy rules are framed by considering both product feature and customer expectation. Final recommendation is done based on the optimal combination.

Yuanchun Jiang et al. [7] provides the importance of capturing the end user feedback after purchase of the product for recommendation system. Simply buying the product would not mean that it can be recommended for other end users. Instead recommendation system supposes the capture the real feedback or review comments after they used the product. This approach improves the customer satisfaction level by considering the user profile and review comments. Sung-Shun Weng et al. [8] propose new multidimensional recommendation model which improves the accuracy of recommendation system. The approach is applied to the movie recommendation system, it identifies the users key influence preference factors and compute rating by considering all this multidimensional input factor. User profile and product rating are considered in this approach.

Ruben Gonzelez Crespo et al. [9] done detail analysis on challenge in the education sector for both students and teachers in getting the needed information and shown the importance of recommendation system to extract the relevant information and topics to the knowledge seekers. Using recommendation system also improves the satisfaction of the student's expectations. Another approach rule based recommendation system is proposed by Vicente Arturo Romero Zaldivar et al. [10] for education applications. In this approach Meta based rule recommendation is used to understand the profile, the interest of the user and helps to recommend the details for similar kind of users.

Mohanraj et al. [11] proposed Ontology driven bees foraging technique to understand the user navigation pattern and predict the user interest. The score is calculated based on user navigation or selection in web pages. By learning this user profile and interest, proposed approach capture the navigation pattern and helps to recommend the relevant information in the web pages. Sunita et al. [12] proposed a course recommendation system to recommend some interested course to a new student. Clustering technique is used to group the related courses and then apriori algorithm is then applied with associate rule to find the best combination and recommend courses that is interested by the new student.

Jian-Wu Bi et al. [13] used type 2 Fuzzy numbers for sentimental analysis, which helps to improve the accuracy rate and also predict the relevant sub set of products from the large value of the product set for the recommendation. Chao Yang et al. [14] proposed neural network based recommendation method which considers both user item interaction and also user list interaction. This improves the accuracy of the recommendation list and also found more relevant products to recommend.

Aminu Dau et al. [15] used deep learning method for product recommendation. In this method sentiment score is calculated based on each aspect and then finally overall product score is computed and used for product recommendation. Xiaofeng Zhang et al. [16] proposed deep recommendation system by considering user rating and also user textual information or feedback.

Karthik et al. [17] proposed feature and product ranking approach to predict most suitable product for recommendation based on user preference and interested feature set. The proposed approach is an extension of this feature and product ranking approach. By considering the demographic information like user age group and delivery location, the recommendation can be further improved. For that purpose, this paper introduces a new recommendation system for recommending the right products to the customers in e-commerce.

3 Proposed Work

Proposed multi demographic approach contains following main modules. 1) Preprocessing the customer review 2) Identify the End user age group 3) Compute product rating 4)

Apply Fuzzy logic and computes the recommended level. All the modules are explained in the following chapters. Section 3.1 describes the preprocessing technique applied for the customer review. User reviews play important role in identifying the real end user of the product. By processing each customer review, rating of the product can be computed which gives better value when compared to the overall rating.

Section 3.2 explains the method to identify the end user age group. In order to recommend the relevant product to the right customer. It is important to focus the end user to whom the product is likely to purchase. Section 3.3 explains the technique to identify the product rating. In order to provide better recommendation, it is important to rank the product based on the user reviews and comments. Section 3.4 provides the fuzzy logic details to compute the recommended level. Fuzzy rule is framed by considering the product overall rating, end user age group and customer location or delivery location.

Figure 1 explains the overall proposed model. It takes customer reviews as input and provides the recommendation list based on customer present interest.

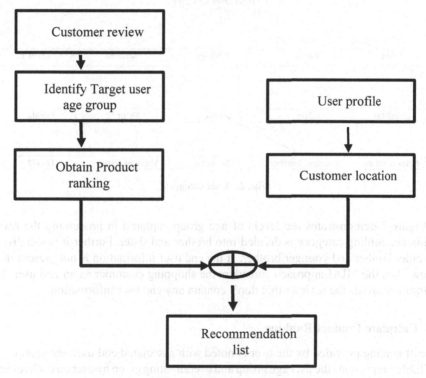

Fig. 1. Proposed model

3.1 Pre Processing

In most of the ecommerce sites, customer review section is a free text area. Customers text their user experience and feedbacks. These text reviews are required to preprocess it, remove all unnecessary information's, stop words etc.

In preprocessing phase, stop words are removed from the review text and POS tagging is applied. Applying the preprocessing reduce the computation time and also improve the accuracy of the recommendation list.

3.2 Identify End User Age Group

In most of the customer reviews, real end users are mentioned by the purchased customers. By capturing this user age group helps recommendation system to propose the liked product to the similar kind of age group customers.

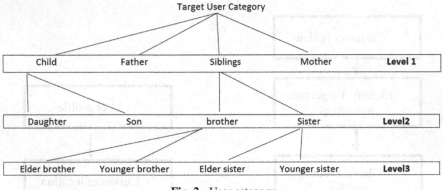

Fig. 2. User category

Figure 2 demonstrates the levels of age group captured in processing the review comments. Sibling category is divided into brother and sister. Further it is sub divided into elder brother and younger brother. If the end user information is not present in the review, then the MDH approach considers the shopping customer as an end user. This assumption avoids the reviews that don't contain any end user information.

3.3 Calculate Product Ranking

Overall ranking provided by the user is noted with associated end user age group.

Table 1 represents the user age group and overall rating given by user on each reviews.

The average ranking value for each age group is computed and noted as product ranking for respective age group.

$$ranking_{ageGroup} = \frac{\sum_{k=1}^{n} review_{ranking}}{Total\ number\ of\ reviews\ of\ same\ age\ gorup}$$

Overall product ranking for each group helps to recommend the product based on the current customer interest.

Table 1. Product ranking

Item Id	User age group	Review overall rating (Out of 5)
5834CS01	Son	4
5834CS01	Daughter	2
5834CS01	Son	4
5834CS01	Son	4

3.4 Fuzzy Logic

Fuzzy logic helps to obtain the recommendation level by taking computed product ranking and customer delivery location. The MDH recommendation system gets the age group to whom like to purchase as input from the shopping user. Fuzzy rule is written to decide the recommendation level based on the rating of the product and the location of the customer (Table 2).

Table 2. Fuzzy rule

Rule	Fuzzy rule description	Outcome
Rule 1	If age group exactly matches and rating is high and location exactly matches	Highly recommended
Rule 2	If age group exactly matches and rating is high and location matches	Highly recommended
Rule 3	If age group exactly matches and rating is medium and location matches	Highly recommended
Rule 4	If age group matches and rating is medium and location exactly matches	Recommended
Rule 5	If age group matches and rating is medium and location somewhat matches	Recommended
Rule 6	If age group not matches and rating is medium and location not matches	Not recommended
Rule 7	If age group not matches and rating is low and location not matches	Not recommended

Algorithm (MDH Approach)

Input: Online review comments of each product and age group interested to purchase
Output: Recommended product for the requested target user
Step 1: Process all the reviews for each product

Step 2: Apply preprocess technique and remove unwanted data from the reviews
Step 3: Identify the target end user category for each customer review
Step 4: Compute overall product rating score for each associated target user category
Step 5: Apply Fuzzy rule and determine the recommend level
Step 6: Share the highly recommended product based on target user
Overall product ranking computed for each age group looks like in Table 3.

Table 3. Product recommendation

Item	Overall rating	End user age group recommended	Overall rating (Age group 1)	Overall rating (Age group 2)
5834CS01	3	Brother	3.2	4.2
5834CS01	4	Sister	3.5	3.1

4   Results and Discussion

Dataset: Amazon review dataset is considered for this work. The data set contains more than 10,000 of products and 25,000 customer reviews. Each entry in the dataset contains the details present in the Table 4. Customer id is part of the user profile. Review comments and user rating are available for each customer reviews.

Table 4. Dataset details

S. No.	Dataset items	Example
1	Customer id	239845
2	Overall rating	4
3	Product id	5834CS01
4	Review comments	"Awesome product, worth for money"
5	User rating	3.5
6	Purchased date	April/2016

4.1   Performance Metrics

In the evaluation, most widely used content based and collaborative filtering recommendation system approach is considered. In addition to these conventional recommendation systems, Hybrid recommendation system also taken in to account for comparison. Hybrid recommendation system is combination of both content and collaborative filtering technique. Hybrid approach outstands when compare to the individual recommendation system. Precision and Recall metrics are evaluated with all these recommendation techniques. Recommendation system specific metrics Item, User and Catalog coverage are also considered for evaluation.

$$Precision = \frac{Number\ of\ relvant\ recommendation\ product}{Total\ number\ of\ recommended\ products}$$

$$Recall = \frac{Number\ of\ relavant\ recommended\ prodcut}{Total\ number\ of\ product\ that\ to\ be\ recall}$$

Item Coverage: This metric helps to know the percentage of products available in recommendation list over the number of potential available products.

$$Coverage_{item} = \frac{n_r}{N} * 100$$

Where, n_r represents the number of recommended products and N represents the number of potential available products.

User Coverage: This metric helps to obtain the percentage of users or customer that the system able to provide the recommendation list over the number of potential users.

$$Coverage_{user} = \frac{n_u}{N} * 100$$

Where, n_u represents the number of users that system can capable to generate recommended products and N represents the number of potential users.

Catalog Coverage: This metric helps to obtain the percentage of user – product pair over the total number of possible user product pair.

$$Coverage_{catalog} = \frac{n_{up}}{N * U} * 100$$

Where, n_{up} represents number of user pair product considered in product recommended products and N represents the number of possible user product pair.

Table 5. Experiment sets

Experiment	Number of products	Number of users
Exp-1	2000	1200
Exp-2	4500	2000
Exp-3	6000	2800
Exp-4	8000	3500
Exp-5	10000	5000

4.2 Experimental Results

The dataset contains more than 80,000 reviews and 1000 products. Customers purchased more than 4 times are grouped separately and used for predicting the accuracy of the product recommendation. For example, if the customer done shopping ten times in the history, then First 4 transactions are considered as test data to understand the behavior of the user. Based on these initial transactions, recommendation list is computed by the MDH approach and checked how it's relevant to the reaming transaction that were really purchased by the customer X. This method helps to identify the accuracy of the proposed MDH recommendation system.

Totally 5 experiment results are recorded by considering a different number of products and user list. Table 5 represents the number of product and users included in each experiment.

Inference: By referring all the experiment outcomes, Hybrid approach result are better when compare to the individual content based or collaborative filtering techniques. Whereas Feature and product based ranking performs betters on comparing the Hybrid approach. Enhancement done on top of Feature and product ranking approach – MDH approach is further shown better result when compared to feature and product ranking feature. MDH approach improves 8% of accuracy of predicting relevant product for recommendation when compare to the existing recommendation system. MDH approach also shows better results when considering the coverage evaluation outcome. All three User, item and catalog coverage shows better results when compare to the other existing recommendation system.

Figure 3 and Fig. 4 gives the comparative results of precision and recall evaluation. Both precision and recall evaluation outcome shows MDH approach shows better results when compare to other existing recommendation approach.

Figure 5 and Fig. 6 gives the comparative results of user and item coverage evaluation. Coverage is recommendation system specific metrics. Both user and item coverage results imply MDH approach shows better results when compare to other existing recommendation approach.

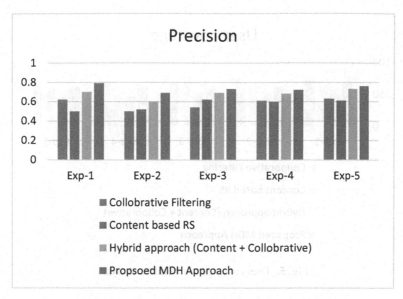

Fig. 3. Precision value comparative analysis

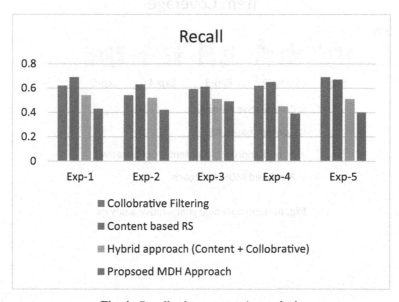

Fig. 4. Recall value comparative analysis

Figure 7 gives the comparative results of catalog coverage evaluation. The outcome of the of all 5 experiment set, shows better results with MDH approach when compared to other existing recommendation approach.

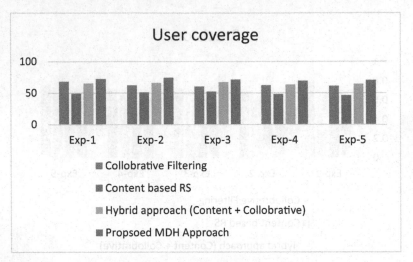

Fig. 5. User coverage comparative analysis

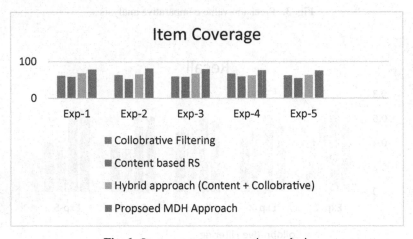

Fig. 6. Item coverage comparative analysis

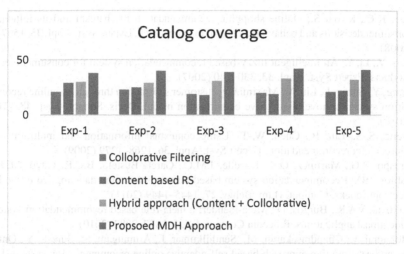

Fig. 7. Catalog coverage comparative analysis

5 Conclusion and Future Work

In order to provide interesting and relevant products to the customer, new Multi demo graphic hybrid based product recommendation system is developed. Enhancing the recommendation system fuzzy approach with Demographic information, gives better results. Fuzzy rule is written to consider the product overall ranking and also the delivery location of the customer. Considering demo graphic information end user age and location is further improving the accuracy of the recommendation system. In the MDH approach, the first end user group is determined and product rank is computed. Later product list is further filtered based on the delivery location. Experimental results show MDH approach better results prediction accuracy is improved by 8% when considering the user location and age group. In addition to accuracy, coverage also evaluated and found product and user coverage of MDH gives better results.

In future, planned to improve the preprocessing technique and apply neuro fuzzy to improve the accuracy further. In the neural implementation, negative product list and outcome of the earlier recommendation list can be considered, which will improve the result of the product recommendation.

References

1. Yuan, S.-T., Cheng, C.: Ontology-based personalized couple clustering for heterogeneous product recommendation in mobile marketing. Expert Syst. Appl. **26**, 461–476 (2004)
2. Kim, J.K., Cho, Y.H., Kim, W.J., Kim, J.R., Suh, J.H.: A personalized recommendation procedure for Internet shopping support. Electron. Commer. Res. Appl. **1**, 301–313 (2002)
3. Liu, D.-R., Shih, Y.-Y.: Hybrid approaches to product recommendation based on customer lifetime value and purchase preferences. J. Syst. Softw. **77**, 181–191 (2005)
4. Choi, S.H., Kang, S., Jeon, Y.J.: Personalized recommendation system based on product specification values. Expert Syst. Appl. **31**, 607–616 (2006)

5. Lee, K.C., Kwon, S.: Online shopping recommendation mechanism and its influence on consumer decisions and behaviors: a causal map approach. Expert Syst. Appl. **35**, 1567–1574 (2008)
6. Cao, Y., Li, Y.: An intelligent fuzzy-based recommendation system for consumer electronic products. Expert Syst. Appl. **33**, 230–240 (2007)
7. Jiang, Y., Shang, J., Liu, Y.: Maximizing customer satisfaction through an online recommendation system: a novel associative classification model. Decis. Support Syst. **48**, 470–479 (2010)
8. Weng, S.-S., Lin, B., Chen, W.-T.: Using contextual information and multidimensional approach for recommendation. Expert Syst. Appl. **36**, 1268–1279 (2009)
9. Crespo, R.G., Martínez, O.S., Lovelle, J.M.C., García-Bustelo, B.C.P., Gayo, J.E.L., De Pablos, P.O.: Recommendation system based on user interaction data applied to intelligent electronic books. Comput. Hum. Behav. **27**, 1445–1449 (2011)
10. Zaldivar, V.A.R., Burgos, D.: Meta-mender: a meta-rule based recommendation system for educational applications. Procedia Comput. Sci. **1**, 2877–2882 (2010)
11. Mohanraj, V., Chandrasekaran, M., Senthilkumar, J., Arumugam, S., Suresh, Y.: Ontology driven bee's foraging approach based self-adaptive online recommendation system. J. Syst. Softw. **85**, 2439–2450 (2012)
12. Aher, S.B., Lobo, L.M.R.J.: Combination of machine learning algorithms for recommendation of courses in E-Learning System based on historical data. Knowl. Based Syst. **51**, 1–14 (2013)
13. Bi, J.-W., Liu, Y., Fan, Z.P.: Representing sentiment analysis results of online reviews using interval type-2 fuzzy numbers and its application to product ranking. Inf. Sci. **504**, 293–307 (2019)
14. Dau, A., Salim, N., Rabiu, I., Osman, A.: Recommendation system exploiting aspect-based opinion mining with deep learning method. Inf. Sci. **512**, 1279–1292 (2020)
15. Zhang, X., Liu, H., Chen, X., Zhong, J., Wang, D.: A novel hybrid deep recommendation system to differentiate user's preference and item's attractiveness. Inf. Sci. **519**, 306–316 (2020)
16. Karthik, R.V., Ganapathy, S., Kannan, A.: A recommendation system for online purchase using feature and product ranking. In: Eleventh International Conference on Contemporary Computing (IC3) (2018)
17. Yang, C., Miao, L., Jiang, B., Li, D., Cao, D.: Gated and attentive neural collaborative filtering for user generated list recommendation. Knowl. Based Syst. **187**, 104839 (2020)

Simulation of Path Planning Algorithms Using Commercially Available Road Datasets with Multi-modal Sensory Data

R. Senthilnathan, Arjun Venugopal(✉), and K. S. Vishnu

Department of Mechatronics Engineering, SRM Institute of Science and Technology,
Kattankulathur, India
arjunvenugopal07@gmail.com

Abstract. Road datasets for computer vision tasks involved in advanced driver assist systems and autonomous driving are publicly available for the technical community for the development of machine learning aided scene understanding using computer vision systems. All the perceived data from multiple sensors mounted on the vehicle must be fused to generate an accurate state of the vehicle and its surroundings. The paper presents details of the simulation implementation of local path planning for an autonomous vehicle based on multi-sensory information. The simulation is carried out with sensory inputs from RGB camera, LIDAR and GPS. The data is obtained from the KITTI dataset. A variant of the D-star algorithm is utilized to demonstrate global and local path-planning capabilities in the simulation environment.

Keywords: Multi-sensory data fusion · Object detection · Range detection · RGB-D data · Path planning

1 Introduction

Advanced Driver Assist Systems (ADAS) and autonomous driving technology have greatly contributed to the explosive growth of the field of deep learning which was fueled by the massive collection and availability of road datasets for public usage. Such tasks use multiple sensors such as cameras, inertial measurement units (IMU), global positioning systems (GPS), range finders such as LIDAR sensor and RADAR sensor. An autonomous driving vehicle utilizes all these sensors for tasks such as localization of the vehicle in a metric map, mapping of surroundings with reference to the vehicle, etc. The sensors fundamentally aid in the perception stage of the autonomous vehicle's control loop. The role of perception is to aid the autonomous vehicle to plan trajectories and execute graceful navigation to the destination. Hence it may be appreciated that all the perceived information must be ending up utilized in path planning stages especially for the purpose of local path planning which often is an obstacle avoidance problem. The paper presents the details of the simulation of local path planning algorithms based on the sensory dataset of road scenes available as part of the KITTI benchmarking suite [1, 2]. The task of path

© IFIP International Federation for Information Processing 2020
Published by Springer Nature Switzerland AG 2020
A. Chandrabose et al. (Eds.): ICCIDS 2020, IFIP AICT 578, pp. 261–275, 2020.
https://doi.org/10.1007/978-3-030-63467-4_21

planning generally manifests in at least two ways namely global and local. Global path planning is carried out from source to destination of an autonomous driving mission based on GPS data with reference to a map of the environment. Local path planning requires an accurate estimate of the objects in the road such as vehicles, pedestrians, civil structures, etc. In autonomous vehicles, utilizing multiple sensors is generally the norm. This is required to address the inherent sensor noise and aliasing present in the data obtained from a completely unstructured environment. The current work utilizes the sensory data of the sample driving scenario from the KITTI dataset for simulation studies. Simulation studies refer to the offline processing of real-world data acquired from sensors such as RGB-camera, LIDAR, GPS, etc. This multi-modal sensory information is acquired from the real world and can be further used to implement intelligent behavior in many systems. The intelligence of a system depends on the interpretation of sensory data for the perception of the environment. Also, the accuracy of a system corresponds to the amount of data the system is able to collect or process within the given span of time. Simulation facilitates analysis and visualization of the performance of the system in a virtual environment in which the errors can be corrected then and there. When the simulation results are up to the mark, the model can be used for deployment in a real-time environment. Many simulation-based studies related to ADAS and autonomous driving had been proposed.

1.1 Elements of Simulation Studies

The simulation studies are based on images, range data (LIDAR), and GPS information. The various tasks using such information are addressed below.

Object Classification. It is used to classify the set of similar objects in a scenario as a group. Finally, the scene will consist of several such groups.

Object Detection. To detect objects, if any, which are of interest present in a scenario at any point in time and to generate a bounding box for such objects. The object is segmented by a bounding box generated around the detected objects.

Range Detection. The distance between the ego vehicle with respect to the objects is found out from the given scenario and to generate a bounding box for such objects.

Registration of Range Data on RGB Image. The object is segmented by a boundary which is with respect to the distance from the ego vehicle.

Collision Avoidance. Similar to range detection, but the focus is not just to detect the object, rather use such information to avoid the collision of the object with the ego vehicle.

Path Planning. To generate a path, the ego vehicle must take for the given start and goal point which ensures collision-free traverse.

1.2 Limitations of Simulation Studies

The simulation studies are performed in an environment that assumes the system or its parameters be static in nature which is not the case, and any changes in the system affect the overall performance. Some constraints of such systems are presented as follows:

Environmental Constraints. Any change in the behavior of the environment with respect to the environment it was trained result in an unpredictable solution or failure.

Real-time Constraints. The simulation studies do not work as efficiently as it does in a simulated environment when compared to the real-world environment.

Data availability Constraints. In a simulated environment, the data is already available for the computational purpose, which is not the case in a real-time environment, since the collection of data from a real-world environment in a real-time manner is a crucial task. The real-world data acquisition may not exactly meet the fidelity.

2 Details of Dataset

To find the suitable dataset for simulation studies, the datasets available for road data were referred, and among them, the three most suited datasets which contain RGB, range and GPS information are filtered for final selection. Some of the datasets considered for the purpose include the Ford Campus Vision and LIDAR dataset [3], The KITTI Vision dataset, and Oxford RobotCar Dataset [4]. Among these, the KITTI dataset was chosen based on the merit of the abundance of supporting materials for development and popularity in benchmarking for various computer vision tasks. The other advantage of the KITTI dataset is the content-based grouping. The various groupings include stereo, flow, depth, odometry, tracking and road semantics. For all the work reported in this paper, the KITTI tracking dataset acquired during the year 2012 is utilized [2]. The tracking section consists of different categories of files which include left and right color images, LIDAR point clouds, GPS/IMU data, camera calibration matrices [5], training labels, L-SVM reference detections, Regionlet reference detections, tracking development kit many more. Among these, the simulation specific files are utilized from the tracking dataset which is explained in the following sections.

Since in the current work the RGB images are used only for semantic understanding of the scene the left camera images are utilized for the same. This file consists of RGB information taken from the left camera mounted on top of the ego vehicle. The RGB information is packed as Portable Network Graphics (PNG) file format and has a resolution of 1242 × 375 pixels.

The second sensory data used in the work is the LIDAR point cloud data. This file consists of laser information taken from the Velodyne laser scanner mounted on top of the ego vehicle and the specifications are listed in Table 1.

The third sensory data utilized is the GPS/IMU combo. This file consists of GPS/IMU information taken at each instance the frame from the camera was captured, the GPS/IMU sensor is also mounted on top of the ego vehicle. The various information contained in the data is listed in Table 2.

Table 1. Specifications of LIDAR point cloud data.

Parameter	Specification
File format	Text (TXT)
Point information utilized in the current work	'x' – coordinate in m 'y' – coordinate in m 'z' – coordinate in m

Table 2. Specifications of GPS/IMU data.

Parameter	Specification
File format	Text (TXT)
GPS information utilized in the current work	Latitude in deg Longitude in deg

One of the important tasks involved in utilizing multiple sensors is the requirement for registration of correspondence of information recorded by the sensors with respect to the one frame of reference. KITTI dataset provides accurate measurements of the physical locations of the sensors with reference to each other in terms of positions and orientations. This information is generally part of system calibration which enables tasks such as sensor fusion. In KITTI dataset a set of calibration matrices that defines the transformation between various sensors located on the vehicle, which are listed in Table 3.

Table 3. KITTI calibration matrices.

Parameter	Specification
Matrix information utilized in the current work	Projection matrix of left color image Rotation matrix of the camera Transformation matrix from LIDAR to camera

3 Deep Learning Inferences on RGB Images

When multiple objects are present in the image, the task is to localize the objects in the image, also known as object detection. From the term object detection, it is evident that it predicts an output or detects objects which are of interest based on application. When an input is fed to the system (i.e., images), the computer runs a suitable algorithm and bounding boxes are generated for the objects present in the input image. Since images are large data and consist of several features, it is impossible for shallow networks to predict the output with high accuracy. Therefore, a deep learning network which consists of a greater number of layers is chosen. For the purpose of object detection, a pre-trained model of Faster RCNN with Resnet 101 [6] is utilized. The training specification of the pre-trained model is listed in Table 4.

Table 4. Training specifications of Faster R-CNN Resnet 101.

First Stage	Second Stage
Anchor generator:	Bounding box predictor:
Height stride: 16	Regularizer:
Width stride: 16	Regularizer: 12
Scales: 0.25, 0.5, 1.0 and 2.0	Weight: 0.0
Aspect ratios: 0.5, 1.0 and 2.0	Initializer:
Bounding box predictor:	Variant scaling
Regularizer:	Factor: 1
Regularizer: 12	Uniform: True
Weight: 0.0	Dropout:
Initializer:	Usage: False
Truncated normal	Keep probability: 1.0
Stddev: 0.00999999977648	Post processing
Non max suppression:	Batch non max suppression:
score threshold: 0.0	score threshold: 0.300000011921
iou threshold: 0.699999988079	iou threshold: 0.600000023842
max proposals: 100	max detections per class: 100
Localization loss weight: 2.0	max total detections: 100
Objectness loss weight: 1.0	Localization loss weight: 2.0
Initial crop size: 14	Objectness loss weight: 1.0
Maxpool kernel size: 2	Score converter: SOFTMAX
Maxpool stride: 2	
Training Configuration	**Evaluation Configuration**
Batch size: 1	Number of examples: 8000
Learning rate	Maximum evaluations: 10
Step: 0	Moving average: False
Learning rate: 0.000300000014249	Shuffle: False
Step: 900000	Number of readers: 1
Learning rate: 2.99999992421e-05	
Step: 120000	
Learning rate: 3.00000010611e-06	
Momentum: 0.899999976158	
Moving average: False	
Gradient clipping: True	

The model uses Faster R-CNN algorithm for object detection when an input image is fed to the model and classifier used is a type of Residual network with 101 layers called

ResNet-101, The dataset was trained on MS Common Objects in Context (COCO) [7] which is a dataset with 330K images and 80 object categories of which more than 200K of them are labeled. An illustration of the Faster R-CNN network architecture is presented in Fig. 1.

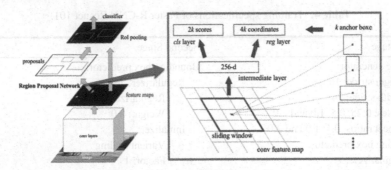

Fig. 1. Faster RCNN architecture [6].

For the pre-trained network implementation, TensorFlow API is used. The files of a pre-trained model include those listed in Table 5.

Table 5. Files in a pre-trained model [8].

File	Description
checkpoint	This file contains all the checkpoint name and its path.
model.ckpt.data-00000-of-00001	This file contains the values of variables i.e., weights, biases, placeholders, gradients, hyper-parameters, etc.
model.ckpt.index	This file contains a table of each tensor and its values
model.ckpt.meta	This file contains the complete graph, separately from the variables
frozen_inference_graph.pb	This file contains the model architecture and weights in a single file
pipeline.config	This file contains the training specification of the network, training configuration, evaluation configuration, etc.

Inferencing on a pre-trained model is advantageous since it does not require to train the model which is again a time-consuming process and moreover for a simulation task based on popular datasets pre-trained models are more than adequate. Figure 2 shows a sample result of network inference. The inferencing was done for 154 frames which are continuously captured image sequence and in all cases the model was able to detect the objects which are of interest. The pre-trained model was set to detect 4 classes of labels namely *"person"*, *"two-wheeler"*, *"light vehicle"*, and *"heavy vehicle"*, but the actual model was trained to detect 90 classes of labels. Therefor the classes which are not required for the scope of this task are omitted and are represented as *"N/A"*. All the detectable classes may be noted in the figure.

Fig. 2. Object detection results

4   Range Data Processing

The LIDAR point clouds are also referred to as range data. This data consists of the distance of the surrounding objects scanned with reference to a home position which in this case corresponds to ego vehicle. This data gives a description of the position of each object as a collection of points. The detailed description of the data structure and its plot visualization are explained in the following sub-sections.

4.1   Description of Data Container

The primary data structure used for data storage is arrays. Arrays give the unique feature of storing the same type of data under the same name and the data can be an n-dimensional vector. The programming language used is *Python 3* and since python does not contain an inbuilt array data structure, the *NumPy* library is used for support. The LIDAR point cloud is stored as a binary file and contains a minimum of 1 lakh points per file where each point has its respective *x, y, z* and *reflectance* in meters. This binary file is unpacked using the *fromfile* functionality in *NumPy*.

4.2   LIDAR Data Visualization

Visualization is the only way to observe, the correctness of obtained data or the data on which different processing techniques are used, with reference to the desired behavior. Visualization involves plotting of points in 2-D and 3-D space to understand the output image or any other data. Here, *Matplotlib* is used to handle all kinds of visualization tasks. The LIDAR data is a collection of points in 3-D space, in which each point is represented by the three-coordinate system. This information is unpacked and stored as an array of $(n \times m)$ where, n corresponds to rows and m corresponds to columns, in this case, m is taken as 3. Figure 3 shows the plot visualization of LIDAR data.

4.3   Correspondence Between RGB and Range Data

In an RGB image, the three layers are stacked above one another to visually produce real-world information based on its actual color. The RGB image is the scaled two-dimensional version of the real-world information which does not provide any information with regard to the actual distance of the object, actual height of the object, or about the actual width of the object. Whereas, the data obtained from LIDAR gives the distance of the object with respect to the LIDAR and this information is on the metric scale. To establish the correspondence between camera-RGB data and range data, both the data has to be in pixels. This is done by performing matrix multiplication with four different matrices in which the LIDAR coordinated from metric scale are converted to pixel information, similar to how the camera captures the image. The first two matrices are the function of internal and external parameters of the camera which obtained from camera modeling, the third matrix is used to transform the LIDAR frame with respect to the camera frame and the final matrix corresponds to the LIDAR points.

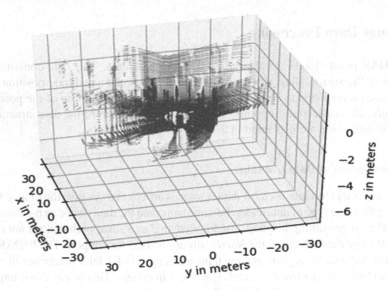

Fig. 3. 3-D plot of LIDAR data

In order to establish a correspondence between LIDAR data and RGB image, the necessary transformation equation [9] must be formulated which are provided as part of the dataset. The image points i.e., x and y in pixels are obtained by performing the matrix multiplication in the order presented in Eq. 1. The projection matrix is also called as the intrinsic parameters of the camera and the rotation matrix is also called as the extrinsic parameters of the camera. The transformation matrix is to relate the LIDAR frame with respect to the camera frame.

$$s \begin{bmatrix} x \\ y \\ z \end{bmatrix} = \begin{bmatrix} fx & 0 & cx & dx \\ 0 & -fy & cy & dy \\ 0 & 0 & 1 & dz \end{bmatrix} \times \begin{bmatrix} r11 & r12 & r13 & 0 \\ r21 & r22 & r23 & 0 \\ r31 & r32 & r33 & 0 \\ 0 & 0 & 0 & 1 \end{bmatrix} \times \begin{bmatrix} t11 & t12 & t13 & tx \\ t21 & t22 & t23 & ty \\ t31 & t32 & t33 & tz \\ 0 & 0 & 0 & 1 \end{bmatrix} \times \begin{bmatrix} lx \\ ly \\ lz \\ 1 \end{bmatrix} \quad (1)$$

Where,

- 'x' and 'y', the 'x' and 'y' of the LIDAR point in pixels respectively.
- 'fx' and 'fy', the focal length in 'x' and 'y' respectively.
- 'cx' and 'cy', the principal point offset in 'x' and 'y' respectively.
- 'dx', 'dy', and 'dz', the distortion in 'x', 'y', and 'z' respectively.
- '$r11$'–'$r33$', the rotation with respect to the camera frame.
- '$tr11$'–'$tr33$', the rotation of LIDAR frame with respect to camera frame.
- 'tx', 'ty', and 'tz', the translation of LIDAR frame with respect to camera frame.
- 'lx', 'ly', and 'lz', the 'x', 'y', and 'z' of the LIDAR points in metric scale respectively.

By performing the matrix multiplication, the correspondence between RGB and range can be obtained. Hence, by plotting the points of LIDAR which lies in the image resolution, the RGB-D data is obtained. Figure 4 shows an RGB-D image in which the LIDAR range information are indicated by green, blue, red and yellow colored points in the increasing order of range.

Fig. 4. RGB-D data (Color figure online)

The RGB-D data obtained is segmented based on the requirement i.e., the objects of interest. Here, the segmentation is based on the number of range points received back from the LIDAR sensor for all objects of interest. Figure 5 shows the range segmented RGB-D image in which the LIDAR range information are indicated by green, blue, and red colored points in the increasing order of range.

Fig. 5. Segmented RGB-D data (Color figure online)

5 Path Planning

The work reported in this paper demonstrates the simulation of path planning based on multi-sensory data acquired from the real-world. In order to demonstrate the local path planning capabilities based on the proposed method, the vehicle must first have an intended path to connect any given source point to the destination. This task in mobile robotics literature is referred to as global path planning. Since it is a simulation of the path planning algorithm that is intended, the actual path traveled by the vehicle used for acquiring the dataset is considered as the intended path to travel in the simulation environment. It must be noted that the path is the actual path traveled by the vehicle. This information may be obtained from the GPS data in the dataset. The GPS coordinates from the KITTI dataset are in the form of latitude and longitude measured degrees. In order to have a path planning performed on a map, it is convenient to have the map coordinates recorded in linear dimensions in meters. The python library '*utm*' is used to convert the coordinates to meters and the plot obtained is in the easting-northing convention. To convert this convention into the *x-y* right-handed orthogonal convention, the subtraction of every point with the first point is carried out. Hence, the first point will be (0,0) and the rest of the points take this as reference. Table 6 shows the test case results obtained on different sample video scenes from the same dataset. The figures illustrated in this paper are results obtained from Test Case – 1.

5.1 GPS-Based Global Map

The GPS information of the original acquired dataset contained 15 parameters and out of which the 2 parameters utilized were latitude and longitude. By using these parameters and performing appropriate mathematical conversions, the path travelled by the ego vehicle was obtained which in this case is considered as the global map. The objects detected from the deep learning network's inference on the RGB image and the corresponding point cloud data were also plotted in this graph as shown in Fig. 6. Objects are represented as blue colored blobs and red-colored points represent the GPS path taken by the ego vehicle.

Table 6. Specification of test cases

Test Case - 1	Specification
Number of frames from video scene	154
Number of GPS points from scene	154
Objects detected using pre-trained network	9
Objects under 20m LIDAR range	7
Detected classes of labels:	
Person	2
Two-wheeler	4
Light vehicle	1
Heavy vehicle	0
Test Case - 2	Specification
Number of frames from video scene	374
Number of GPS points from scene	374
Objects detected using pre-trained network	12
Objects under 20m LIDAR range	8
Detected classes of labels:	
Person	0
Two-wheeler	0
Light vehicle	8
Heavy vehicle	0

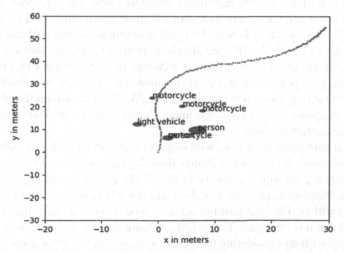

Fig. 6. GPS data-based global intended path (Color figure online)

5.2   Global Map-Based Local Path Planning

Path planning algorithms are basically graph traversal algorithms that estimates the best and least possible distance from a start position to a goal position using numerous

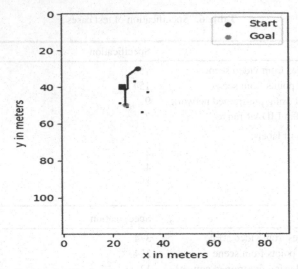

Fig. 7. Local path planning on a single frame (Color figure online)

theorems. The nature of the required task involved direct traversal from start to goal. The possible path planning algorithms referred were algorithms which had the capability to change the cost function and re-plan frequently. The selection list came down to Lifelong Planning A Star (LPA*) path planning algorithm [10] and D Star (D*) path planning algorithm [11]. LPA* is the modified form of A Star (A*) path planning algorithm [12] and is used to find the shortest path under changing costs. The start position and goal position are static and the direction of search is from start to goal. D* Lite implements the same behavior as that of D Star (D*) path planning algorithm [13] and is based on LPA*. Similar to LPA*, the D* Lite algorithm is used to find the shortest path under changing costs except that when there is a change in cost, the algorithm re-plans the costs from the goal position to the current position. The direction of search is from goal to start and the start position is not static unlike LPA*. D* Lite was preferred over LPA* because of its ability to take decisions under changing environmental conditions with respect to the current position.

The local path planning is done with respect to this GPS data-based global intended path obtained using the number of frames from the video scene and by performing D* Lite path planning algorithm as shown in Fig. 7. The position of the ego vehicle in the first frame is taken an origin, which is also taken initially as the start position in the path planning algorithm. The goal position is taken to be 20 m away from the start position along the global intended path. This is just to ensure that any local planning should not be carried out without considering the global outcomes. A slight variation is applied to the way in which the D* Lite algorithm is applied. In original D* Lite algorithm, the goal position is considered to be static while the start position may vary. In this paper, the D* Lite algorithm is performed from frame to frame. Therefore, both the start position and goal positions vary in every frame. After a path is generated in a frame, the goal position becomes the start position and a new goal position is set using the same criteria for rest of the frames. This criterion is applied to ensure the dynamic nature of the environment is considered while performing the path planning. In Fig. 7 the green point represents the start and red represents a goal, the black blobs are objects present in the current frame as detected by the deep learning network and registered against the LIDAR data for range from, the ego-vehicle. The black line represents the path generated from start to goal on the particular frame.

6 Conclusion

The evaluation of the entire studies was based on software simulation and process flow can be understood from Fig. 8. The simulation studies were performed mainly using the acquired LIDAR data, camera data and GPS data. The task involved the usage of different sensory data which contained different information for understanding the scene as captured by the sensors. Therefore, all the sensory information as a package is referred to as multi-modal sensory data. This data was used in order to achieve certain tasks within the scope of this paper which includes object detection, range detection and path planning. The objects are detected from the image obtained using the camera, the range of the objects are taken from the LIDAR sensor. Correspondence is established between the objects detected from different sensors and are projected to a 2-D global map. This 2-D global map is further used as reference for local path planning by using D* Lite path planning algorithm to generate a path from a given start position to a given goal position. The proposed system was able to generate a local path for every frame in the sample video scene for different test cases.

The studies presented in this paper was able to meet the expected outcome and performance metrics of the proposed system. Such metrics include reproducibility of the results under similar sensory data which proved to be valid from different test cases both obtained from KITTI dataset. The metrics of the system can also be universally repeated for any dataset other than KITTI dataset provided the sensory data obtained from those consist of information from camera, LIDAR and GPS taken simultaneously which was the case in KITTI dataset. The time and space complexity of the system is linear as when the number of inputs grows, the proposed system takes longer time to complete and longer memory space.

The test case usage of the multi-modal sensory data proved to be true in simulation. The same can be implemented on a real time situation, collecting real time data provided the global map is known, also considering factors such as the dynamics of the system and environment to provide better assistive capabilities to the user at any given point of time. By the usage of dynamic path planning algorithms, the system can continuously monitor the change in position of the obstacles, which in turn helps collision avoidance.

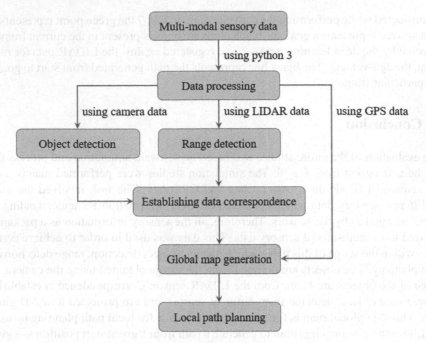

Fig. 8. Block diagram of the proposed system

References

1. The KITTI Vision Benchmark Suite Homepage. http://www.cvlibs.net/datasets/kitti/. Accessed 22 Apr 2020
2. Geiger, A., Lenz, P., Urtasun, R.: Are we ready for Autonomous Driving? The KITTI Vision Benchmark Suite. In: Proceedings of the IEEE Conference on Computer Vision and Pattern Recognition, pp. 3354–3361. IEEE, USA (2012)
3. Pandey, G., McBride, J.R., Eustice, R.M.: Ford campus vision and lidar data set. Int. J. Robot. Res. **30**(13), 1543–1552 (2011)
4. Maddern, W., Pascoe, G., Linegar, C., Newman, P.: 1 year, 1000 km: the Oxford RobotCar dataset. Int. J. Robot. Res. **36**(1), 3–15 (2017)
5. Object Tracking Evaluation 2012. http://www.cvlibs.net/datasets/kitti/eval_tracking.php. Accessed 22 Apr 2020
6. Ren, S., He, K., Girshick, R., Sun, J.: Faster R-CNN: towards real-time object detection with region proposal networks. IEEE Trans. Pattern Anal. Mach. Intell. **39**(6), 1137–1149 (2017)
7. Lin, T.-Y., et al.: Microsoft COCO: common objects in context. In: Fleet, D., Pajdla, T., Schiele, B., Tuytelaars, T. (eds.) ECCV 2014. LNCS, vol. 8693, pp. 740–755. Springer, Cham (2014). https://doi.org/10.1007/978-3-319-10602-1_48
8. Huang, J., et al.: Speed/accuracy trade-offs for modern convolutional object detectors. In: Proceedings of the IEEE Conference on Computer Vision and Pattern Recognition, pp. 3296–3297. IEEE Computer Society, USA (2017)
9. Geiger, A., Lenz, P., Stiller, C., Urtasun, R.: Vision meets robotics: the KITTI dataset. Int. J. Robot. Res. **32**(11), 1231–1237 (2013)
10. Koenig, S., Likhachev, M., Furcy, D.: Lifelong planning A*. Artif. Intell. **155**(1–2), 93–146 (2004)

11. Koenig, S., Likhachev, M.: D* lite. In: Proceedings of the AAAI Conference of Artificial Intelligence, pp. 476–483. AAAI Press, USA (2002)
12. Hart, P.E., Nilsson, N.J., Raphael, B.: A formal basis for the heuristic determination of minimum cost paths. IEEE Trans. Syst. Sci. Cybern. **4**(2), 100–107 (1968)
13. Stentz, A.: Optimal and efficient path planning for unknown and dynamic environments. In: Proceedings of the IEEE International Conference on Robotics and Automation, pp. 3310–3317. IEEE, USA (1994)

Implementation of Blockchain-Based Blood Donation Framework

Sivakamy Lakshminarayanan$^{(\boxtimes)}$ ⓘ, P. N. Kumar ⓘ, and N. M. Dhanya ⓘ

Department of Computer Science and Engineering, Amrita School of Engineering,
Amrita Vishwa Vidyapeetham, Coimbatore, India
sivakamy.lak@gmail.com, {pn_kumar,nm_dhanya}@cb.amrita.edu

Abstract. Existing blood management systems in India function as
Information Management systems that lack dynamic updates of blood
usage and detailed blood trail information, starting from donation to
consumption. There exists no communication platform for surplus blood
in one region to be requested from another region where blood is scarce,
leading to wastage of blood. Lack of transparency and proper blood
quality checks have led to several cases of blood infected with diseases
such as HIV being used for transfusion. This paper aims at mitigating
these issues using a blockchain-based blood management system. The
issue of tracking the blood trail is modelled as a supply-chain management issue. The proposed system, implemented in the Hyperledger Fabric
framework, brings more transparency to the blood donation process by
tracking the blood trail and also helps to curb unwarranted wastage of
blood by providing a unified platform for the exchange of blood and its
derivatives between blood banks. For ease of use, a web application is
also built for accessing the system.

Keywords: Blood donation framework · Blockchain · Hyperledger ·
Hyperledger Fabric

1 Introduction

Blood is an existential requirement for all living beings. Every few seconds, someone, somewhere needs blood. Since blood cannot be manufactured, the demand
for blood to save lives can be satiated only by donations from humans. Millions
of lives are saved every year by the transfusion of blood and its derivatives to
patients during complex medical and surgical procedures. An adequate supply
of blood is also needed for patients suffering from prolonged, life-endangering
illnesses such as leukemia.

Based on the data obtained through the WHO Global Database on Blood
Safety (GDBS) for the year 2015, for every sample of 1000 people, the blood
donation rate is 32.6 in high-income countries, 15.1 in upper-middle-income
countries, 8.1 in lower-middle-income countries and 4.4 in low-income countries.

A. Chandrabose et al. (Eds.): ICCIDS 2020, IFIP AICT 578, pp. 276–290, 2020.
https://doi.org/10.1007/978-3-030-63467-4_22

Although 112.5 million units of blood are collected every year globally, many patients who require blood transfusion do not have timely access to safe blood, especially in middle-income and upper-income countries [17].

1.1 Blood Transfusion Service in India

In India, the National Blood Transfusion Council (NBTC), established in 1996, is the federal body that oversees and manages the Blood Transfusion Service. The NBTC along with the State Blood Transfusion Councils (SBTCs) is tasked with reviewing the status of the blood transfusion services and conducting annual quality checks in blood banks. NBTC and National AIDS Control Organisation (NACO) are the technical bodies that are tasked with framing guidelines to ensure safe blood transfusion, provide infrastructure to blood centres, develop human resources and formulate and implement the blood policy in India [18].

In 2016, it was reported by the Ministry of Health and Family Welfare that only 10.9 million units of blood were donated while the requirement was 12 million units [19]. To address the huge gap in the demand and supply for blood in the country, several governmental and non-governmental blood donation organizations were established in different parts of the country. These organizations conduct frequent blood donation camps and also organize several awareness programmes to emphasize on the importance of blood donation. Most of the organizations have an online portal where willing donors register themselves and receive notifications when camps are organized. Realizing the importance of a unified portal, the Indian Government launched eRaktKosh [22], an initiative to connect, digitize and streamline the workflow of blood banks across the nation in 2016. The portal functions as a unified Management Information System (MIS). However, it can not track the blood chain completely.

1.2 Issues in Blood Donation

It is ironic to observe that in the country that proclaims a shortage of about 1.1 million (11 lakh) units of blood, over 28 lakh units of blood and its components have been wasted in five years [23]. Wastage of blood refers to the discarding of blood or any of its components rather than administering it to a patient. The main reason behind such a disparity is the lack of communication between regions where surplus blood stock is available and those where there is absolute scarcity. Surprisingly, even comparatively more developed states like Maharashtra, Andra Pradesh, Karnataka, and Tamilnadu reported wastage of not just whole blood, but also critical life-saving derivatives of blood such as RBCs and plasma (which have longer shelf-life). It can be observed clearly that India's blood scarcity woes can be reduced by a huge margin with just voluntary blood donations, provided there is an umbrella framework that supports communication and transfer of blood supplies across regions. For this, an efficient blood cold chain [24] management system that monitors the storage and transport of blood from donors to final transfusion sites is required.

It has to be noted that the inter-blood bank transfers have been encouraged by the Government. Despite this, some blood banks refuse to accept units from other banks with the reasoning that they may not be safe. There is also no way to track the flow of information from donation to consumption. This highlights the need for transparency in the blood inspection process. The benefits of this two-fold. Apart from providing a trust mechanism for encouraging blood stock exchange between blood banks, it also helps in preventing accidental transfusion of blood infected with diseases such as HIV, hepatitis, and syphilis. The latter holds extreme importance as over 14,474 cases of HIV contraction through transfusion of infected blood have been reported from 2010–2017 [25].

Under the National Blood Transfusion Services Act 2007, people who are found guilty of donating or selling blood in exchange for money may be imprisoned up to three months with a fine. But there have been many cases of forced blood extraction over the years. Though paid donations were banned by a Supreme Court ruling in 1996, reports of the practice continuing have surfaced. There have also been several cases where patients were forced to find replacement blood donors for stock replenishment even during emergencies despite efforts from the government to insulate patients [19]. These are major pitfalls in the Blood Transfusion Service which need attention.

To address these issues, the paper proposes to implement a unified, secure, end-to-end, permissioned blockchain-based blood chain management system that curbs unwarranted wastage of blood, ensures transparency in the donation process and ensures track-ability of donated blood. Such a system would also prevent illegal sales of blood/plasma and ensure that donors do not donate too frequently by monitoring their donation history.

2 Literature Survey

Most literature related to Blood Donation and Transfusion delved into providing a platform for people in need of blood to contact blood banks and donors in their vicinity ("replacement" donors) to request for blood. [12] proposed an Android mobile application that acts as a portal where users can view details about blood banks, donors and request blood from them. This system also consisted of a reporting mechanism using which the blood bank admins could monitor the inventory. T. Hilda Jenipha and R. Backiyalakshmi [13] implemented a cloud-based Android application that tracks the current location of registered donors and finds potential donors in the vicinity of where the blood request originated. [14] suggested the use of CART Decision Tree to identify whether a registered donor is likely to donate blood in case the need arises. This system is also modelled for "replacement" donations. The aforementioned solutions do not address the issues of blood cold chain management or the wastage of blood and also have no control over the health of the donors and the integrity of the donation process.

Research pertaining to the Blood Supply chain has focused on challenges such as transportation time and ensuring blood supply in the event of disasters. [7]

proposed a model that determines optimal locations for setting up blood facilities and blood inventory levels that will satiate the demand. [11] and [10] proposed an RFID-based information management system for the blood issue chain. But the major pitfall of such a system is security, as data could be modified.

Kim, Seungeun, and Dongsoo Kim [2] proposed the high-level design of a blockchain-based blood cold chain management using the shared-ledger concept. The architecture, however, does not track donor details, facilitate donor matching or track the wastage of by-products of blood. [3] proposes the design of KanChain, a new Ethereum-based permissioned framework for blood distribution. The proposal is to track the exchange of tokens, called KanCoins, between stakeholders to trace the blood chain. Peltoniemi, Teijo, and Jarkko Ihalainen, in [1], conducted an exploratory study on how distributed ledger technologies could be used within the supply chain of plasma derivatives. It also elucidates the concept of providing incentives to active donors to encourage more voluntary donation.

This paper aims to implement a unified framework for monitoring and managing donated blood by harnessing the potential of open-source blockchain infrastructure that provides better transparency, end-to-end tracking and reliability when compared to existing solutions.

3   Blockchain Technology

Blockchain is essentially a decentralized, public ledger in which transactions are accepted only when they are validated by every node in the network, making it immutable. It can be thought of as a time-stamped chain of data records, called blocks, linked together using cryptographic hashes and stored in a cluster of nodes. Each block/record consists of a set of transactions (defined in Smart Contracts) with a pointer to the next block thereby forming a linked-list of blocks [15,16]. Hence the name "blockchain".

Blockchain framework stores a digital footprint of all transactions in a block in a binary tree called "Merkle Tree". Leaf nodes in a Merkle tree are hashes of transactions while interior nodes are hashes of their child nodes [6]. Constructed bottom-up, nodes are repeated hashed until only one node is left, called the Merkle Root, which is stored in the Block Header. SHA-2 is used as the default hash function. Since Merkle Root is a hash, it can be used to test if a specific transaction is present in the block or not.

There are three classes of blockchain-based on the access levels of nodes that are allowed in the network - public, private and consortium [6]. Public blockchains are open to all with no central authority for control. Nodes do not require permissions to access data or transact i.e. it a permission-less blockchain. Private blockchains on the other hand, place restrictions and access controls on who can join the network and transact. Since users require permissions to join the network, private blockchains are permissioned in nature. Private blockchain frameworks are either shipped with Identity Management tools or allow third-party plug-ins that offer Membership Service Providers.

In consortium blockchains, the rights to authorize transactions are held only with specific nodes while other nodes can only initiate transactions and review transaction history. Due to the restricted number of nodes, private and consortium blockchains are more susceptible to data tampering than public blockchains. However, the former is more efficient and scalable.

The nodes in blockchain use Public Key Cryptography for digitally signing transactions. A node wishing to transact executes it and floods it in the network after digitally signing it with its private key. The transaction is validated by other nodes in the network. A consensus algorithm is executed to ensure that the transactions are validated by an adequate number of nodes. Once validated, the block is updated as a part of the shared ledger by all nodes in the network. It is worthy to note that a node in a blockchain may be member node which can only initiate transactions or a validator node, which can initiate and validate transactions.

4   Proposed System

The Blood Management System is designed as a private, permissioned blockchain allowing access only to specific stakeholders as illustrated in Fig. 1. The main reason for choosing permissioned blockchain is to prevent public access to sensitive information such as donor and donation details. The use of blockchain adds more transparency to the blood trail as it provides a trust-able tracking mechanism. It also enhances the security of the system as the fundamental design of a blockchain prevents unauthorized/illegal data modification.

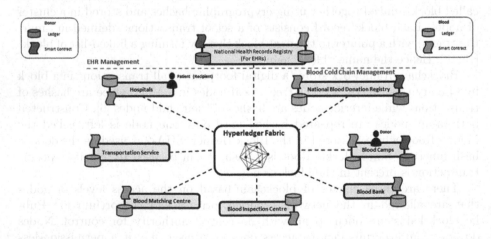

Fig. 1. Conceptual architecture of the proposed Blood Donation System.

4.1 Entities in the Proposed System

There are two main requirements to ensure that the proposed closed-loop system functions securely as intended, without any scope for malpractice. Firstly, it is assumed that there exists a National Health Records (NHR) Registry which maintains Electronic Health Records (EHRs) [8,9] of all Indian Residents. The EHRs are used to monitor the medical history of every person in the country. The design of EHR is out of the scope of this paper. For implementation, a database consisting of basic details of people along with a flag indicating their eligibility to donate blood was used. Secondly, anyone wishing to donate blood must be registered with the National Blood Donation Registry (NBDR). A unique Blood Donor ID, linked to the EHR ID, is allocated to each donor by this Registry. The advantage of this is two-fold. The donor's medical history can be readily verified while collecting blood samples using the EHR ID. The Blood Donor ID facilitates tracking donors and their donation history. This way, the frequency of donation by every individual can be monitored. This will help in curbing frequent donations for money and forced extractions. Also, donors are notified once their blood has been used to save lives, thereby offering a gratifying experience. If the donated blood is found unfit for usage during an inspection, the framework is designed to notify the donor immediately. This will enable them to take necessary medical action against any possible illnesses they may have contracted.

The other entities in the system are the Blood Camps where blood is collected, Blood Inspection Centres (BICs) where the collected blood is tested, Blood Banks where tested blood is stored, Hospitals, Blood Matching Centres (BMCs) and the Transportation Services.

4.2 The Workflow of the Proposed System

This section describes the overall work-flow required in a Blood Donation System as illustrated in Fig. 2. The Donor registers with the National Blood Donation Registry using their unique EHR ID. The medical details of the person are verified with the EHR managed by the National Health Records Registry using the EHR ID. If the person is found to be eligible to donate blood, a Donor ID is allocated to them. The Donor must use a valid Donor ID to donate blood. Before donation, the Donor's medical parameters such as temperature, blood pressure, and heart rate are verified. The Donor's eligibility to donate blood is verified using the EHR ID. It is also verified whether the donor's last donation was over 3 months ago. If the donor is found eligible for donation, their blood is collected and a unique Blood ID is allocated to the collected sample. Else, blood is not collected. Blood collection can happen in Blood Camps as illustrated in Fig. 2 or directly in Blood Banks. The collected blood sample is then transported to the Blood Inspection Centre (BIC) by the Transportation Service provider in a vehicle whose details are tagged to the Blood ID in the blockchain. The blood is then inspected at the BIC based on guidelines by NBTC [28] and the report is generated. The proof of inspection along with its HASH is logged into the blockchain. If the blood is found unfit for usage, the donor is notified and the

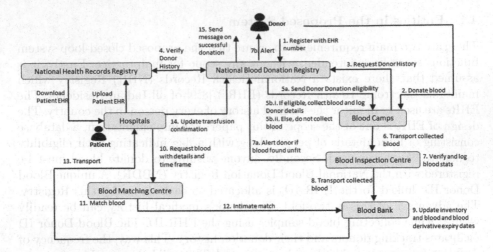

Fig. 2. Conceptual Workflow the proposed Blood Donation System.

blood is discarded. If the blood is found fit, it is then transported to the blood bank and stored. In case of any requirement, the hospital posts a request to the Blood Matching Centre (BMC) for blood. The BMC finds the best blood match based on factors such as blood group, request location and expiry date. Once a match is found, the blood is transported from the blood bank to the hospital. For ensuring transparency, the request for blood and its consumption are tagged to the recipient. This is done to ensure the patient's privacy and also prevent requested blood from being used for another patient. Once the blood is consumed, the donor is notified regarding the same.

4.3   Privacy and Access Controls

It has to be noted that the EHR, Donor and Blood records hold extremely sensitive information and thus must be privacy protected. For this, access controls are enforced by configuring Access Control List (ACL) files. For instance, the entity collecting blood from Donors can only READ the EHR ID and the flag variable mentioning eligibility to donate blood from the EHR records. Similarly, all Donor information is abstracted from transport service providers; only the Blood IDs and sender and receiver details are made available. Donors are given READ ONLY access to their donation data i.e. blood IDs and dates and statuses of donations. Hospitals are only allowed to do two transactions - requesting blood and updating blood usage post allocation. Donor information is thus completely hidden (DENY ALL ACCESS) from hospitals. In the current deployment, the Network Administrator is given ALLOW ALL access. However, in future deployments, we aim at encrypting information using pluggable cryptographic packages to ensure that only entities with valid keys will be able to decrypt and therefore access the data.

4.4 Blockchain Framework

The architecture of the proposed system revolves around "Hyperledger Fabric", an open-source platform for distributed ledger solutions [5,6]. Hyperledger Fabric is preferred over other enterprise-grade blockchain platforms like Ethereum because it is tailor-made for B2B solutions and allows access controls to be enforced for data confidentiality and privacy through the use of membership services. Also, the proposed solution does not require the crypto-currency stack that Ethereum provides. Hyperledger Fabric also allows the different modules and components to be customized as per requirements. It supports the deployment of permissioned blockchains in which the participating entities have known identities. This is well-suited for our use-case as only specific stakeholders must be allowed to access the network. Other permissioned blockchain frameworks like Quorum were designed primarily for financial use-cases. For managing user authentication and authorization, Hyperledger Fabric uses X.509 certificates. It also offers the flexibility of configuring authentication services to suit the requirements [27]. The framework supports the concept of Channels for data privacy - data in a Channel is only made available to those entities which subscribe to it. Hyperledger Fabric also supports smart contract logic using the concept of chaincodes. In addition, it provides the option of including access controls as a part of the chaincode to restrict data visibility. Though Hyperledger Fabric itself does not provide data encryption, it supports the use of pluggable cryptographic interfacing packages like the BlockChain Cryptographic Service Provider (BCCSP). However, such additional configurations are not included in the scope of this paper.

5 Implementation Details

The initial development of the system was done in Hyperledger Composer [20], a toolset used to model the back-end business logic that can be run on the Hyperledger Fabric blockchain infrastructure. The Network module of Hyperledger Composer consists of definitions of Assets, Participants, and Transactions. In our system, Blood, Donor, EHRs, and transport vehicles are considered as Assets while the rest of the stakeholders mentioned in Sect. 4.1 are considered as Participants. The functional definition or logic of the Transaction is called the Business Logic. The Access Control Lists consist of information regarding access privileges of Assets, Participants and Users. Hyperledger Composer also provides a way to create optional Query files in which SQL-like queries can be defined for the system. These components are packaged into a BNA file (Business Network Archive), which can be used to create the Fabric network and deploy the system. The installation and deployment of the Blood Donation System (Project BloodLine) are shown in Fig. 3. We used Fabric version 1.2 deployed in Ubuntu 18.04 in a machine with Intel Core i7 processor and 8 GB RAM.

It is worthy to note that the Business Logic defined in Hyperledger Composer forms the basis of Smart Contracts (termed as Chaincodes) in Hyperleder

Fig. 3. Installation and deployment of the Blood Donation System in Hyperledger Fabric using the BNA file.

Fabric. This logic consists of functional definitions of all transactions in the network. Such transactions were identified and developed using Javascript. Some of the core transactions include the procedure for logging donated blood, booking transport vehicle for transferring blood from one place (say a Blood Bank) to another (a Hospital), requesting blood and allocating blood based on the request.

As this is the first attempt at implementing a blockchain-based Blood Donation System, we have developed a baseline algorithm for blood allocation when a request for blood is issued by a hospital. For simplicity, a scenario with two blood banks and ten hospitals within a 5 km radius was considered. A priority queue of the blood stock is maintained for blood allocation - the blood of the required blood group, with the earliest expiry date, is allocated first. This algorithm can be extended further to include factors such as distance between hospital and blood bank, transportation time and urgency of the requirement. The design of allocating transport vehicle to transfer blood across locations is a study by itself and shall not be discussed as a part of this paper.

Another important contribution of the proposed system is to notify the donor when donated blood has been used and also alert the donor when the blood is found to carry any reactive disease as a part of the inspection process. The former brings a sense of gratification for the donors and will encourage them to donate regularly as the appreciation of donors' contributions is a vital catalyst for improved donor retention [4]. The latter stems out of social and ethical responsibility which would help the donor take timely and appropriate medical care.

6   Results and Discussion

The implemented system was tested with sample EHR and Donor datasets created using volunteer data. For simplicity, a scenario with two blood banks and ten hospitals within a 5 km radius was considered. The blockchain was initialised with the genesis blocks including those for Blood and Donor. The system was

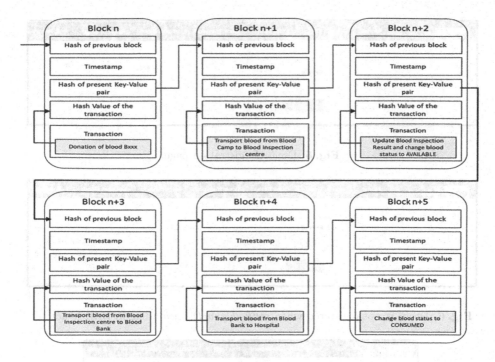

Fig. 4. Block structure for Blood Donation System.

configured with a block size of 10 transactions per block. For testing the implementation, 100 blood requests were simulated, some of which are discussed in this Section. Each transaction results in the creation of a new block, which is linked to the previous block as shown in Fig. 4. The average transaction throughput was found to be *80 tps*. More stress tests must be conducted on the system with parallel requests and heavier loads in the future. Since the system is based on Hyperledger Fabric, the blood trail (transaction results) are recorded in the Key Value Store (KVS). For ease of use of the proposed system, a web application is designed. The user in the screenshots is logged in as an Admin and has privileges to initiate all transactions.

Figure 5 illustrates the Donor Registration page. Donors are registered using their EHR Id and are allowed to register only when the *isEligibleForDonation* flag is set in their EHR. In case they are not eligible, the person is not allowed to register as shown in Fig. 6. Also, donors without a valid EHR Id are not allowed to register with the network as shown in Fig. 7.

For recording the collection of blood, the form in Fig. 8 is used. If the medical parameters such as temperature, blood pressure and pulse rate are abnormal, blood is not collected. Instead, a notification is displayed as shown. If the previous donation by the donor fell within three months, a notification to not collect blood is issued as illustrated in Fig. 9. Also, the EHR of the donor is looked-up again to verify that the donor is still eligible to donate. This is done to ensure

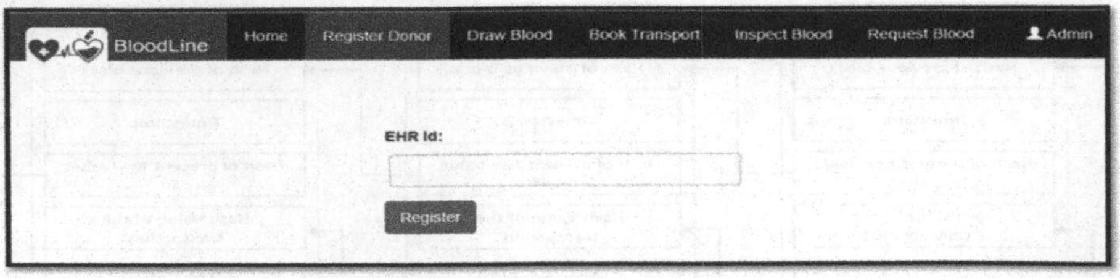

Fig. 5. Donor Registration page.

Fig. 6. Donor is not registered when *isEligibleForDonation* flag in EHR is not set.

Fig. 7. Donor is not registered when EHR Id is not found.

that even the latest medical updates are considered while choosing donors. In the future, we also plan to automatically flag and notify a donor in case the *isEligibleForDonation* flag in EHR is updated based on medical records.

The web app also contains a portal to record blood inspection results. The HASH of the inspection report helps in verifying its authenticity and increases the trustworthiness of the inspection process - any small modification of the report will lead to a different HASH. A sample report was used to show the generation and storage of the file's HASH.

To help hospitals to request blood on behalf of the patient, the page in Fig. 10 is used. The matched blood details are then shared with the hospital. The proposed blood matching algorithm was successful in allocating blood from the second blood bank even though the first blood bank did not have stock. The advantages of this are two-fold. Firstly, it ensures that blood is available to regions where the stock may not be available. Also, by ensuring that donated

Fig. 8. Example of preventing blood donation when the donor's medical parameters are off-range.

Fig. 9. Example of preventing blood donation when the previous donation by the donor was less than 3 months ago.

blood is put to use irrespective of location constraints, it prevents unnecessary wastage of blood.

The proposed solution ensures that donated blood is traced right from the time of donation to the time of consumption. It also provides a mechanism to track the donation eligibility of donors in terms of medical fitness and time elapsed since the previous donation. The proof of quality checks on the blood is also included as a part of the system for increased trust while sharing blood across regions. The blood matching module ensures that the allocation of blood is done fairly and transparently. Overall, the proposed system can be used as an end-to-end blood trail monitoring solution that is capable of increasing the effectiveness of blood utilization and preventing malpractices.

Fig. 10. Interface for the hospitals to request blood on behalf of patients.

7 Conclusion

In this paper, we have successfully implemented a Blockchain-based Blood Donation System. To ensure ease of use, we have also implemented a Web Application as the front-end for the system. This App can be used to track donated blood across the entire process of blood transfusion thereby ensuring complete transparency. The proposed system can also be enhanced for similar areas like organ donation on a national and international scale.

8 Future Work

Since this was the first attempt at implementing the blockchain framework for Blood Donation Management, only tracking of blood was implemented. As a part of the next phase, we intend to add a tracking facility for derivatives of blood for a more rounded framework. This is important as different derivatives of blood have different shelf-life. The proposed framework for blood cold-chain management is in its nascence and there is a need to come up with performance metrics dedicated to it. Bench-marking and design of performance metrics are required for further analysis. Also, blood matching was done based only on the closest stock expiry date. However, additional factors such as location, real-time traffic, the severity of the requirement, stock availability and predicted demand in Blood Banks could be incorporated as a part of the blood matching algorithm to make it more efficient. We intend to implement this as a part of the next phase.

References

1. Peltoniemi, T., Ihalainen, J.: Evaluating Blockchain for the Governance of the Plasma Derivatives Supply Chain: How Distributed Ledger Technology Can Mitigate Plasma Supply Chain Risks. Blockchain in Healthcare Today (2019)
2. Kim, S., Kim, D.: Design of an innovative blood cold chain management system using blockchain technologies. ICIC Express Lett. Part B, Appl. Int. J. Res. Surv. 9(10), 1067–1073 (2018)

3. Çağlıyangil, M., Erdem, S., Özdağoğlu, G.: A blockchain based framework for blood distribution. In: Hacioglu, U. (ed.) Digital Business Strategies in Blockchain Ecosystems. CMS, pp. 63–82. Springer, Cham (2020). https://doi.org/10.1007/978-3-030-29739-8_4

4. Naskrent, J., Siebelt, P.: The influence of commitment, trust, satisfaction, and involvement on donor retention. VOLUNTAS: Int. J. Voluntary Nonprofit Organ. **22**(4), 757–778 (2011)

5. Androulaki, E., et al.: Hyperledger fabric: a distributed operating system for permissioned blockchains. In: Proceedings of the Thirteenth EuroSys Conference, pp. 1–15. ACM (2018)

6. Sajana, P., Sindhu, M., Sethumadhavan, M.: On blockchain application: hyperledger fabric and ethereum. Int. J. Pure Appl. Math. **118**(18), 2965–2970 (2018)

7. Jabbarzadeh, A., Fahimnia, B., Seuring, S.: Dynamic supply chain network design for the supply of blood in disasters: a robust model with real world application. Transp. Res. Part E: Logist. Transp. Rev. **70**, 225–244 (2018)

8. Ekblaw, A., Azaria, A., Halamka, J.D., Lippman, A.: A case study for blockchain in healthcare:"MedRec" prototype for electronic health records and medical research data. In: Proceedings of IEEE Open and Big Data Conference, vol. 13, p. 13 (2016)

9. Atherton, J.: Development of the electronic health record. AMA J. Ethics **13**(3), 186–189 (2011)

10. Davis, R., Geiger, B., Gutierrez, A., Heaser, J., Veeramani, D.: Tracking blood products in blood centres using radio frequency identification: a comprehensive assessment. Vox Sanguinis **97**(1), 50–60 (2009)

11. Dusseljee-Peute, L.W., Van der Togt, R., Jansen, B., Jaspers, M.W.: The value of radio frequency identification in quality management of the blood transfusion chain in an academic hospital setting. JMIR Med. Inform. **7**(3), e9510 (2019)

12. Casabuena, A., et al.: BloodBank PH: a framework for an android-based application for the facilitation of blood services in the Philippines. In: TENCON 2018–2018 IEEE Region 10 Conference, pp. 1637–1641. IEEE (2018)

13. Jenipha, T.H., Backiyalakshmi, R.: Android blood donor life saving application in cloud computing. Am. J. Eng. Res. (AJER) **3**(02), 105–108 (2014)

14. Santhanam, T., Sundaram, S.: Application of CART algorithm in blood donors classification. J. Comput. Sci. **6**(5), 548 (2010)

15. Zheng, Z., Xie, S., Dai, H., Chen, X., Wang, H.: An overview of blockchain technology: architecture, consensus, and future trends. In: 2017 IEEE International Congress on Big Data (BigData Congress), pp. 557–564. IEEE (2017)

16. Nofer, M., Gomber, P., Hinz, O., Schiereck, D.: Blockchain. Bus. Inf. Syst. Eng. **59**(3), 183–187 (2017). https://doi.org/10.1007/s12599-017-0467-3

17. Who.int. n.d. Blood Safety And Availability. https://www.who.int/news-room/fact-sheets/detail/blood-safety-and-availability. Accessed 30 November 2019

18. Nbtc.naco.gov.in. n.d. National Blood Transfusion Council (NBTC) - Mohfw, India. http://nbtc.naco.gov.in/. Accessed 30 November 2019

19. En.wikipedia.org. 2016. Blood Donation In India. https://en.wikipedia.org/wiki/Blood_donation_in_India. Accessed 30 November 2019

20. Hyperledger. n.d. Hyperledger Composer - Hyperledger. https://www.hyperledger.org/projects/composer. Accessed 30 November 2019

21. Hyperledger.github.io. n.d. Using Playground — Hyperledger Composer. https://hyperledger.github.io/composer/v0.19/playground/playground-index. Accessed 30 November 2019

22. Eraktkosh.in. n.d. E-Rakt Kosh: Centralized Blood Bank Management System. https://www.eraktkosh.in/BLDAHIMS/bloodbank/transactions/bbpublicindex. html. Accessed 30 November 2019
23. The Times of India. n.d. Blood Collection: Blood Banks Waste 2.8 Million Units In 5 Years — Mumbai News - Times Of India. https://timesofindia.indiatimes. com/city/mumbai/blood-banks-waste-2-8m-units-in-5-yrs/articleshow/58333394. cms. Accessed 30 November 2019
24. Who.int. n.d. WHO — Blood Cold Chain. https://www.who.int/bloodsafety/ processing/cold_chain/en/. Accessed 30 November 2019
25. Business-standard.com. n.d. 14,474 Cases Of HIV Through Blood Transfusion, But Govt Denies Crisis. https://www.business-standard.com/article/current-affairs/14-474-cases-of-hiv-through-blood-transfusion-but-govt-denies-crisis-116120200521_1.html. Accessed 30 November 2019
26. Smtpjs.Com - Send Email From Javascript. Smtpjs.com: https://smtpjs.com. Accessed 30 November 2019
27. Hyperledger-fabric.readthedocs.io. n.d. Membership Service Providers (MSP) - Hyperledger-Fabricdocs Master Documentation. https://hyperledger-fabric. readthedocs.io/en/release-1.4/msp.html. Accessed 30 November 2019
28. Naco.gov.in. n.d. Guidelines for Blood Donor Selection and Blood Donor Referrals. http://naco.gov.in/blood-transfusion-services-publications. Accessed 30 November 2019

IoT Based Crop-Field Monitoring and Precise Irrigation System Using Crop Water Requirement

Kanchana Rajaram$^{(\boxtimes)}$ (iD) and R. Sundareswaran (iD)

Department of Computer Science and Engineering, Sri Sivasubramaniya Nadar
College of Engineering, Anna University, Chennai 603110, Tamil Nadu, India
rkanch@ssn.edu.in, sundareswaran1616@cse.ssn.edu.in

Abstract. Existing practices of crop irrigation is manual and based on generic traditional recommendations. Crops when provided lesser water, shows reduced growth and reduced uptake of calcium. Excessive irrigation leads to root death and water wastage. Hence, irrigating crops with precise water becomes an important problem. Towards this objective, an IoT based crop field monitoring and precise irrigation system is proposed that monitors crop-field and computes precise crop water requirement based on its life cycle and climatic conditions. Using this computed crop water requirement, a pump motor is operated automatically whenever soil moisture decreases below permanent welting point. The motor is shut down once the required water is pumped out to crops. The proposed system is installed in a crop-field of brindle plant and the crop is irrigated for 6 months. It is observed that 53% of water has been saved from wastage.

Keywords: IoT · Precision irrigation · Crop water requirement ·
Gateway · Wireless sensor network · Amazon web service ·
Evapotranspiration

1 Introduction

The Internet of Things (IoT) is a system of interrelated computing devices, mechanical and digital machines, objects, animals or people that are provided with unique identifiers and the ability to transfer data over a network without requiring human-to-human or human-to-computer interaction The Internet of Things extends internet connectivity beyond traditional devices like desktop and laptop computers, smart phones and tablets to a diverse range of devices and things that utilize embedded technology to communicate and interact with the external environment, all via the Internet. Examples include connected security systems, thermostats, cars, electronic appliances, lights in household and commercial environments, alarm clocks, speaker systems, vending machines and more.

© IFIP International Federation for Information Processing 2020
Published by Springer Nature Switzerland AG 2020
A. Chandrabose et al. (Eds.): ICCIDS 2020, IFIP AICT 578, pp. 291–304, 2020.
https://doi.org/10.1007/978-3-030-63467-4_23

IoT is the expansion of communication network and internet application, which is a technique to sense the physical world by sensing technology and the intelligent devices through the interconnection, calculation, processing and knowledge mining. It achieves information exchange and seamless communication among people and objects or among the things. It controls the physical world in real-time enabling accurate management and scientific decision-making.

Smart irrigation refers to the supply of water to the agricultural field by controlling the pump motor automatically by using IoT technique. Excessive irrigation of crops results in wastage of water and human resources leading to water scarcity. Water requirement of a particular crop varies based on soil moisture and climatic parameters such as temperature, humidity and wind speed. In addition, it varies according to the crop's life cycle. None of the existing methods for automated irrigation have attempted to precisely compute water requirement of crops and irrigate them accordingly. Hence, an IoT based system has been proposed that monitors crop-field and automatically irrigates based on crop water requirement.

The rest of this paper is organized as follows: Sect. 2 surveys the existing literature and Sect. 3 introduces background concepts. Section 4 illustrates the proposed IoT framework and Sect. 5 explains the tested results of the proposed work. Section 6 concludes the work and points out future directions.

2   Literature Survey

The existing works on automated irrigation have been surveyed and presented below: Nisha et al. [1], Rao et al. [9] and Banumathy et al. [10] have proposed an automated crop-field monitoring and irrigation system using wireless sensor networks and gateways that irrigates plants automatically for fixed duration of time. However, it does not prevent wastage of water. Morais et al. [2] uses TDMA and CSMA protocol to acquire data which results in slowed down communication with frequent disconnections.

Tanveer et al.'s work [3] collects sensed data as and when climatic parameters cross a safety threshold and controls irrigation. Jiao et al. [4] proposed a three layered system for precise irrigation with remote monitoring. These two works have been tested for a controlled environment at green house and not for large natural fields. In Pavithra et al.'s work [5], when the moisture sensor senses low moisture content of the soil, it signals a mobile phone via a microcontroller which activates an alert/buzzer. In this work, bluetooth based communication is proposed. However, bluetooth has a low range of communication and it requires repeaters/routers to send the data precisely. Tensiometric and volumetric techniques proposed in [6], monitors soil moisture and temperature for irrigating crops. However, this work may not provide precise irrigation as other climatic parameters are ignored. Pavankumar et al.'s work [7] considered humidity, temperature and soil moisture to compute water required to irrigate the crops. Verma et al. [8] proposed crop field monitoring system that irrigates crops and administers water-soluable fertilizers. However, the water requirement of crops has been ignored. Bharathi et al.'s work [11] used SCADA and PLC for automated irrigation. This work does not sense any data and focused mainly on water storage. In summary, none of the automated drip irrigation [13] and automated plant irrigation systems [12] have considered water requirement of crops.

3   Background

Computation of Crop Water Requirement (CWR) [15] is based on the following factors:

- The effect of climate on crop water requirements.
- Crop characteristics.
- Local conditions for cultivation and irrigation such as distance, altitude, size of fields, and salinity.

Evaporation [14] refers to conversion of water into water vapour directly from surfaces like river, soil etc. **Transpiration** implies conversion of water into water vapour through plant cells and plant tissue. **Evapotranspiration** [15] is a combination of evaporation and transpiration. It can be computed using the following methods:

- Blaney-Criddle Method
- Radiation Method
- Pan Evaporation Method
- Penman Method

Penman Method is used for computation of Evapotranspiration (ET_o) based on temperature, humidity, windspeed and radiation/sunshine duration. It has been proved that, compared to other methods, Penman method provides better and accurate results [15]. Hence, crop water requirement is computed using Penman method in our work.

4   Proposed Work

4.1   Architecture Design

The proposed work consists of soil moisture sensor, temperature sensor, humidity sensor and a wind speed sensor. These sensors collect the real time data and the sensed values are sent to the gateway through RF communication medium as depicted in Fig. 1. From gateway, the values are sent to the amazon's AWS cloud server through GSM communication medium, where the CWR is calculated. The sensed data values are stored in the logger database in MySQL server at the cloud. Based on the sensed data, the pump motor is operated by controlling the relay through the motor controller connected to the gateway via RF communication medium.

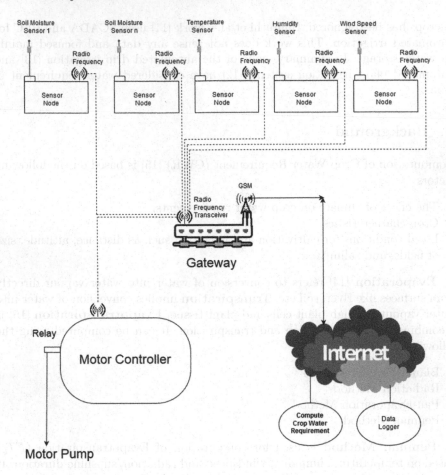

Fig. 1. Crop field monitoring and precise irrigation system - architecture

The proposed system comprises of the following four modules:

- Setting up of Wireless Sensor Network.
- Sensing and logging of data.
- Computation of CWR.
- Automated control of pump motor.

4.2 Setting up of Wireless Sensor Network

The soil moisture sensor is connected to a sensor node/mote as shown in Fig. 2. The mote is battery operated and is equipped with zigbee transmitter to transmit the soil moisture values to a gateway. Several motes are installed in the crop-field forming a WSN. The humidity sensor, temperature sensor and windspeed sensor are also connected to the gateway as shown in Fig. 3. The sensed values

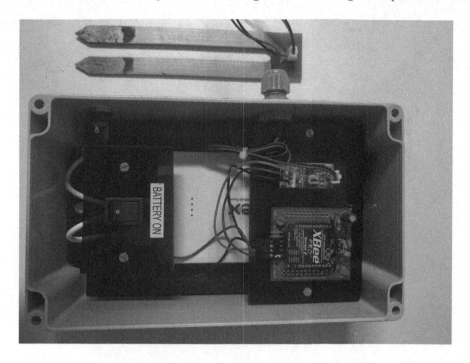

Fig. 2. Mote connected with soil moisture sensor

are communicated to amazon AWS cloud server through GSM module as shown in Fig. 4. A snapshot of AWS server is shown in Fig. 5. The gateway shown in Fig. 4. Consists of a relay to control the pump motor, on receiving a signal from CWR computation module in the AWS server. The AWS cloud server consists of an EC2 instance and RDS Database to store and retrieve the sensed values for the calculation of crop water requirement.

4.3 Sensing and Logging of Data

The environmental conditions are sensed and logged in the database available in the cloud server.

Mote for Soil Moisture Sensor. The sensed soil moisture value is in HEX format and it is caliberated into % value for the computation purpose. The following table is used to store the sensed values of different soil moisture sensors in % along with their IDs (Table 1).

Table 1. Schema for data logging from soil moisture sensors

Timestamp	Soil moisture sensor ID	Soil moisture (%)
...	...	...

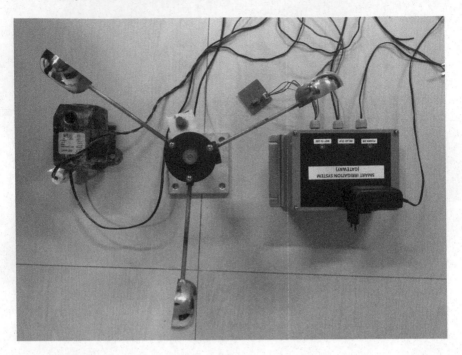

Fig. 3. Gateway connected with sensors and a pump motor

Mote for Humidity Sensor. The humidity sensor senses the humidity value in the environment. The sensed value is calibrated using DHT.lib in arduino controller and displayed in %. Figure 6 depicts the DHT11 Humidity sensor that is interfaced in the gateway.

The following table is used to log the humidity value in % along with timestamp (Table 2).

Table 2. Schema for data logging from humidity sensor

Timestamp	Humidity (%)
...	...

Mote for Temperature Sensor. The temperature sensor senses the temperature value in the environment and displays the output in °C which is calibrated in arduino controller. Figure 6 depicts the LM35 Temperature sensor that is interfaced with the gateway. The following table is used to store the temperature value in °C along with timestamp (Table 3).

Table 3. Schema for data logging from temperature sensors

Timestamp	Temperature (°C)
...	...

Fig. 4. Gateway comprising GSM module

Mote for Windspeed Sensor. The windspeed sensor as shown in Fig. 7 senses the speed of the wind in the environment giving the output in pulses that is calibrated into KMPH (Kilometer Per Hour) in arduino controller. This windspeed sensor is connected to the gateway.

The following table shows the schema to store the windspeed value in KMPH along with timestamp (Table 4).

Table 4. Schema for data logging from windspeed sensor

Timestamp	Windspeed (KMPH)
...	...

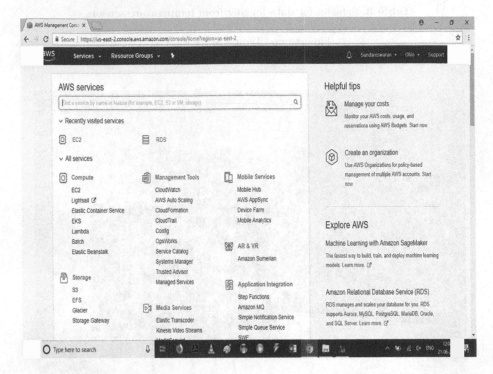

Fig. 5. Snapshot of Amazon Web Services (AWS) server

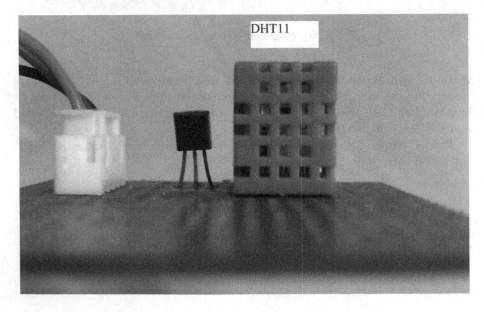

Fig. 6. Temperature sensor mote

Fig. 7. Windspeed sensor mote

4.4   Computation of Crop Water Requirement

The water requirement for a crop depends upon its life cycle and the climatic conditions. Crop Water Requirement (ET_{crop}) is calculated as

$$ET_{crop} = K_c.ET_o mm/day \qquad (1)$$

where,
K_c= Crop Coefficient
ET_o = Reference crop evapotranspiration

Evapotranspiration (ET_o) is computed using Penman method as given below:

$$ET_o = c[W.R_n + (1 - W).f(u).(e_a - e_d)]mm/day \qquad (2)$$

where,
ET_o = reference crop evapotranspiration in mm/day
W = temperature-related weighting factor
R_n = net radiation in equivalent evaporation in mm/day
f(u) = wind-related function
e_a = saturation vapor pressure at daily mean air temperature in millibars
e_d = mean actual vapour pressure of the air, in millibars
c = adjustment factor to compensate for the effect of day and night weather conditions

ET_o is calculated using the sensed values of temperature, humidity and wind speed. The mean of minimum and maximum sensed temperature is used to look-up weighting factor from the table given in [15]. Similarly, according to minimum sensed humidity value and mean temperature values, e_a is looked up from the table in [15]. e_d is computed using the formula:

$$e_d = e_a * RH_{mean}/100 \tag{3}$$

Wind related function, f(u) is computed using sensed wind speed value

$$f(u) = 0.27(1 + U/100) \tag{4}$$

where U = 24 h wind run in km/day.
Adjustment factor c and radiation values R_n are looked up from the corresponding tables available in [15].
In the AWS server, the calculated CWR provides the output in mm/day which is converted into litres as follows:

$$CWR = ET_o * A * 1000 \text{ L} \tag{5}$$

where A is the area of crop field in square meters.

4.5   Automated Control of Pump Motor

The gateway receives CWR in litres from the CWR computation module residing in AWS cloud server. The time duration for which the pump motor must be operated is calculated from the HP (Horse Power) of the motor. For example, a pump motor of 1HP or 750 W, pumps at the rate of 56 L/min. The time duration has been calculated in the locally in the gateway for running the motor to pump calculated CWR. The pump motor is operated to supply the calculated CWR by running it for a pre-computed time duration. The relays that are connected to the pump motor are switched ON and OFF by sending RELAY10T and RELAY00T commands through GSM in the gateway.

5   Experimentation

The proposed system for crop-field monitoring and automated irrigation has been tested in a crop field of Solanum melonjena (Brinjal plant) cultivated in an area of 100 m^2 near our college campus. Four motes that are connected to soil moisture sensors were installed in the crop field as shown in Fig. 8. A gateway connecting these sensors wirelessly is shown in Fig. 9. Other sensors such as temperature, humidity, and windspeed and the pump motor which are also installed in the field and are connected directly to the gateway. The system was observed for a period of 6 months. The microcontroller in the gateway monitored soil moisture 4 times a day. The gateway senses other climatic values when soil moisture falls below permanent welting point; sends the sensed values to AWS server where

Fig. 8. Soil moisture sensor mote placed in the crop field

Fig. 9. Gateway connected with motor and placed in the crop field

Date	Timestamp	Temperature in C	Humidity in %	Windspeed in KMPH	Crop water requirement for 1 Sq.M in Litres
21-12-2017	09:30:35	26	60	5	6.814
21-12-2017	02:10:20	28	55	2	7.112
21-12-2017	06:30:42	27	59	4	6.834
21-12-2017	09:45:26	25	65	8	5.165
26-01-2018	08:30:19	22	75	4	4.153
26-01-2018	12:39:52	25	63	7	5.012
26-01-2018	05:20:57	24	66	6	6.741
26-01-2018	08:50:22	21	70	9	3.853
12-02-2018	09:06:35	24	59	8	5.654
12-02-2018	12:58:53	27	55	4	6.952
12-02-2018	05:38:23	25	58	6	6.521
12-02-2018	09:25:39	23	65	7	5.865
07-03-2018	10:00:25	26	60	6	6.842
07-03-2018	01:59:36	29	50	4	7.685
07-03-2018	06:15:43	27	51	3	6.894
07-03-2018	09:55:12	25	53	7	6.231
27-04-2018	08:50:25	27	56	4	6.753
27-04-2018	01:05:35	30	45	2	7.992
27-04-2018	05:50:55	28	48	3	7.234
27-04-2018	09:02:52	25	52	7	6.529
23-05-2018	09:25:24	28	49	4	6.991
23-05-2018	01:45:30	36	40	2	9.451
23-05-2018	05:45:15	30	42	3	7.254
23-05-2018	10:29:39	27	52	7	7.101

Fig. 10. CWR for a brinjal plant for 6 months

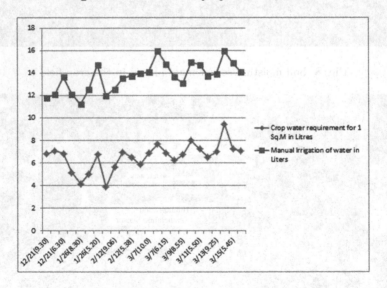

Fig. 11. Smart irrigation system versus traditional irrigation system

CWR is computed; the motor is operated by the gateway for a precomputed time duration based on CWR.

The snapshot of the table shown in Fig. 10 lists CWR values for 6 months, sensed 4 times everyday. In a traditional irrigation system, the farmer of the crop field watered the plants by operating the motor for 30 min. Over a period of six months, the amount of water used for irrigating the crops by the farmer as well as through our proposed system were observed. Total amount of water used by traditional irrigation system over a period of 6 months is 320.193 L while with our smart irrigation system is 157.633 L. Hence, with our smart irrigation system, 162.56 L of water was saved. The graph shown in Fig. 11 shows the comparison of water consumed with the proposed system and the traditional system. It is evident that smart irrigation system saves around 53% of water from wastage.

The overall cost involved in setting up the proposed system in the crop field is approximately Rs. 40,000. In order to cover more acres of crop field, the same setup can be used with an addition of six soil moisture sensors per acre. The additional cost incurred for extending the proposed system is only 15% of the initial setup cost. Since the pump motor is switched ON and OFF according to CWR, it is found that the proposed system saves around 30% of electrical energy as compared to the manual irrigation system.

6 Conclusion

Soil moisture is monitored using a sensor and when it decreases below permanent welting point, temperature and humidity sensor readings are obtained and the specific crop water requirement has been computed. The pump motor controller module takes the crop water requirement in litres as input and operates a pump motor via RF transceiver. It switches off the motor after it runs for a computed time duration to pump the required amount of water.

A wireless sensor network of the sensors has been set up and the proposed approach has been tested with the prototype setup in the lab. In addition, the proposed smart irrigation system was tested in the crop field of brinjal plant for six months. With the proposed approach for smart irrigation based on CWR and climatic parameters, it is proved that 53% of water can be saved when compared to traditional irrigation methods using generic recommendations. Moreover, this system proves to be cost effective and proficient in conserving water and reducing its wastage.

Scenarios such as dry motor, absence of power supply and shortage of water supply must be considered and are to be developed in future.

References

1. Nisha, G., Megala, J.: Wireless sensor network based automated irrigation and crop field monitoring system. In: Sixth International Conference on Advanced Computing (ICoAC), pp. 189–194 (2014)
2. Morais, R., Valente, A., Serdio, C.: A Wireless sensor network for smart irrigation and environmental monitoring: a position article. In: 5th European Federation for Information Technology in Agriculture, Food and Environment and 3rd World Congress on Computers in Agriculture and Natural Resources (EFITA/WCCA), pp. 845–850 (2005)
3. Tanveer, A., Choudhary, A., Pal, D., Gupta, R., Husain, F.: Automated farming using microcontroller and sensors. Int. J. Sci. Res. Manag. Stud. (IJSRMS) 2(1), 21–30 (2015)

4. Jiao, J., Ma, H.M., Qiao, Y., Du, Y.L., Kong, W., Wu, Z.C.: Design of Farm Environmental Monitoring System Based on the Internet of Things. Advance Journal of Food Science and Technology **6**(3), 368–373 (2014)
5. Pavithra, D.S., Srinath, M.S.: GSM based automatic irrigation control system for efficient use of resources and crop planning by using an android mobile. IOSR J. Mech. Civil Eng. (IOSR-JMCE) **11**(4), 49–55 (2014)
6. Darshna, S., Sangavi, T., Mohan, S., Soundharya, A., Desikan, S.: Smart irrigation system. IOSR J. Electron. Commun. Eng. (IOSR-JECE) **10**(3), 32–36 (2015)
7. Naik, P., et al.: Arduino based automatic irrigation system using IoT. Int. J. Sci. Res. Comput. Sci. Eng. Inf. Technol. IJSRCSEIT **2**(3), 881–886 (2017)
8. Ankit Kumar, V., Bhagavan, K., Akhil, V., Amrita, S.: Wireless network based smart irrigation system using IOT. Int. J. Eng. Technol. **7**(1.1), 342–345 (2018)
9. Rao, R.N., Sridhar, B.: IoT based smart crop-field monitoring and automation irrigation system. In: Second International Conference on Inventive Systems and Control ICISC, pp. 478–483 (2018)
10. Banumathi, P., Saravanan, D., Sathiyapriya, M., Saranya, V.: An android based automatic irrigation system using bayesian network with SMS and voice alert. Int. J. Sci. Res. Comput. Sci. Eng. Inf. Technol. IJSRCSEIT **2**(2), 573–578 (2017)
11. Bharathi, G., Prasunamba, C.G.: Automatic irrigation system for smart city using PLC AND SCADA. Int. J. Sci. Res. Comput. Sci. Eng. Inf. Technol. IJSRCSEIT **2**(4), 309–314 (2017)
12. Kumar, B.D., Srivastava, P., Agrawal, R., Tiwari, V.: Microcontroller based automatic plant irrigation system. Int. Res. J. Eng. Technol. (IRJET) **4**(5), 1436–1439 (2017)
13. Pooja, P., Pranali, D., Asmabi, S., Priyanka, N.: Future of the drip irrigation system: a proposed approach. Multidiscip. J. Res. Eng. Technol. **4**(1), 1055–1060 (2017)
14. Allen, R.G., Pereira, L.S., Raes, D., Smith, M.: Crop evapotranspiration-Guidelines for computing crop water requirements-FAO Irrigation and drainage. International Commission for Irrigation and Drainage **300**(9), (1998)
15. Doorenbos, J., Pruitt, W.O.: FAO irrigation and drainage. In: Food and Agriculture Organization of the United Nations Rome (1977)

A Machine Learning Study of Comorbidity of Dyslexia and Attention Deficiency Hyperactivity Disorder

Junaita Davakumar(✉) [iD] and Arul Siromoney [iD]

College of Engineering Guindy, Anna University, Chennai, India
junaitadavakumar@gmail.com, arul.siromoney@gmail.com

Abstract. Neurodevelopmental disorders in children like dyslexia and ADHD must be diagnosed at earlier stages as the children need to be provided with necessary aid. Comorbidity of dyslexia and ADHD is very high. Children with comorbidity of dyslexia and ADHD face comparatively more difficulty than children with just one of the disorders. Since all the three, dyslexia, ADHD and comorbid cases share many similar characteristics, it is hard to distinguish between cases which have only dyslexia or ADHD and those which have both. Manual analysis to differentiate based on standard scores of the psycho analysis tests provided inconsistent results. In this paper, we have applied standard machine learning techniques Random Forest, Support Vector Machine and Multilayer Perceptron to the diagnosis test results to classify between ADHD and comorbid cases, and dyslexia and comorbid cases. Analysis using the different individual psycho analysis tests is also done. Application of machine learning techniques provides better classification than the manual analysis.

Keywords: Attention deficiency hyperactivity disorder · Dyslexia · Comorbidity · Machine learning · Classification

1 Introduction

Dyslexia and Attention Deficiency Hyperactivity Disorder (ADHD) are both neurodevelopmental disorders which have a high rate of comorbidity or combined occurrence [1]. These neurodevelopmental disorders are chronic and should be diagnosed at childhood, to provide support [2]. These neurodevelopmental disorders make the children depressed as they are unable to perform well academically like their peers with no ailments. This is mainly because of the lack of guidance and proper diagnosis of the disorders during childhood. Dyslexia is the inability to read or spell though the person is highly capable and has adequate intelligence [3]. ADHD is the lack of attention, presence of high impulsivity in individuals [4]. Children with comorbid issues face more difficulties than the children with just one of the disorders [5, 6]. Though the rates of comorbidity or combined occurrences of ADHD and dyslexia are high, they are individual diseases and do not cause each other. Although they are different diseases, individuals with dyslexia

A. Chandrabose et al. (Eds.): ICCIDS 2020, IFIP AICT 578, pp. 305–311, 2020.
https://doi.org/10.1007/978-3-030-63467-4_24

and individuals with ADHD share certain similar characteristic traits. It has been found through various studies of families and twins that both ADHD and dyslexia are inherited [7–10]. ADHD and Dyslexia have been routinely diagnosed using various psycho analysis tests which do not provide precise outcomes. Diagnosis of comorbid cases of dyslexia and ADHD is even more strenuous. When a case is not correctly diagnosed as being comorbid, it means that necessary treatment is not provided for one of the two illnesses.

In this paper we have used standard machine learning techniques Random Forest (RF), Support Vector Machine (SVM) and Multilayer Perceptron (MLP) for classifying and analyzing dyslexia and comorbid cases, ADHD and comorbid cases to achieve better outcomes. The impact of various psycho analysis tests on the classification is also analyzed. To our knowledge, there is no prior work on the analysis of comorbidity of dyslexia and ADHD, and its classification, using machine learning techniques.

2   Related Work

Although there are many genetic studies of dyslexia and ADHD, there are not many conclusive results that link the two disorders. Work has been done to identify genetic, cognitive and neural overlap between the two disorders. Sanchez et al. have studied the association of genes with dyslexia, ADHD and comorbid samples as well [11]. Marino et al. [12] have investigated a strategy to clarify which genes are important for dyslexia. Eva et al. [1] have analyzed the link between dyslexia and ADHD from epidemiological, genetic, neurofunctional, neuropsychological and therapeutic perspectives.

Structural and functional magnetic resonance imaging combined with comprehensive behavioural testing has been used to characterize the behavior of comorbid dyslexia and reading disability (RD) [13]. Lauren et al. have done an analysis of voxel-based morphometry studies to find whether there is any overlap in the gray matter correlates of dyslexia and ADHD [14]. Comorbidity between dyslexia and ADHD has also been clarified by investigating cognitive endophenotypes [15].

3   Materials and Methods

3.1   Methods Used

SVM [16] is a machine learning algorithm that uses supervised learning for classification. Using the training samples, a hyperplane is constructed to separate the samples into two classes. This hyperplane is used to classify the new samples. Random Forest [17] is also a machine learning algorithm that is used for classification and regression. Multiple decision trees are grown using the training samples. To classify a new sample, each tree gives a class and that class is assumed to be given a vote. The class that gets the maximum number of votes is decided as the class for the new sample. MLP [18], a type of feedforward artificial neural networks with at least three layers of nodes and nonlinear activations is used for classification. MLP uses a supervised learning technique called backpropagation for training.

3.2 Dataset Description

In this study a public domain dataset [19, 20] has been used. The dataset includes the results of psycho analysis tests on 26 children with only dyslexia, 27 children with only ADHD and 27 children with comorbid ADHD and dyslexia. The dataset attributes are in four sections namely, demographics, dyslexia tests, ADHD tests and motor skills tests. All the children took up the tests along with their legal guardian and have given a written informed consent [20].

Demographics. This section of the dataset attributes includes age, sex, TONI 4 test and handedness. TONI 4 (Test of non-verbal intelligence, Fourth Edition) tests the non-linguistic and non-motor skills of the examinee, limiting to the analysis of the general intelligence. Abstract reasoning is tested and cognitive abilities such as reading, writing, speaking and listening are avoided. Handedness questionnaire includes eighteen questions based on certain actions testing the hand preference of the examinee.

Dyslexia Tests. This section of the dataset attributes includes pseudo-word decoding and spelling tests (Wechsler Individual Achievement Test). Wechsler Individual Achievement Test, Second Edition, is a standard academic based test that is used for the identification of learning disabilities. In Pseudo-Word Decoding subtest, the phonetic knowledge is tested by making the examinee read out meaningless words aloud. In the spelling subtest, words, letters and combination of words are dictated, for which the examinees are expected to write spellings.

ADHD Tests. Conners 3-Parent test is a behavioral questionnaire which is a commonly used test for the diagnosis of ADHD is included in this section.

Motor Skills Tests. The motor skills tests include the attributes Grooved pegboard (GPB) and the Leonard Tapping Task (LTT). Grooved pegboard test is for the analysis of fine motor skills using dexterity and Leonard Tapping Test is for the analysis of simple and complex motor skills.

3.3 Experiments

We have classified between dyslexia and comorbid cases, and ADHD and comorbid cases using the machine learning techniques RF, SVM and MLP. Machine learning classifiers are applied on the results of ADHD tests, dyslexia tests, motor skills tests and demographics, of children with just ADHD, just dyslexia and combined occurrence of dyslexia and ADHD. For the classification using the entire dataset, overall accuracy, receiver operation characteristic and f-measure are found. For the analysis using the individual and combination of psycho analysis tests, overall accuracy alone is found. 3-fold cross validation is done for each classifier. The public domain tool WEKA is used.

4 Results and Analysis

The standard score for the dyslexia tests is 80 and children scoring below 80 are diagnosed with dyslexia. Using the standard score, only 3 out of 26 (11%) children were classified correctly. For the ADHD tests, the standard score is 60 and children scoring above 60 are diagnosed with ADHD. Using this, 19 out of 27 (70%) children were classified correctly. For a child to be diagnosed with comorbidity of dyslexia and ADHD it should pass the criteria of both dyslexia and ADHD tests. Using this, only 6 out of 27 (25%) children were classified correctly. True condition in all the subtests was considered for the classification.

4.1 Comparison of the Results of Dyslexic and Comorbid Cases

In the overall analysis of the results of the psycho analysis tests of dyslexic and comorbid cases using machine learning techniques, it is seen that random forest classifier performs better when compared to support vector machine and multilayer perceptron. The entire dataset has been used for the classification (Table 1).

Table 1. Overall dataset analysis using RF, MLP and SVM (dyslexia vs. comorbid)

Classifiers	Overall prediction accuracy	Receiver operation characteristic	F-measure
Random forest	71.6%	0.761	0.717
Support vector machine	64.1%	0.642	0.642
Multilayer perceptron	66.0%	0.660	0.748

In the results shown in Table 2, it is seen that ADHD tests provide better classification than any other psycho analysis test. It is seen that dyslexia tests provide bad results. This is due to the fact that, the difference between the comorbid cases and the dyslexic cases would be the lack of ADHD characteristics in the dyslexic cases.

Table 2. Prediction accuracy using different individual tests (dyslexia vs. comorbid)

Classifiers tests	RF	SVM	MLP
Demographics	62.2%	50.9%	58.4%
ADHD tests	73.5%	81.1%	84.9%
Dyslexia tests	49.0%	50.9%	43.3%
Motor skills test	60.3%	56.6%	50.9%

4.2   Comparison of the Results of ADHD and Comorbid Cases

In the overall analysis of the results of the tests of ADHD and comorbid cases using machine learning techniques, it is seen that support vector machine classifier performs better when compared to random forest and multilayer perceptron. The entire dataset has been used for the classification (Table 3).

Table 3. Overall dataset analysis using RF, MLP and SVM (ADHD vs. comorbid)

Classifiers	Overall prediction accuracy	Receiver operation characteristic	F-measure
Random forest	53.7%	0.533	0.597
Support vector machine	66.6%	0.667	0.665
Multilayer perceptron	62.9%	0.656	0.625

In the results shown in Table 4, it is seen that dyslexia tests provide better classification than any other psycho analysis test. It is seen that ADHD tests provide bad results. This is due to the fact that, the difference between the comorbid cases and the ADHD cases would be the lack of dyslexia characteristics in the ADHD cases.

Table 4. Prediction accuracy using different individual tests (ADHD vs. comorbid)

Classifiers tests	RF	SVM	MLP
Demographics	44.4%	42.5%	40.7%
ADHD tests	37.0%	46.2%	44.4%
Dyslexia tests	66.6%	70.3%	66.6%
Motor skills test	55.5%	57.4%	50%

5   Conclusion

Differentiating between ADHD and comorbid cases as well as dyslexia and comorbid cases is a difficult task as they share common characteristics. From the above analysis it is seen that, applying machine learning techniques to the results of the psycho analysis tests rather than using standard cut off scores for the diagnosis has proved to classify better. Doctors can thus use machine learning techniques for the preliminary diagnosis of comorbid cases of dyslexia and ADHD.

References

1. Eva, G., Antonella, G.M.D., Paolo, C.: Comorbidity of ADHD and dyslexia. Dev. Neuropsychol. **35**(5), 475–493 (2010). https://doi.org/10.1080/87565641.2010.494748
2. Tager-Flusberg, H.: Neurodevelopmental Disorders. The MIT Press, Cambridge (1999)
3. Vellutino, F.R.: Dyslexia: Theory and Research. MIT Press, Cambridge (1979)
4. American Psychiatric Association: Diagnostic and Statistical Manual of Mental Diseases (DSM- V), 5th edn. American Psychiatric Publishing, Washington, DC (2013)
5. Willcutt, E.G., Pennington, B.F., Boada, R., Ogline, J.S., Tunick, R.A., Chhabildas, N.A., et al.: A comparison of the cognitive deficits in reading disability and attention deficit/hyperactivity disorder. J. Abnorm. Psychol. **110**, 157–172 (2001)
6. Mental health problems in neurodevelopmental disorders. http://www.kcl.ac.uk/ioppn/depts/cap/research/MentalHealthProblemsinNeurodevelopmentalDisorders. Accessed 05 Feb 2020
7. Del'homme, M., Kim, T.S., Loo, S.K., Yang, M.H., Smalley, S.L.: Familiar association and frequency of learning disabilities in ADHD sibling pair families. J. Abnorm. Child Psychol. **35**, 55–62 (2007)
8. Friedman, M.C., Chhabildas, N., Budhiraja, N., Willcutt, E.G., Pennington, B.F.: Etiology of the comorbidity between RD and ADHD: exploration of the non-random mating hypothesis. Am. J. Med. Genet. Part B, Neuropsychiatric Genet. **120**, 109–115 (2003)
9. Gayan, J., Olson, R.K.: Genetic and environmental influences on orthographic and phonological skills in children with reading disabilities. Dev. Neuropsychol. **20**, 483–507 (2001)
10. Willcutt, E.G., Pennington, B.F., DeFries, J.C.: A twin study of comorbidity between attention-deficit/hyper-activity disorder and reading disability. Am. J. Med. Genet. (Neuropsychiatrics Genetics) **96**, 293–301 (2000)
11. Sanchez-Moran, M., Hernandez, J.A., Dunabeitia, J.A., Estevez, A., Barcena, L., Gonzalez-Lahera, A., et al.: Correction: Genetic association study of dyslexia and ADHD candidate genes in a Spanish cohort: implications of comorbid samples. PLoS ONE **13**(12), e0209718 (2018). https://doi.org/10.1371/journal.pone.0209718
12. Marino, C., Giorda, R., Vanzin, L., Molteni, M., Lorusso, M.L., Nobile, M., et al.: No evidence for association and linkage disequilibrium between dyslexia and markers of four dopamine-related genes. Eur. Child Adolesc. Psychiatry **12**, 198–202 (2003)
13. Langer, N., Benjamin, C., Becker, B.K.C., Gaab, N.: Comorbidity of reading disabilities and ADHD: Structural and functional brain characteristics. Hum. Brain Mapp. **40**(9), 2677–2698 (2019). https://doi.org/10.1002/hbm.24552
14. McGrath, L.M., Stoodley, C.J.: Are there shared neural correlates between dyslexia and ADHD? A meta-analysis of voxel-based morphometry studies. J. Neurodevelopmental Disord. **11**(31) (2019)
15. Debbie, G., Margaret, S., Charles, H.: Time perception, phonological skills and executive function in children with dyslexia and/or ADHD symptoms. J. Child Psychol. Psychiatry **52**(2), 195–203 (2011)
16. Cortes, C., Vapnik, V.: Support-vector networks. Mach. Learn. **20**(3), 273–297 (1995). https://doi.org/10.1007/BF00994018
17. Breiman, L.: Random forests. Mach. Learn. **45**(5) (2001). https://doi.org/10.1023/a:1010933404324
18. Rumelhart, D.E., Hinton, G.E., Williams, R.J.: Learning internal representations by error propagation. In: Rumelhart, D.E., McClelland, J.L. (eds.), Parallel Distributed Processing: Explorations in the Microstructure of Cognition, vol. 1. MIT Press (1986)

19. Marchand-Krynski, M-E., Morin-Moncet, O., Belanger, A-M., Beauchamp, M.H., Leonard, G.: Shared and differentiated motor skill impairments in children with dyslexia and/or attention deficit disorder: From simple to complex sequential coordination. PLoS ONE **12**(5) (2017). https://doi.org/10.1371/journal.pone.0177490
20. [dataset] https://doi.org/10.1371/journal.pone.0177490.s001

Program Synthesis: Synthesizing Operators for Integer Manipulation

Jayasurya Seenuvasan(ID), Shalini Sai Prasad$^{(\boxtimes)}$(ID), and N. S. Kumar(ID)

PES University, Bangalore, Karnataka 560085, India
suryaseenu@gmail.com, shalinisaiprasad@gmail.com, nskumar@pes.edu

Abstract. We describe a language to synthesize a linear sequence of arithmetic operations for integer manipulation. Given an input-output example, our language synthesizes a set of operators to be applied to the input integers to obtain the given output. The sequence is generated by using Microsoft Prose, a program synthesis framework and the Genetic Algorithm. Our approach generates a set of ranked solutions that can be made unique on additional input-output examples that are consistent.

Keywords: Program synthesis · Genetic Algorithm · PROSE · Arithmetic operations

1 Introduction

Program synthesis is the generation of programs in some underlying Domain Specific Language (DSL) from a high-level specification. The concept of program synthesis was theorized in the 1950s, where Alonzo Church defined the problem to synthesize a circuit from mathematical requirements [1]. Since then, various domains such as spreadsheet manipulation [2], data wrangling [3], and generation of competitive coding programs [4] have made use of the concept of synthesizing programs.

In this paper, we explore various techniques of synthesizing programs and develop a language to generate formulae consisting of arithmetic operations. We implemented inductive programming, which is the generation of a program from input-output specifications, through the use of Microsoft PROSE [5] and have used the Genetic Algorithm [6] in synthesizing a combination of expressions for a given list of integers.

The synthesizer was able to learn various operators to be linearly applied to different kinds of inputs.

2 Literature Survey

There are several ways to approach the task of synthesizing programs. The synthesis problem is reduced to a search problem that involves searching over a

Supported by PES University.

candidate space of possible programs to find an expression or combination of expressions that satisfies a set of constraints given by a user.

According to a study by Sumit Gulwani et al. [7], there are four methodologies of program synthesis: enumerative search, stochastic search, constraint solving, and deduction based program synthesis.

Oracle Guided Component-Based Synthesis [8] by Susmit Jha et al. delves into the synthesis of loop-free programs for bit manipulation. The approach involves the use of a combination of an oracle to guide learning from examples and synthesis of components from constraints using satisfiability modulo theory (SMT) solvers [9]. Given a set of bitwise operators and arithmetic operators, the tool they developed, Brahma synthesizes loop-free programs to obtain a one-bit vector from the other. The tool also helped in the deobfuscation of programs. The program is synthesized from a set of base components and a set of input-output examples. It uses an I/O oracle, which returns an output when given an input and a validation oracle, which verifies the correctness of a synthesized program. We have provided an extension of this concept with our synthesizer for integer manipulation.

Automatic String Processing in Spreadsheets Using Input-Output Examples, by Sumit Gulwani [2], is a benchmark of program synthesis. Here, a language to perform string manipulation tasks was developed, which synthesized a program from input-output examples. The DSL uses regular expressions to add constraints to the synthesized operators. The algorithm used also allows ranking in the case of multiple synthesized programs. This algorithm has been used as an add-in feature in Microsoft Excel, called Flashfill. The algorithm uses version space algebra and a Directed Acyclic Graph (DAG) representation of the code to be synthesized. The shortcomings of Flashfill include not being able to handle semantic synthesis, i.e., converting a date to a day as well as being limited to strings. Our DSL overcomes this data type dependency by implementing this feature for integers.

DeepCoder by Matej Balog et al. [4] adopted a machine learning technique to solving the task of program synthesis for competitive style coding problems by using a neural network to predict the program synthesized for various outputs given an input. They solve the problem of inductive program synthesis by defining a DSL with high-level operations. A recurrent neural network predicts how well an operation will solve a given problem and then searches over the program space using the predicted probabilities to find the solution. The limitations of DeepCoder include the inability to synthesize complex programs that require algorithms like dynamic programming as their DSL is limited. Another limitation is the number of examples to be given to the tool has to be five or more, which is not reflective of actual competitive coding examples that provide only two or three input-output examples. Our approach requires one example to synthesize multiple formulae, which are narrowed down if provided with additional examples.

3   Overview

Microsoft PROSE [5] is a program synthesis framework for creating custom Domain-Specific Languages. The framework requires that a user specifies four features of the DSL.

i) The grammar consisting of operations provided by the DSL
ii) The semantic meaning of these operations
iii) The constraints on the operations, i.e., witness functions
iv) The ranking of features in the case of multiple synthesized programs

PROSE takes an input-output example, generates the operator(s) to be learned, and applies the synthesized operators on a new input to produce an output. We have incrementally created the synthesizer using four approaches to tackle more complex inputs. We define our DSL as the list of arithmetic operations: Addition, Subtraction, Multiplication, and Division.

The types of syntheses provided by our DSL are:

i) Synthesis of a binary operator and an integer.
ii) Synthesis of a binary operator.
iii) Synthesis of a single operator to reduce a list of integers to an aggregate.
iv) Synthesis of a sequence of operators to reduce a list of integers to an aggregate.

3.1   Approach 1: Synthesizing an Operation and an Integer

For this type of program, the input to the PROSE framework is two integers: x and y, such that x is the input to the synthesized program, and y is the expected output. The synthesizer generates an operator op and an integer k, such that $op(x, k)$ results in y. The grammar is as follows:

$$S \rightarrow Add(x, k)|Subtract(x, k)|Multiply(x, k)|Divide(x, k)$$

If the input-output example given is

$$x = 5, \qquad y = 20$$

The programs synthesized are

$$Multiply(x, 4)$$

$$Add(x, 15)$$

$$Subtract(x, -15)$$

If a second example is given where,

$$x = 2, \qquad y = 8$$

The synthesized programs narrow down to a program that satisfies both input-output examples, i.e.,

$$Multiply(x, 4)$$

The synthesizer generates multiple operators to obtain the final solution, and these solutions are ranked by assigning a score to each synthesized program. The programs are listed such that the absolute value of the smallest operand synthesized would get a higher rank. Thus, in the above example, the first input-output example would rank $Multiply(x, 4)$ first.

3.2   Approach 2: Synthesizing a Binary Operation

This type of program takes a pair of integers x and y as input and an integer j as the expected output. The synthesizer generates as operator op such that $op(x, y)$ results in j. The grammar for this synthesis required the use of delegates. The grammar is as follows:

$$S \rightarrow Eval(x, y, Func)$$

where $Func$ is synthesized based on the output given along with the witness functions provided. The synthesizer compares the expected output j and the outputs obtained by applying all the operators on the given inputs x and y. If the outputs match, the synthesizer learns that the operator must be used for subsequent inputs. Multiple operators may be learned in this program and we use arithmetic precedence to rank the learned operators.

If the input-output example given is

$$x = 2, \qquad y = 2, \qquad j = 4$$

The programs synthesized are

$$Eval(x, y, Multiply)$$

$$Eval(x, y, Add)$$

If a second example is given where,

$$x = 3, \qquad y = 3, \qquad j = 9$$

The synthesized programs narrow down to a program that satisfies both input-output examples, i.e.,

$$Eval(x, y, Multiply)$$

3.3   Approach 3: Synthesis of a Single Operator to Reduce a List of Integers to an Aggregate

This type is an extension of the previous approach, but the synthesized operator is applied to a list of integers $l = [x1, x2, x3, ..., xn]$ and an expected output integer j. The synthesizer applies the operator sequentially between each element in the list and checks if the aggregate is equal to the expected result. The grammar for this type also uses delegates.

$$S \rightarrow Eval(<list>, Func)$$

The input-output example given is

$$l = [1, 2, 3], \qquad j = 6$$

The programs synthesized are

$$Eval(l, Multiply)$$

$$Eval(l, Add)$$

If a second example is given where,

$$l = [1, 2, 3, 4], \qquad j = 10$$

The synthesized programs narrow down to a program that satisfies both input-output examples, i.e.,

$$Eval(l, Add)$$

3.4   Approach 4: Synthesis of a Sequence of Operators to Reduce a List of Integers to an Aggregate Using Genetic Algorithm

The last type is where a sequence of operators is synthesized. The input to the synthesizer is a list of integers $[x1, x2, x3, ..., xn]$ and an expected output j. The synthesizer generates a sequence of operations to be applied sequentially on each list element to obtain j.

If the input-output example given is

$$l = [1, 2, 3, 4], \qquad j = 4$$

The synthesizer returns the following sequences of operators:

$$[+, /, *]$$

$$[+, -, +]$$

If a second example is given where,

$$l = [4, 4, 3, 5], \qquad j = 10$$

The synthesized programs narrow down to a program that satisfies both input-output examples, i.e.,

$$[+, -, +]$$

We have used the Genetic Algorithm [12] to help us generate the possible sequence of operators that can be applied to the input list of integers. The algorithm initially generates a random list of sequences to form the initial population. The size of the initial population is 25% of the candidate space of all series of operators possible. Each sequence of operators is applied to the input list to compute the output.

During subsequent iterations of the algorithm, the fitness score for each series is calculated as the absolute difference between the generated output and expected output. The lesser the absolute difference between the outputs, the fitter is the individual. 40% of the fittest individuals pass onto the next generation. The remaining individuals for the population for the next generation are computed by performing crossover and mutation on randomly picked pairs of individuals as parents in the current population. Each parent produces two children for the next population. We have used single-point crossover and coin-flip mutation for this approach.

The algorithm runs for a certain number of iterations that were varied during testing, and finally returns combinations of operators that give the expected output.

4 Results

We illustrate the working of the genetic algorithm with an increasing number of operations to learn. We test the algorithm by providing simple examples where one or more known solutions exist and tabulating the results, as in Table 1.

Table 1. Tests run on Genetic Algorithm

Iterations	Input	Output	No. of sequences learned	Time (ms)
100	1 2 3	6	2	957
150	1 2 3 4	4	2	932
200	8 4 6 2 1	13	5	982
250	9 2 3 3 7 2	60	1	1286
300	5 3 4 6 5 8 3	15	19	2122
350	4 6 6 4 9 5 3 6	53	29	6934
400	6 5 4 2 7 9 5 8 3	36	222	27771
450	1 2 3 4 5 6 7 8 9 1	46	176	128429

The algorithm runs until convergence and generates sequences of operators that result in the output. The algorithm takes exponential time to converge as the input size increases linearly, as shown in the graph in Fig. 1.

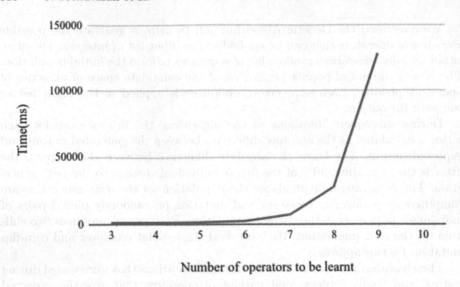

Fig. 1. Graph showing convergence time based on input size

It is observed that the genetic algorithm may result in the same sequence to be generated multiple times as a result of crossover and mutation. The algorithm may not generate some possible sequences because of the randomness of the initial population. These are the limitations of choosing such an approach as compared to an exhaustive search.

5   Future Work

We are attempting to strengthen the implementation of the genetic algorithm by eliminating duplicate sequences that could potentially reduce the time taken by it to converge. Compared to an exhaustive search the number of sequences tested are lesser but due to the crossover and mutation overhead, the genetic algorithm is much slower than the exhaustive search for this small subset of inputs.

Our program's DSL is limited to simple arithmetic operations, namely Addition, Subtraction, Multiplication, and Division. The DSL can be extended to include relational operators, and the grammar could be modified to include statements such as branches and loops to generate complex programs.

References

1. Friedman, J.: Application of recursive arithmetic to the problem of circuit synthesis. J. Symb. Logic **28**, 289–290 (1963). Review: Alonzo Church. Summaries of talks presented at the Summer Institute for Symbolic Logic Cornell University, 1957, 2nd edn., Communications Research Division, Institute for Defense Analyses, Princeton, N. J., 1960, pp. 3–50. 3a–45a

2. Gulwani, S.: Automating string processing in spreadsheets using input-output examples. In: Proceedings of the 38th Annual ACM SIGPLAN-SIGACT Symposium on Principles of Programming Languages (POPL11), 26–28 January 2011, Austin, Texas, USA. Association for Computing Machinery, New York (2011)

3. Gulwani, S., Jain, P.: Programming by examples: PL meets ML. In: Chang, B.-Y.E. (ed.) APLAS 2017. LNCS, vol. 10695, pp. 3–20. Springer, Cham (2017). https://doi.org/10.1007/978-3-319-71237-6_1

4. Balog, M., Gaunt, A.L., Brockschmidt, M.L., Nowozin, S.L., Tarlow, D.L.: Deep-Coder: learning to write programs. In: International Conference on Learning Representations, Toulon, France, 24–26 April (2017)

5. Microsoft Program Synthesis using Examples SDK. https://microsoft.github.io/prose/

6. Mallawaarachchi, V.: Introduction to Genetic Algorithms - Including Example Code. https://towardsdatascience.com/introduction-to-genetic-algorithms-including-example-code-e396e98d8bf3

7. Gulwani, S., Polozov, O., Singh, R.: Program Synthesis. Now Publishers Inc., Boston (2017)

8. Jha, S.A., Gulwani, S.A., Seshia, S.A., Tiwari, A.A.: Oracle-guided component-based program synthesis. In: Proceedings of the 32nd ACM/IEEE International Conference on Software Engineering - ICSE 2010, vol. 1, pp. 215–224 (2010)

9. Barrett, C., Sebastiani, R., Seshia, S., Tinelli, C.: Satisfiability modulo theories. In: Handbook of Satisfiability, vol. 185, pp. 825–885 (2009)

Location Based Recommender Systems (LBRS) – A Review

R. Sujithra @ Kanmani[✉] and B. Surendiran

Department of Computer Science and Engineering, National Institute of Technology Puducherry, Karaikal, India
sujithrakanmani@gmail.com, surendiran@nitpy.ac.in

Abstract. Recommender system has a vital role in everyday life with newer advancement. Location based recommender system is the current trend involved in mobile devices by providing the user with their timely needs in an effective and efficient manner. The services provided by the location based recommender system are Geo-tagged data based services containing the Global Positioning System and sensors incorporated to accumulate user information. Bayesian network model is widely used in geo-tagged based services to provide solution to the cold start problem. Point Location based services considers user check-in and auxiliary information to provide recommendation. Regional based recommendation can be considered for improving accuracy in this Point location based service. Trajectory based services uses the travel paths of the user and finds place of interest along with the similar user behaviours. Context based information can be incorporated with these services to provide better recommendation. Thus this article provides an overview of the Geo-tagged media based services and Point Location based services and discusses about the possible research issues and future work that can be implemented.

Keywords: Geotagged media · Point location · Trajectory · Location based recommender systems (LBRS)

1 Introduction

The "ranking" or "preference" a consumer will give to a product is foreseen by a recommender system. The basic classification of the recommendation system consists of content based filtering, collaborative filtering and hybrid filtering (see Fig. 1).

Content based filtering method contains providing items, which are acquainted to users. It is also known as cognitive filtering, which suggest items grounded on a contrast between the item content and a user profile. Collaborative Filtering depends on similar user ratings. It is capable of predicting the future preference of the user with top-k recommendations. It is categorised as model based filtering and memory based filtering. In Model based systems, models are created initially and system studies algorithms for doing procedures based on the data trained. Then using this models prediction is being made for definite data. Memory based filtering make use of rating information to

A. Chandrabose et al. (Eds.): ICCIDS 2020, IFIP AICT 578, pp. 320–328, 2020.
https://doi.org/10.1007/978-3-030-63467-4_27

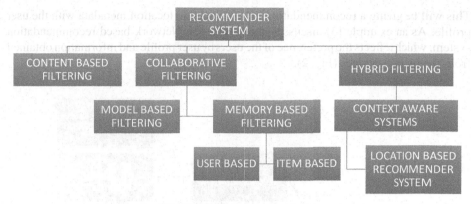

Fig. 1. Classification of recommendation system

compute the similarity among items or users and provides recommendation based on similarity measure. It is of two types calculating the similarity based on user ratings and item ratings. Hybrid filtering involves the combination of content and collaborative filtering. Context-Aware systems comes under this category, which collects both user ratings of items and their properties. Location based recommender system is an example of context aware system as it uses location, time, etc. for providing recommendations.

The involvement of Global Positioning System (GPS) in mobiles made the location based services trendy today. By accessing the location co-ordinates, the user activities can be analysed, and their preferences can be given in a better way. Thus the LBRS offers user with the appealing materials in an effectual way, such that the user can label their location media stuffing like text, photos, videos, etc. Recommender Systems is a vital applications that can be straight fitted into Location Based Services (LBS) field. Providing suitable and adapted recommendation built on the position of the user or location is a salient exploration trick [1].

2   Location Based Recommendation System (LBRS)

The Location based systems can be separated into classes based on the goal of the recommendation such as Stand- alone location recommender systems, and Sequential location recommender systems. Stand-alone location recommends a user with entity locations that matches their favourites and Sequential location recommender system provides a chain of locations to a user based on their favourites and constraints (time and cost) [2].

2.1   Standalone Recommendation System

It can be divided into User profile based recommendation, Location histories based recommendation and User trajectories based recommendation.

User Profile Based Recommendation

This will be giving a recommendation by matching the location metadata with the user profile. As an example [3] discusses about Bayesian Network based recommendation system, which reflects the preference of the users by user profile and information obtained from the mobile devices (Fig. 2).

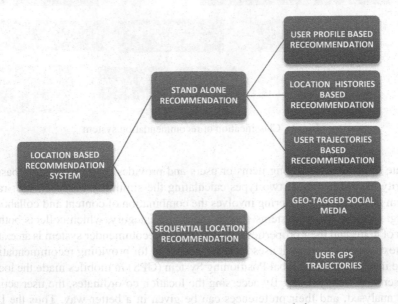

Fig. 2. Classification of location based recommendation system

Location Histories Based Recommendation
Location histories based recommendation system considers the rating history and check-in histories thereby it improves the quality of recommendation. Collaborative filtering technique is mostly used for finding the similar user's rating. By considering the rating history a user is capable of rating the location categories [4]. When a user is interested in location recommendation, that particular user's ratings are considered and higher predicted location matching with their ratings are recommended. Therefore, deployment of location data involving rating history improves the recommendation accuracy.

User Trajectories Based Recommendation
It contains a richer set of geographical information such as location co-ordinates (latitude, longitude and time). Thus using these trajectory logs and data the user's movement and behaviour in the real world can be collected.

2.2 Sequential Location Recommendation

It can be divided into Geo-tagged social media based, and User GPS trajectories based.

Geo-tagged Social Media
It relies upon the data revealed from geo-tagged photos, which may offer a custom-made trip arrange for a holiday maker, i.e., the fashionable destinations to go to, the order of

visiting destinations, the arrangement of time in every destination, and also the typical path travelled at intervals every destination [5]. Users can state personal preference like location visited, travel period, time/season visited, Associate in Nursing destination vogue in an interactive manner. Mining techniques can be applied for providing effective results in case of geo-tagged based recommendation.

User GPS Trajectories
It includes the length used up at a location and therefore, the sequence of visited locations, which will advance serial location recommendations. GPS trajectories spawned by multiple users, fascinating locations and classical travel sequences at a given geospatial region [6]. Such info will facilitate the link among users and locations, and modify recommendation on travel effectively.

3   Services Provided by Location Based Recommender Systems (LBRS)

Existing LBRS services will be classified as geo tagged media based, point location based, and trajectory-based (see Fig. 3). In this paper we are going to discuss briefly about Geo-tagged and Point based services and their future work.

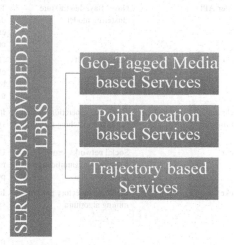

Fig. 3. Services provided by LBRS

3.1   Geo Tagged Media Based Services

This services permit users to feature location with user's broadcasting contents like text, photos and videos that were formed within the real world. Passive tagging happens once a user expressly makes and increases the contents of location [7]. Geo-tagged media-based services permit a user to look at alternative users content during a geographical content by using digital maps incorporated in phones [8]. Common applications that offer LBRS services embrace Flickr, Twitter and Panoramio [9] (Table 1).

3.2 Point Location Based Services

These services make users to share user's locations like restaurants, cinemas or mall [19]. The foremost common applications of this services include Facebook, Instagram and Foursquare that promote users to share their current location. Users of such application's area unit abounding with choices to achieve arrival at totally different locations that users visited area in their regular routine to share knowledges and information by providing a feedback [20]. One among those services that enable period location track of users is they will find their friends round their locations that ease in improving a user's social activities. As an example, when discovering a friend's location from their social network, we can give the opinion for lunch or searching activity. The use of feedback permits users to share their suggestions. Such feedbacks can help in accumulating recommendation. Unlike geo tagged media, some extent location (venue) is that the main element related to the purpose location [9, 21]. The following explains the various Point Location based services works and the Table 2 explains various Point Location based services.

Table 1. Survey of Geo-tagged media based services

Related work	Dataset used	Techniques used	Inference
Zhang et al. [10]	Twitter API	Novel Bayesian mixture clustering model	TrioVecEvent methodology that leverages multimodal embeddings to attain correct on- line native event detection
Yang et al. 2017 [11]	Flickr	Network motifs	Geo-Infa novel motif-based clustering framework using the behaviour patterns of the tourists is proposed
Sun et al. 2018 [12]	Flickr and other contextual information	Candidate generation and candidate matching techniques	It provides a recommendation system that matches with user's preferences
Bujari et al. 2018 [13]	Flickr	Social networks, crowd sourcing and gamification method	User feedback based recommendation system was proposed
Cai et al. 2018 [14]	Flickr	Semantic trajectory pattern mining algorithm	It proposed a recommendation system by considering the semantic trajectory pattern of other user's sequences and preferences
Mor et al. 2018 [15]	Flickr	Cell-based clustering and bi-directional constrained pathfinder Nearest Neighbour route calculation algorithm	Popular tourist's destinations are recommended using photographer's activities and tourism context features
Wu et al. 2018 [16]	Flickr	Clustering by Fast Search and Find of Density Peaks clustering algorithm	This work proposes a Spatially - installed hotspot system utilizing web based life information (geotagged)
Barros et al. 2019 [17]	Flickr	Analysis of geotagged data	This work proposes the application of geotagged data in finding behaviour of the visitors in endangered area

(continued)

Table 1. (*continued*)

Related work	Dataset used	Techniques used	Inference
Sun et al. 2019 [18]	Flickr	Candidate generation, candidate ranking, SVM and gradient boosting algorithm	This work provides personalized recommendation by resolving cold start and long tail problem

4   Highlights for Improvement

The geotagged media based recommendation service can be given in a better way by incorporating with GPS, sensors to accumulate info and interaction among diverse users

Table 2. Survey of Point Location based services

Related work	Dataset used	Techniques used	Inference
Wang et al. 2018 [22]	Foursquare and gowalla	Multinomial distribution	Proposed method on modelling users' past check-ins and supporting information to facilitate POI recommendation
Ding et al. 2018 [23]	Foursquare.	A Matrix Factorization method	It is utilized to display connections among highlights and acquire include vector portrayals of clients and POIs
Guoqiong et al. 2018 [24]	Real check –in dataset from WW	Latent Dirichlet Allocation (LDA) topic model and Higher Order Singular Value Decomposition (HOSVD) Algorithm	A POI recommendation system for solving data sparsity by mining time aware topic preference of the user is used
Baral et al. 2018 [25]	Yelp Dataset	Dense subgraph extraction and ranking-based techniques	A ReEL (Review aware Explanation of Location Recommendation) using deep neural network. It can be used for providing group recommendations
Ri et al. 2018 [26]	Foursquare and Yelp	Ranking model (GSBPR)	Geo-Social Bayesian Personalized Ranking model (GSB-PR), which is based on the pairwise ranking
Cheng et al. 2018 [27]	Foursquare and Gowalla dataset	Context Graph Attention (CGA) Model	POI recommendation system supporting heterogeneous context information
Zhu et al. 2018 [28]	Review website.	Distance-based pre- filtering and distance-based ranking adjustment	POI based group recommendation is proposed. Context factors for group Recommendation can be considered

(*continued*)

Table 2. (*continued*)

Related work	Dataset used	Techniques used	Inference
Xu et al. 2019 [29]	Foursquare and Gowalla dataset	This model concentrates on preference and geographic factors of the user in providing recommendation	This system addresses two issues such as city guidelines for traveller categories and recommendations for a single traveller inside city POI

in movable environment. The context aware attributes like time, weather can be added to the system for providing better recommendation. The cold start problem and data sparsity problem can be overcome by proposing some new hybrid recommendation system, which is the amalgamation of content and knowledge based system. Even Bayesian Network model also gives an effective response to cold start and data sparsity problems.

Point Location media based recommendation system can be incorporated with chronological effect, geological influence, temporal cyclic effect and semantic effect (Graph based embedding model) for making ease of data sparsity and cold start problem. Clustering around the user by their fondness on categories and context factors such as (time aware scenarios, traffic, and consumption level) is useful for providing group recommendations and regional based recommendation. Geo-diversification concept can also be integrated for providing better recommendation.

5   Conclusion

Recently, the usage of Location has been a growing trend in the recommendation system. This paper covers all Location based recommender systems and the services provided by them, such as Geo-tagged media based services, Point location media based services. Geo-tagged media offers travel recommendations based on travel preferences combined with social geo-tagged images linked to a user's visit context. The key goal behind the Point location based recommendation is to help users accurately in locating Point Of Interests with overall positive reviews thereby providing Geographical, temporal, categorical and social recommendation. It is used to find the similarity of user behaviour through the travel paths. Group identification can be carried out as it identifies group similarities apart from clustering. The datasets and techniques used in these services are discussed and arrived at a conclusion that this Location based recommendation system which is a hybrid method involves in the improvement of the prediction quality for various social media services like Flickr, Foursquare and Twitter etc. We can conclude that the location details also provides added advantage to a recommendation system.

References

1. Tiwari, S., Kaushik, S., Jagwani, P.: Location based recommender systems: architecture, trends and research areas. In: IET International Conference on Wireless Communications and Applications (ICWCA 2012), pp. 1–8. IET (2012). https://doi.org/10.1049/cp.2012.2096

2. Bao, J., Zheng, Y., Wilkie, D., Mokbel, M.: Recommendations in location-based social networks: a survey. GeoInformatica **19**(3), 525–565 (2015). https://doi.org/10.1007/s10707-014-0220-8

3. Park, M., Hong, J., Cho, S.: Location-based recommendation system using Bayesian User' s preference model in mobile devices. Expert Syst. Appl. **39**(11), 1130–1139 (2012). https://doi.org/10.1016/j.eswa.2012.02.038

4. Logesh, R., Subramaniyaswamy, V., Vijayakumar, V., Li, X.: Efficient user profiling based intelligent travel recommender system for individual and group of users. Mob. Netw. Appl. **24**(3), 1018–1033 (2018). https://doi.org/10.1007/s11036-018-1059-2

5. Lu, X., Wang, C., Yang, J., Pang, Y., Zhang, L.: Photo2Trip : generating travel routes from geo-tagged photos for trip planning. In: Proceedings of the 18th ACM International Conference on Multimedia, pp. 143–152. (2010). https://doi.org/10.1145/1873951.1873972

6. Zheng, Y., Zhang, L., Xie, X., Ma, W.: Mining interesting locations and travel sequences from GPS trajectories. In: Proceedings of the 18th International Conference on World Wide Web, pp. 791–800 (2009). https://doi.org/10.1145/1526709.1526816

7. Majid, A., Chen, L., Chen, G., Mirza, H.T., Hussain, I., Woodward, J.: A context-aware personalized travel recommendation system based on geotagged social media data mining. Int. J. Geogr. Inf. Sci. 37–41 (2012). https://doi.org/10.1080/13658816.2012.696649

8. Chon, Y., Cha, H.: LifeMap: a smartphone-based context provider for location-based services. IEEE Perv. Comput. **10**(2), 58–67 (2011). https://doi.org/10.1109/mprv.2011.13

9. Rehman, F., Khalid, O., Madani, S.: A comparative study of location based recommendation systems. Knowl. Eng. Rev. **31**, 1–30 (2017). https://doi.org/10.1017/S0269888916000308

10. Zhang, C., Liu, L., Lei, D., Zhuang, H., Hanra, T., Han, J.: TrioVecEvent: embedding-based online local event detection in geo-tagged tweet streams. Knowl. Data Discov. 595–605 (2017). https://doi.org/10.1145/3097983.3098027

11. Yang, L., Wu, L., Liu, Y., Kang, C.: Quantifying tourist behavior patterns by travel motifs and geo-tagged photos from Flickr. SPRS Int. J. Geo-Informatica **6**(11), 1–18 (2017). Article number 345. https://doi.org/10.3390/ijgi6110345

12. Sun, X., Huang, Z., Peng, X., Chen, Y., Liu, Y.: Building a model-based personalised recommendation approach for tourist attractions from geotagged social media data. Int. J. Digit. Earth 1–18 (2018). https://doi.org/10.1080/17538947.2018.1471104

13. Bujari, A., Ciman, M., Gaggi, O., Palazzi, C.E.: Using gamification to discover cultural heritage locations from geo-tagged photos. Pers. Ubiquit. Comput. **21**(2), 235–252 (2017). https://doi.org/10.1007/s00779-016-0989-6

14. Cai, G., Lee, K., Lee, I.: Itinerary recommender system with semantic trajectory pattern mining from geo-tagged photos. Expert Syst. Appl. **94**, 32–40 (2017). https://doi.org/10.1016/j.eswa.2017.10.049

15. Mor, M., Dalyot, S.: Computing touristic walking routes using geotagged photographs from Flickr. ETH Zurich Research Collection, pp. 1–7 (2018). https://doi.org/10.3929/ethz-b-000225591

16. Wu, X., Huang, Z., Peng, X.I.A., Chen, Y.: Building a spatially-embedded network of tourism hotspots from geotagged social media data. Cyber-Phys.-Soc. Comput. Netw. **6**, 21945–21955 (2018). https://doi.org/10.1109/ACCESS.2018.2828032

17. Barros, C., Moya-Gómez, B., Gutiérrez, J.: Using geotagged photographs and GPS tracks from social networks to analyse visitor behaviour in national parks. Curr. Issues Tour. 1–20 (2019). https://doi.org/10.1080/13683500.2019.1619674

18. Sun, X., Huang, Z., Peng, X., Chen, Y., Liu, Y.: Building a model-based personalised recommendation approach for tourist attractions from geotagged social media data. Int. J. Digit. Earth **12**(6), 661–678 (2019). https://doi.org/10.1080/17538947.2018.1471104

19. Wang, H., Ouyang, W., Shen, H., Cheng, X.: ULE: learning user and location embeddings for POI recommendation. In: Proceedings of 2018 IEEE 3rd International Conference on Data Science Cyberspace, DSC 2018, pp. 99–106 (2018). https://doi.org/10.1109/dsc.2018.00023

20. Sarwat, M., Levandoski, J.J., Eldawy, A., Mokbel, M.F.: LARS*: an efficient and scalable location-aware recommender system. IEEE Trans. Knowl. Data Eng. **26**(6), 1384–1399 (2014). https://doi.org/10.1109/TKDE.2013.29

21. Zheng, V.W., Zheng, Y., Xie, X., Yang, Q.: Towards mobile intelligence: learning from GPS history data for collaborative recommendation. Artif. Intell. **184–185**, 17–37 (2012). https://doi.org/10.1016/j.artint.2012.02.002

22. Wang, H., Ouyang, W., Shen, H., Cheng, X.: ULE : learning user and location embeddings for POI recommendation. In: 2018 IEEE Third International Conference on Data Science Cyberspace, pp. 99–106 (2018). https://doi.org/10.1109/dsc.2018.00023

23. Ding, R., Chen, Z.: RecNet: a deep neural network for personalized POI recommendation in location-based social networks. Int. J. Geogr. Inf. Sci. **32**(8), 1–18 (2018). https://doi.org/10.1080/13658816.2018.1447671

24. Guoqiong, L., Zhou, Z., Changxuan, W., Xiping, L.: POI recommendation of location-based social networks using tensor factorization. In: 2018 19th IEEE International Conference on Mobile Data Management, pp. 116–124 (2018). http://doi.ieeecomputersociety.org/10.1109/MDM.2018.00028

25. Baral, R., Zhu, X., Iyengar, S.S., Li, T.: ReEL : review aware explanation of location recommendation. In: Proceedings of the 26th Conference on User Modeling, Adaptation and Personalization, pp. 23–32 (2013). https://doi.org/10.1145/2509230.2509237

26. Gao, R., et al.: Exploiting geo-social correlations to improve pairwise ranking for point-of-interest recommendation. China Commun. **15**(7), 180–201 (2018). https://doi.org/10.1109/cc.2018.8424613

27. Zhang, S., Cheng, H.: Exploiting context graph attention for POI recommendation in location-based social networks. In: Pei, J., Manolopoulos, Y., Sadiq, S., Li, J. (eds.) DASFAA 2018. LNCS, vol. 10827, pp. 83–99. Springer, Cham (2018). https://doi.org/10.1007/978-3-319-91452-7_6

28. Zhu, Q., et al.: Context-aware group recommendation for point-of-interests. IEEE Access **6**, 12129–12144 (2018). https://doi.org/10.1109/ACCESS.2018.2805701

29. Jiao, X., Xiao, Y., Zheng, W., Wang, H., Hsu, C.H.: A novel next new point-of-interest recommendation system based on simulated user travel decision-making process. Futur. Gener. Comput. Syst. **100**, 982–993 (2019). https://doi.org/10.1016/j.future.2019.05.065

Author Index

Printed in the United States
by Baker & Taylor Publisher Services